THE PATH TO SINGULARITY

HOW TECHNOLOGY WILL CHALLENGE
THE FUTURE OF HUMANITY

J. CRAIG WHEELER

FOREWORD BY
NEIL DEGRASSE TYSON

Prometheus Books

Essex, Connecticut

(PB) Prometheus Books

An imprint of The Globe Pequot Publishing Group, Inc.
64 South Main Street
Essex, CT 06426
www.globepequot.com

Distributed by NATIONAL BOOK NETWORK

British Library Cataloguing in Publication Information Available

Library of Congress Cataloging-in-Publication Data

Names: Wheeler, J. Craig, author.
Title: The path to singularity : how technology will challenge the future of humanity /
    J. Craig Wheeler ; foreword by Neil deGrasse Tyson.
Description: Lanham, MD : Prometheus, [2024] | Includes bibliographical references.
    | Summary: "Technology is developing at an increasingly swift pace, and there is
    likely to be a tipping point when change occurs at a pace so rapid that humans are
    not able to adjust, either individually or as a society. J. Craig Wheeler argues that
    we must take charge of our technology now, before we lose the ability to control
    it"— Provided by publisher.
Identifiers: LCCN 2024010275 (print) | LCCN 2024010276 (ebook) | ISBN
    9781493085439 (cloth) | ISBN 9781493085446 (epub)
Subjects: LCSH: Artificial intelligence—Forecasting. | Artificial intelligence—Social
    aspects. | Singularities (Artificial intelligence)
Classification: LCC Q334.7 .W45 2024 (print) | LCC Q334.7 (ebook) | DDC
    303.48/34—dc23/eng/20240509
LC record available at https://lccn.loc.gov/2024010275
LC ebook record available at https://lccn.loc.gov/2024010276

To Andreas and Karen, who will have
to live the future outlined here.

CONTENTS

ACKNOWLEDGMENTS

I would first like to express my thanks to the students in my class AST 321 The Future of Humanity at the University of Texas at Austin, who engaged in exploring the future with me. They indulged both my quirks as I learned along with them and my untutored attempts to improve their use of the English language. I would especially like to note two students: Rebecca Larson, who introduced herself and her nontraditional path into astronomy to me before the first class in 2013 and brought her enthusiasm to every class; and Vincent Lau, whose sharp insight changed the way I thought and taught about exponential change.

I'm grateful to Daniel Jaffe who, as department chair, recognized my enthusiasm and gave me license to teach this interdisciplinary topic that was only barely about astronomy. Later as vice-president for research, Dan developed the remarkable, campus-wide, interdisciplinary program Bridging Barriers that gave me ground floor access to the organization of the Good Systems team that was exploring ethical artificial intelligence. My interaction with Good Systems yielded more grist for this book.

Special thanks go to two people who provided insights into the topics here long before I had any notion that a class and a book were in store. One is my cousin Bruce Campbell (http://airplanehome.com/), who quintessentially marches to his own drummer. Not only have we recently had long conversations about the likelihood of sentient artificial intelligence, but I also realized as I was writing this book that we have had related exchanges stretching back perhaps four decades. Bruce helped plant the seeds.

The other person is my friend and colleague John Scalo. John introduced me to the concepts of astrobiology, was my partner in our attempts to get funding for a multi-institutional NASA Astrobiology Center, and endlessly impressed and entertained me with his deep and broad insights on a vast array of topics. Listening to John is like attending an improvisational word jazz session on the sciences of life, and much more. It would be difficult for me to map all the ways John influenced this book over the decades since that trip to Greece when we got lost on the Isle of Spetses while searching for the mansion featured in John Fowles's *The Magus*.

More recently, I have been stimulated by reconnecting with my friend Jay Boisseau, who was my PhD student some time ago. Jay has had a distinguished career in supercomputing and established a dynamic institution here in Austin, the Austin Forum on Technology and Society (www.austin forum.org/). The Austin Forum sponsors a variety of venues designed to bring together technologists to discuss the means to have a positive impact on society. Jay's depth of knowledge of the field and its practitioners and his incisive ability to lead complex discussion on multiple topics continue to impress me. Several portions of this book draw directly from my participation in Austin Forum presentations over the last several years.

Many family members supported me on the long road from first glimmer to final product. J. Robinson Wheeler and H. L. Wheeler provided valuable help with proof reading and comments on style, coherence, and impact. Diek W. Wheeler introduced me to elements of neural networks, brain electrochemistry, and the role of the hippocampus associated with the work he did in neurophysics for his PhD at the University of Texas at Austin, postdoc in Munich, and current research at George Mason University. Greta Ann Herin gave me a wonderful, patient, introduction to the structure of the cortex at a key juncture.

I am especially grateful to my friend, fellow writer, and raconteur Wayne Bowen, who provided feedback on the manuscript and moral support through the whole process.

I met and befriended Neil Tyson when he was in graduate school here in Austin. His charisma was evident even then. It has been a pleasure to watch and wonder from afar as, through hard work and talent, he emerged as the public face of American astronomy and prolific author, Neil DeGrasse Tyson. Neil responded to my inquiry about writing a foreword with warmth and grace. He also read the manuscript and provided detailed suggestions on typos, style, presentation, science, and insight into the book business. The responsibility for remaining flaws is, of course, mine.

I want to give thanks to my agent, Regina Ryan, who took a flyer on an unsolicited email and signed me on as a client. Regina gave me careful, intensely professional guidance, shepherded me though the process of constructing a proper book proposal, provided sound advice on the structure and content of the book itself, and then stuck with me on the turbulent seas of pitch making and contract signing. I have been enriched by the experience of working with her.

Finally, I want to thank my editor Jonathan Kurtz at Prometheus, who responded enthusiastically to my attempt to cover a sprawl of interrelated topics in this book. I am also grateful to my production editors, David Bailey and Nicole Carty, who moved the project forward with both rigor and flexibility in response to this author's quirks, and to Brianna Soubannarath and Emily Jeffers, who supported the outreach and publicity.

FOREWORD

When tuned into the day's news, we're treated to the perspectives of full-time commentators opining on culture, domestic politics, geo-politics, and technology. They tend to have backgrounds in journalism or in communication. Others might also be trained in history, politics, or law, and maybe business. On social media platforms, strongly voiced opinions flow from everyone no matter their background—including the neighbors who think and vote just the way you do and the neighbors who don't.

The scarcity of scientists in this 24/7 news cycle might leave the viewer to suppose that scientists have nothing to say about world events. Or they might suppose the views of scientists are not as important as the views of media pundits. In the United States, science and engineering professions employ about 6 million people. That's one in fifty Americans, but almost all of them cluster in academia and in private and government labs. Statistically that means most people cannot claim a scientist or engineer as a friend, and some people will live their entire lives having never met a scientist.

Meanwhile, practically everything we value about modern civilization—our health, wealth, and security as well as our systems of transportation, communication, and information technologies—all derive from innovations in science, technology, engineering, and mathematics.

Scientists are human and generally susceptible to the full range of human emotions and biases just like everybody else. They live and walk among us. They read books, watch TV, and listen to music. Scientists carry opinions and vote. Good scientists are also aware of how bias can influence their conclusions. They invest special effort to minimize that influence, including peer review, which accounts for its high value in the scientific enterprise.

Scientists are also deep thinkers. From the beginning of one's training, especially in the physical sciences, the mind is honed to analyze, interpret, and draw conclusions from relevant data guided by the universal laws of physics and the attendant mathematical tools that wield the power to grant statistical integrity to that which is real in the world. These are pathways to objective truths independent of what feels good or what we need to be true based on our culture, politics, or religion.

Seems to me these are just the people you want commenting on the endless, future-leaning, breaking-news topics we face today such as energy, automation, robotics, quantum computing, artificial intelligence, gene editing, and synthetic biology. And scientists surely wield views on the socioeconomic disruptions these moving frontiers will unleash on civilization.

The world is long overdue for a peek at the state of society and what its future looks like through the lens of a scientist. And when that scientist is also an astrophysicist, you can guarantee the perspectives shared will be as deep and as vast as the Universe itself.

When such an occasion presents itself, as it does in this book, run, don't walk to experience it. The views you'll find here are enlightening, insightful, and occasionally mind-blowing. You will never see the world the same way. And that's a good thing.

Neil deGrasse Tyson

PREFACE

As an astrophysicist, I have spent my career pondering a particular aspect of our wondrous Universe: stellar evolution, and especially what makes stars explode at the end of their lives. I did not think too deeply about biological evolution or the fate of humanity although I had a long-running email discussion with a cousin, a technologist, on the prospects for super-intelligent machines. Perhaps this began to shift around 2000 when, through my colleague John Scalo, I got involved in astrobiology. John had read widely and thought deeply about genetics, the origin of life, and related topics. He and I wrote a few papers on how astrophysical radiation might affect living things on Earth or elsewhere. We also tried to obtain funding from NASA for a multi-institutional, interdisciplinary astrobiology institute. We failed to get funding, but I learned a lot in the process of constructing the proposals and talking to biologists, geologists, and chemists. In hindsight, I probably learned just enough to be dangerous but not enough to be a contributing expert.

From 2003 to 2005, I had the privilege of cochairing the Committee on the Origin and Evolution of Life of the National Academy of Sciences. Jack Szostak of Harvard, the other cochair, was working on how cells might have formed their first membranes to trap the machinery of life inside. He had previously illuminated the existence and role of telemeres, work for which he won the Nobel Prize in Physiology or Medicine in 2009.

Every meeting of this interdisciplinary committee was like a mini workshop on the most stimulating topics in the field. Through it I met many of the leading lights in astrobiology. There were issues of field-specific jargon and different approaches to science. Physicists tend to start thinking about problems by making simplifying assumptions—"All cows are spherically symmetric," as the old joke goes—whereas molecular biology is already vastly too complex to be readily amenable to that process. Physicists nevertheless try to understand biology with considerable success. Astrobiologists are self-selected to be interdisciplinary with the urge to communicate across such barriers of concepts and terminology. The whole process was intensely interesting intellectually. We produced a report titled "The Astrophysical Context of Life."[1]

Sometime in this era I read a statement by a preacher. This learned man declared his conviction about the truth of the biological evolution of life, including mankind. He went on to argue, however, that it was clear to him that *Homo sapiens*, our current breed, was the peak. There would be no more biological evolution of our species. Contemplating this statement, I realized that I had always assumed since first learning about biological evolution (When? Age fourteen?) that humans would continue to evolve. Into what, I surely had no clue.

During this time I also participated in a program called the Reading Roundup that was intended to help incoming freshmen adjust to life at the University of Texas at Austin. The Academy of Distinguished Teachers at the University of Texas at Austin, of which I was honored to be a member, sponsored this annual affair. The idea was to select a book to be read by incoming freshman over the summer. In the morning before fall classes started, faculty would meet for two hours to discuss the book with students interested in that particular book. Some of my colleagues would pick the same book year after year. I liked to choose a different one every year to encourage the breadth of my own reading. Because of the freedom to explore allowed by this program, one particular year I decided to address the notion raised by the preacher. I suggested reading Darwin's *The Ascent of Man* and that we discuss "What next?" I had forgotten that I had read Darwin at some earlier stage in my life, and I found an old personal copy, inscribed with my signature, on my shelves at home. I then led a two-hour conversation on what we knew about evolution, and I introduced the preacher's perspective as one point of discussion.

I am not sure what the students got out of that particular session, but it was a revelation to me. While my thoughts were somewhat diffuse, many of the students were already cognizant of the idea that we were developing a deep understanding of molecular biology—the role of DNA—and associated technology that would allow us to affect our own evolution more rapidly than Darwin's natural selection could accomplish.

This thought continued to roll around in my head and at some point crystallized in the notion that this could be the topic of a semester-long class. In another wonderful program, the university also encourages classes with a substantial writing component. Though untrained in teaching writing, I care deeply about the art and have over the years taught a number of eclectic classes with a substantial writing component. Another advantage was that these classes were small. Unlike my standard lecture course for non-majors with two hundred students in which it is very hard to get acquainted with individual students, a writing course can be limited to thirty students and be much more personal. I also had the notion that I might construct a very discussion-heavy class so that I could lead the discussion but not have to prepare a lecture every class day. Frankly I was trying to write a professional-level book on supernovae[2] and hoped that I might cut some corners and save time for that project.

The upshot was that I proposed to our department chair, Dan Jaffe, that I teach this course that was barely about astronomy. I'm not quite sure why Dan said yes, but it was my good fortune that he sensed my enthusiasm and was willing to experiment.

I taught the course, AST 381 *The Future of Humanity*, for the first time in spring 2013. Although I did little formal lecturing, the course was demanding to prepare and present. It was also the most fun I ever had teaching, especially the first year. There is an old saying in pedagogy that the teacher learns the first year, students learn the second year, and no one learns the third year. I taught *The Future of Humanity* for five years alternating between fall and spring terms.

That first term I was a kid in a toy store exploring new ideas, new connections, new ways of thinking. The students were engaged. I had them write a lot. They might have even learned a little about how to write. As a principal text we used *The Singularity Is Near* by Ray Kurzweil.[3] I cautioned the students not to take Kurzweil's extreme utopianism literally, and many ended up with a very dystopian view, but Kurzweil summarized many of the important themes. He also covered some important technical notions such as the nature and significance of exponential growth in a manner that few other books did. Over the years *The Singularity Is Near* got a little more outdated and I began to see issues that Kurzweil overlooked, but I kept it as the main framework. I also began to see a wealth of other material—books and articles—that were germane to the course. I read and collected reams of articles, some of which found their way into the course, some of which did not. It was all stimulating.

Our technology has gotten very complex. No individual knows how to build an iPhone with its mix of microtechnology, manufacturing sophistication, and software. Our technology buoys us but also threatens to overwhelm us. I do not know the solution to this but I am confident that an important step is to talk about the issues.

At the end I think I have something to bring to the topic from my perspective as an astrophysicist and as a result of my pondering and interacting with engaged students. This book is the result.

1

SURVIVING AN ERA OF ASTONISHINGLY RAPID TECHNOLOGICAL CHANGE

THE PROBLEM

In the past, humans have always, with some turmoil, adapted to new technologies. Technology is now racing ahead under its own momentum. Humans and human organizations tend to lag. Things are currently changing so rapidly that we may not be able to adapt. This is a qualitatively new phase in human existence.

The biggest wave can start as a gentle swell in mid-ocean. Near shore, the wave crests and breaks. Picture a surfer on a gigantic wave. With the right timing and balance, the surfer can ride the wave and stay on top. The alternative is being tumbled within the surging surf or in the worst case pounded onto a coral reef. As we try to ride our technological wave, the tumbling may be unavoidable. We must avoid slamming into the reef.

THE BACKGROUND

How did we get here? Where are we going? Who or what is in charge?

Humanity is the legacy of an incredible history. The observable Universe as we know it began about 14 billion years ago. The first stars and galaxies coalesced into existence a few hundred million years later. The Sun and its retinue of planets formed much later, about 4.5 billion years ago, in a flare of now-fading star formation throughout the Universe. Life on Earth arose relatively quickly—less than a billion years later. Evolution brought forth *Homo sapiens* only a few hundred thousand years ago. Humans then developed increasingly sophisticated technologies that now threaten to overwhelm us.

Technological change naturally develops its own momentum and evolves ever more rapidly. Knowledge builds on knowledge. The more you know, the more you can do. This applies not only to individuals but also to whole societies.

Humans invented tools then made those tools more effective. They sharpened rocks for hide scraping and fashioned knives and spears. They controlled fire, then developed metallurgy and agriculture. They invented the wheel. Later came the printing press, the steam engine, automobiles, airplanes, fearsome weapons, computers. Current technology is now nearly beyond individual human comprehension.

REVOLUTIONARY TECHNOLOGIES

Many areas of modern science and engineering contribute to the rush of technology.

In the realm of medicine, our ever-deeper understanding of our biological functions at a molecular level will allow the curing of diseases that now plague us. Perhaps long-promised cures for cancer will arise. The burgeoning technology of nanobots will advance. These devices will permeate our blood and lymphatic streams, finding glitches in our biology and repairing them.

There is a growing community of researchers and advocates who seek to treat aging as a disease. The notion is that aging is a biological process that could be treated, even cured. Advocates of solving the aging problem, however, rarely directly address massive questions of the resulting effect on society. What will we do with all the old people? What about babies? Do we stop having them or regulate births?

Human evolution may be in for a revolution. Nature's way of evolution is to make many small changes over incredibly large swaths of time. Nature explores many options to find the most effective way to survive and procreate—Darwin's survival of the fittest. Biotechnology has invented techniques that could allow us to take over from nature and direct our own human evolution, bringing a new era with vast implications. Guiding our evolution may be necessary to ride the wave and compete with the advances in artificial intelligence (AI). We may be driven to merge with our machines to survive. If we do so, will we be any longer human?

There is dramatic change in other areas. Progress in neuroscience may allow us to control brains and thought. There is potential for AI-catalyzed mental telepathy, perhaps the development of a single human shared mind. We are in the age of the Anthropocene when the activity of humans on Earth has a major effect on the environment, climate change being a prime example. Our technology is biting us.

Rockets guided by AI are becoming efficiently re-useable. We are heading off planet to the Moon and Mars and perhaps beyond our Solar System. To what effect and to what end?

ARTIFICIAL INTELLIGENCE

Artificial intelligence (AI) may be the dominant revolutionary technology. AI already permeates much of our current technology. We have AI that can:

- outcompete humans in the most sophisticated games: chess; Go; on-line games of *World of Warcraft* and *Diplomacy*
- invent new strategies, some beyond human comprehension
- write credibly in virtually every language
- write code

We also have robots and all sorts of sensors and detectors. These current capacities—AI, robots, sensors—will be put together.

We have robots—from single, multi-jointed arms to full humanoid bodies—that can

- perform precise motions rapidly and repetitively, from assembling autos to performing sophisticated surgery
- coordinate in teams to play soccer
- function independently or in collaboration with humans
- be guided by AI

Modern sensors perceive not just visual light but ranges throughout the electromagnetic spectrum that are beyond human ken:

- gamma rays
- X-rays

- ultraviolet
- infrared
- radar
- radio

We also have sensors for

- sounds of all frequencies
- aromas
- pressure
- temperature
- atomic particles
- molecules

The output of this wide range of sensors can be digitized and fed into AI programs.

The outcome of combining these current technologies will be AI-powered robots with situational awareness that can move about freely and witness the world in diverse ways, many of those ways beyond human capability. Already the *embodied AI* movement is well underway at Alphabet and myriad startups. These machines will synthesize and learn from sensor data, write their own evolving code, invent, strategize, think, make decisions, and act. Ever faster. Inhumanly fast.

AI thus brings the potential for technology that can develop itself with no human intervention or control. The stuff of science fiction is being made real.

CONSCIOUS MACHINES

The origin and nature of consciousness is a huge and controversial topic. Some think consciousness arose naturally. Others believe that consciousness is a special property beyond physics, chemistry, and biology. Humans have it; rocks don't. Animals have it; maybe plants do at a rudimentary level. There is also great disagreement on the question of whether non-biological entities—machines—can ever become conscious. Some argue that machines that could observe their environment and compare current input with their memory of past experience of the world, machines that could imagine the future, would effectively be conscious.

There is a distinct possibility that when current capabilities in AI, robots, and sensors are combined and refined to work together, conscious machines will emerge that supplant humans in ability and capacity. These machines will likely self-evolve, learn to improve their abilities at incomprehensible speeds, and, like a Ferrari in a foot race, leave humans in their dust.

URGENCY

All this—robots, AI, artificial human evolution, brain reading and writing, climate change, space exploration—is happening at one time. Any one of these developments could seem overwhelming to contemplate. People tend to focus on one of these issues, but they are interconnected. AI is a common thread woven into a host of dramatic advances in science and engineering.

The unprecedented rush of technology, the cresting of the wave, is not stopping, only accelerating. The crucial new factor is that change may be coming so rapidly that neither individuals nor societies can adapt. Even some people involved in the technology tend to underestimate the potential dangers and how rapidly they may evolve into crises and overwhelm us.

There will be huge challenges from technology ahead for human organizations, for economic systems, for democracy. These matters are urgent. The threat of these interwoven crises is more important than most topics that currently dominate the headlines and common human concerns: the national debt, social security, health care, pandemics, war, migration.

Technologists have defined the potential tipping point when machines become conscious as the *singularity*.[1] Even people who accept this possibility tend to think the singularity is far enough away not to affect them personally. Maybe a thousand years from now, at least a hundred. This is probably wrong.

Given all the revolutionary technologies currently in play and the ever-accelerating pace of technological development, the singularity could actually be imminent, arriving during the lifetime of most people alive today. Some even fear that current hidden activity in laboratories around the world could lead to a singularity within months. Certainly the next few years or decades will be critical. It took two decades to begin to seriously respond after the climate crisis was enunciated. If we delay as long to respond to other technological challenges, we could be in serious trouble. The future will not stop arriving just because we are struggling with it.

Leonardo da Vinci said, "I have been impressed with the urgency of doing." His words are apt now.

HOW DO WE RESPOND?

What can we do to try to stay on top? To ride the wave?

We cannot order technology to stop. No one is in charge of this accelerating technological wave. Rather, it is driven by thousands of individual choices made by researchers, business leaders, and governments with increments of knowledge building on all previous knowledge ever more rapidly. Google's AlphaFold solved the incomprehensibly complex protein-folding problem. Microsoft-supported ChatGPT sent shock waves throughout the worlds of business, politics, and education with its ability to write credible, inventive prose, poetry, and code. Current startlingly capable computer programs will rapidly evolve as Google, Microsoft, Amazon, Facebook (now Meta), and government-backed Chinese companies compete for dominance in AI. Analogous advances will occur in neuroscience and biotechnology.

My recommendation is to succumb neither to hype nor to panic. Rather, be aware. Develop an awareness mindset for all the technological changes that are happening around us.[2] Once you start noticing them, you will notice more. An awareness mindset will be relevant long after today's technological revolutions have faded into the past.

Understand the implications of our advancing technology. Both facts and reasonable speculations are important in framing possibilities and guiding our responses. What technology should be encouraged, what avoided? What are the potential impacts on society? What ethical issues arise? How do we ensure that AI, biotechnology, and neuroscience remain aligned with human values?

There must be policy decisions in all these areas. Aware individuals can make sound policy choices. What regulations should we avoid in order to maintain healthy progress? What regulations must we adopt to prevent disaster? Choose your representatives wisely. Vote!

LOOKING AHEAD

We humans are right now in the grip of rapid, self-advancing developments in our science and technology that will dominate our future. We are designing computers and software that are more capable than we are; we are peering into our own brains with the goal of emulating, simulating, and controlling them; we are learning to direct our own biological evolution. How are these developments in computer science, neuroscience, and biotech related? What happens now? Our burgeoning technology has tremendous implications for every person on the planet. If we are to have control over these advances, it is imperative that we understand them and pay close attention to their impact on our lives.

My aim for this book is to provide you with an overview, a primer, to facilitate your awareness and to give you a deeper understanding of the issues even as rapid change swirls around us.

2

EXPONENTIAL AND SUPER-EXPONENTIAL GROWTH
The Path to the Singularity

INTRODUCTION TO THE EXPONENTIAL

Chapter 1 introduced the notion of an accelerating wave of technological change that evolves ever more rapidly, knowledge building on knowledge. This understanding was reflected by Sir Arthur Conan Doyle, who wrote, "Knowledge begets knowledge as money bears interest."[1] The more you know, the more you can do. An alternate way of expressing this is to say that we are now under the sway of *exponential* advances of our technology that will dominate our evolution.

"Exponential" is a common colloquialism, but what do we mean by it? Exponential growth has a precise mathematical description. Exponential change happens when the rate of change of some quantity is proportional to the current amount of that quantity. Exponential change is equivalent to doubling in a fixed amount of time. Exponential change is illustrated in the power of compound interest with a fixed percentage return per year; for example, 8 percent interest doubles your money in about ten years. A Texas example is that the population of feral hogs is doubling every five years, disrupting farmers and their crops.

If couples have a set number of children more than two, then the number of people will grow exponentially. That can also lead to problems if the number of people outgrows resources. Things that are good for limiting the growth of population (war, disease, starvation) are bad sociologically.

In some areas—science, technology—the quantity we consider to be growing is the amount of knowledge. The rate at which our knowledge grows depends on the amount of knowledge we already possess. To the extent that the rate of change of our knowledge is strictly proportional to what we already know, that is the mathematical prescription for exponential growth of knowledge.

One oft-quoted example of exponential growth of knowledge and its application in technological prowess is Moore's law: the capacity of computer chips has doubled about every 18 months for 50 years. The exponential growth in power of computers has occurred as we have learned more and more about how to manipulate electrons in silicon chips. Similar advances have occurred in biology since the revelation of the double-strand spiral structure of deoxyribonucleic acid (DNA). These developments rested on past advances in understanding electricity and cells and on back into the past. The exponential growth of knowledge has been going on for a long time.

Not all technology has progressed at the same rate. We have made impressive progress in moving things from the invention of the wheel to the launch of rockets into space, but recent progress in transportation has not been as rapid as the development of silicon chips. The speed attained by machines went up by a factor of 1,000 in 50 years from the Wright brothers to the first satellite in orbit—a doubling time of about six years—but the speed of rockets did not substantially increase in the subsequent fifty years. If the speed had increased at the same rate as the capacity of silicon chips, we would now be flying faster than the speed of light! Steve Earle captured the lag in transportation technology in his song *Twenty-first Century Blues* when he lamented the lack of "startrekian teletransporters." Even with this delay, we can all sense the revolution coming as self-driving cars disrupt our transportation systems and more people launch into space.

I posit this is the situation in which we find ourselves: exponential growth of both the amount of our knowledge and of the rate at which our knowledge is accumulating. Some areas are faster, some slower, but overall our summed knowledge is growing approximately exponentially. That could be for good or bad.

LINEAR VERSUS EXPONENTIAL GROWTH

Despite common use of the term *exponential* to mean "very rapid," few people fully understand the nature and implications of exponential growth.

The problem is that most people think linearly. While everyone understands the vicissitudes of life with its ups and downs, by and large we anticipate that tomorrow is going to be much like yesterday. Change, if it happens, is slow and steady. There is, however, change, and then there is change. The difference is coupling and feedback. If the amount of some quantity merely accumulates with time, we say the resulting amount changes in proportion to the passage of time; that is, the change is linear in time. Alternatively, if the change begets more change, then the results can be phenomenally different.

To illustrate the difference, I would collect some pennies and bring the pile to class. I would then illustrate linear change by putting down a penny. A few seconds later, I would put down a second and after a third short interval, a third penny. That represents linear growth in the number of pennies. The number of pennies accumulated is exactly proportional to the elapsed time: three seconds, three pennies; one hundred seconds, one hundred pennies. Then I would illustrate a rather extreme example of exponential growth. As before, I would first put down a penny, but then in the next step I would double that and put down two. After a third step, I would put down double that, or four pennies, then eight, then sixteen. I would rapidly run out of pennies in my small collection. After five steps, I would have accumulated not five, but $1 + 2 + 4 + 8 + 16 = 31$ pennies. This exercise was an extreme example because the doubling time, the time it took me to fumble and stack a bunch of pennies, was short.

I challenged the students to estimate how long it would take using each of these methods to fill the classroom with pennies. For this exercise, one needs to estimate the volume of the room and the volume of a penny. The number of pennies you would need is the volume of the room divided by the volume of the penny. The answer for our modest size classroom was about a billion pennies. The question, then, was how long it would take to accumulate a billion pennies with each prescription. In the linear case, you imagine putting a penny down in the far corner of the room. Then after a second, you put down a second penny. You continue in this slow, stodgy way until you have filled the room. To get a billion pennies and fill the room, you need a billion steps, roughly a billion seconds, or about thirty years.

If you use the other prescription, you first put down a penny. A step later, you put down two. A step later, you put down four. You now have seven pennies, but they are all nestled in a tiny corner of the room and the process seems pretty insignificant. A rough estimate of the number of steps, and hence the time, to fill the room would be to raise the increment at each step—a factor of two—to a power equal to the number of steps. You want

2^n to be a billion.[2] The result is that you only need to double about thirty times to get to a billion. That means that if you double the number of pennies at each step, you only need thirty seconds to fill the room. Even if you only put a penny down once per year, a seemingly painfully slow process, you would fill the room in thirty years, the same result as for the linear process of putting a penny per second. The difference is dramatic!

Another way to illustrate the difference between these two processes is to picture the end game. How long does it take to fill the last half of the room? The linear process takes just as long to fill the second half, about fifteen years, as it did the first half. In the last year, you would add about 30 million pennies way up in the far corner opposite the corner where you started. By contrast, think of the doubling process. One step, one second, from the end, the room is half full. In that final step, you fill the whole last half of the room. To reprise, in the first second, you have 1 penny. In the next second, you add 2 and have 3. In the third second, you add 4 and have 7. In the fourth second, add 8, for a total of 15. By step 28, you add 1/8 billion. By the 29th step, you add 1/4 billion, and the room is about half full. In the last, 30th step, you add 1/2 billion, and the room is (over) full. In the first few steps, there were 1, then 3, then 7 pennies down in the far corner of the room. With the same process of doubling pennies each step, in the final three steps the pennies go from up to your knees, to up to your nose, to over your head. That is what makes exponential change so extraordinary. It's not the beginning but the end game that can really sneak up on you.

Picturing the piling up of pennies in a room is just a way to capture your imagination and make the process visceral. That is an artificial exercise. Here is another one that is closer to reality. Picture cells in a test tube. The cells divide—double in number—at some fixed time interval set by the biology of the cells. Suppose it takes a minute for a cell to mature and divide. You can start with only a few cells in the test tube. Then in the first interval, the number of cells doubles. You still only have a few cells in the test tube. If nothing interferes with the doubling process, then nothing seems to happen for quite a while. Then, after not much passage of time—tens of minutes—the test tube is 1/4 full, then a minute later 1/2 full, and in another minute it's overflowing with cellular glop. The finale seems like a dramatic change in behavior but the underlying process has not changed: each cell is biologically programmed to divide after a certain time.

DOUBLING TIMES

A linear process consists of adding a constant amount at each step, each interval of time. The addition does not depend on the total amount accumulated before the current step. An exponential process involves multiplying by a constant factor at each interval. This process depends on the amount you are multiplying so it depends on the amount already accumulated. More specifically, an exponential process is one in which the rate of change of the amount is strictly proportional to the current accumulated amount.[3] The more you have, the more rapidly you gain. With linear growth, the time to double gets longer and longer. For exponential growth, the doubling time is a constant. Conversely, if there is a fixed doubling time, the result will be exponential growth.

The exponential process derives its name because its mathematical description has the amount increasing as a constant raised to an exponent power with the exponent itself increasing in proportion to the time elapsed. In the penny room-filling example, we took the constant factor to be 2, representing a doubling, and raised that to the power n—the number of doubling time intervals. In this simple example, the number of doubling times increases in proportion to time. One could have asked about the time to increase by 1/3 or by a factor of 10, but doubling is a popular way of characterizing this type of growth. Dividing cells have a natural doubling time. Your funds in a bank that pays a specified interest rate have a doubling time that depends precisely on that interest rate. The money you make in a given year depends on the total funds you have accumulated and not the funds with which you began nor the amount you add each year. That is the power of compound interest.

Note that the examples of exponential growth given so far—pennies, cells, your bank account—have limits. These limits are not set by an underlying law—the time to double or the interest rate—but on other factors often related to the external environment. In the case of the pennies, one hopes you stop the experiment when the room is full, but in our imagination we could continue. For cells in a test tube, once they overflow the test tube, they no longer have access to the nutrient agent on which they fed. They will wither and die. For your bank account, things could be more complex. You could take funds out to pay bills. There could be a recession and the Federal Reserve could lower interest rates. The general notion is that exponential growth cannot last forever in a finite environment. We will return to that important point below and many times later in this book.

A recent example of exponential growth and its limitations was the spread of the coronavirus that caused COVID-19. In circumstances when a carrier of the disease can give it to more than one person, the disease will spread rapidly. If a single person gave the virus to the same number of people and the time to infect that greater number of people was constant, then the virus would spread exponentially. The rate at which new people were infected would be proportional to the number of people currently infected and spreading the disease. This is the mathematical description of exponential growth. In the early days in Wuhan, it was not that a single person gave the disease to precisely the same number of people at precisely the same rate, but this was approximately true on average. The result was that the number of people infected exploded at an exponential rate: first one hundred, then two hundred, then four hundred, then eight hundred. In those early days, the doubling time was only two or three days. It's an elementary exercise to ask in how many doubling times would every Chinese citizen, all 1.4 billion, have COVID-19. The answer for a billion people is the same as the billion pennies in a room: about 30 doubling steps, or about 60 to 90 days for doubling times of 2 or 3 days. I worked this out in the first week of March 2020. The answer was that if nothing changed, all Chinese would have been infected by about the end of April.

Of course, things did change. China enforced strict separation laws. Wuhan was locked down. These restrictions were attempts both to change the underlying doubling systematics by preventing the infected from infecting the same average number of people and to cap the test tube by disallowing people from moving to other parts of China. These were steps to bend the curve, to truncate the terrible power of exponential growth.

As we know, the same process then played out globally as the disease became a pandemic spread throughout the world. There were phases of exponential growth and then phases of desperate attempts to alter the natural dynamics of exponential growth at great economic cost.

There was another exponential process going on with COVID-19: the growth of knowledge about the disease. There was unprecedented global coordination and cooperation among normally competing research groups. Each group was influenced by the research of many other groups. Inasmuch as the rate of understanding of COVID-19 increased in proportion to the amount of understanding already gained, our understanding grew exponentially, and solutions came that much faster. This led rapidly to vaccines and may yet yield a cure. If nationalistic competition dominates, progress will be much slower.

DOWN ON YOUR KNEE

It is worthwhile to explore some other aspects of exponential growth. Figure 2.1 illustrates the basic nature of exponential growth. This diagram plots the amount of some quantity on the vertical axis as a function of the number of steps on the horizontal axis. The curve was generated by requiring that the rate of change of the amount at a given time be proportional to the amount extant at that time. The number of steps in this particular diagram is related to the number of doubling times. You can also think of the horizontal axis as the passage of time. The diagram provides a visual image of the rapid growth that can accompany an exponential process.

The diagram in figure 2.1 also illustrates a property of exponential growth that is often invoked but frequently misunderstood. When exponential growth is just beginning, figure 2.1 shows us that it gives the deceptive

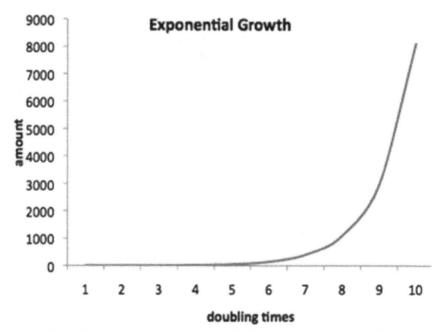

Figure 2.1.   The amount of some quantity that grows exponentially is plotted as a function of the number of steps.

impression that nothing is happening when actually change is occurring. In any short interval of time, exponential growth mathematically looks like linear growth at a constant rate giving the impression that change, if it is occurring, is happening linearly—that is, accumulating at a constant rate with the amount proportional to the time elapsed. This impression is also deceptive. The construction of figure 2.1 thus suggests that nothing much is happening for the first four or five steps, and then the curve surges upward. The diagram seems to indicate that there is a special time when the growth in the amount of the quantity of interest explodes. This apparent change in behavior is often characterized as the knee of the curve. Perhaps you have to picture a person lying on their stomach so that their calf bends upward at the knee in relation to the thigh. This sort of diagram is also referred to as a hockey stick. In figure 2.1, the knee seems to be at about step 8. Something special apparently happens there.

Yes and no. If the growth is exponential, there is no intrinsically meaningful knee. Appendix A shows that if the scales of the axes are expanded or the data plotted in a different way, the knee can be moved around and made to appear to happen after a different number of steps or even disappear. It is true that for exponential growth the increase in the amount is greater in later steps than in earlier steps, but the place where the knee appears shifts depending on how one chooses the axes. The location of the knee is an artifact of how one elects to make the plot. You can put the knee anywhere by adjusting the scale of the vertical axis. A portion of this curve might take on special meaning when compared to external factors. For example, cells can double in a test tube for some time, but there comes a time when the doubling overfills the test tube. That, however, has nothing to do with an intrinsic knee in the curve.

I knew this but still had to learn it in class. The first time I taught the course in the fall of 2013, I was following Ray Kurzweil's excellent illustrations of exponential growth in chapter 2 of *The Singularity Is Near*[4] and expounding on the nature of the knee. A bright young man who habitually sat in the back row, Vincent Lau, held up his hand and pointed out that an exponential law was "scale-free," meaning there was no special place in the curve.[5] Kurzweil makes the point indirectly; I was garbling it. That comment from Mr. Lau changed the way I thought about and taught the material and helped shape my perspectives presented in this chapter. A basic rule of teaching is listen to your students.

WHY EXPONENTIAL?

I establish this somewhat technical background to give us some perspective on what is happening in the technological progress of humanity. No law says technology has to progress exponentially, but there is a sense that the more one knows, the faster one can learn. This rule of thumb applies to individuals but perhaps more appropriately to our collective civilization. No one of us knows all there is to know about technology ranging from microbiology to silicon chips, but collectively our body of knowledge is large and growing. The rate at which collective knowledge grows assuredly depends on the collective amount of knowledge. That implies positive feedback. The more we know, the more we can do. The more we can do, the faster we can learn. Thus it is plausible that the rate of change of knowledge is proportional to our total current reservoir of knowledge. The latter statement—the rate of change of knowledge is proportional to the amount of knowledge—is a verbal statement of the mathematical prescription for the exponential growth of knowledge. Inasmuch as the statement is approximately true on average for the human reservoir of knowledge, then the amount represented in figure 2.1 is the amount of collective technical knowledge. This is growing exponentially with profound implications for our society.

Note from figure 2.1 that this exponential growth could have been going on for a very long time without much seeming to happen. Things did happen, however: tools, control of fire, the wheel, smelting, agriculture, the scientific method, industry, airplanes, DNA, the digital revolution. There has been a steady march of technological development. It starts with the piling up of a few insignificant pennies in the corner and leads to rapidly filling the room. One important question is whether there is a technological room that threatens to be filled.

Erik Brynjolfsson and Andrew McAfee give an interesting, related perspective in their insightful and impactful book *The Second Machine Age*.[6] Brynjolfsson and McAfee start off making their case by looking at the human social development index in a plot like figure 2.1. In this particular case, the human social development index is a specific quantity that represents what I generically called an amount in figure 2.1. In the graph presented by Brynjolfsson and McAfee, it looks as if the development of the steam engine "bent the curve of human history." They do not plot that graph in log space (see appendix) where each step on the vertical axis represents multiplying by a given factor rather than adding a given amount. It is the same data, just plotted in a different way. I do not have the data so I cannot be sure, but I am confident the data would look much less dramatic on a log plot. Perhaps there is a kink with the industrial age, but maybe not. The same argument goes for the concept of the digital age, the second machine age. If the dominant trend is long-term, approximately exponential growth, there is no special epoch. Our total knowledge and the increment of knowledge in a human lifetime are both growing approximately exponentially, but there is no intrinsic knee, industrial or digital.

The essential relation that knowledge begets knowledge has been with us for a very long time. In that sense, our current times are not special. If there is no knee, though, why do we feel that progress is accelerating? Because it is! The underlying intrinsic exponential growth law is always about the same. There is no special knee. To understand the impact, we have to bring in external factors such as the size of the room.

I have pondered how to express why we sense a knee in scale-free, exponential growth. Returning to the metaphor of doubling pennies, if you lived in an epoch corresponding to the fourth step—perhaps corresponding to the industrial revolution—you would see eight pennies added to the pile. You would be justified in thinking, "Whoa! There's a lot going on in my life." In the next step, however, sixteen pennies would appear, and your epoch would seem less special overall.

COMPLEXITY

Another perspective on this issue removes people from a central role. One might also consider the amount in the figure 2.1 to represent a general notion of complexity. To the best of our knowledge, the Universe began in a rather simple state—nearly homogeneous force fields and radiation—and has grown more complex with time. The force fields coalesced into quarks and the quarks into more familiar particles: protons, neutrons, and electrons. The latter formed atoms and the atoms agglomerated into molecules. The molecules formed cells and the cells became multi-celled creatures. The multi-celled creatures developed into plants and animals. Some of the animals became *Homo sapiens*, and *Homo sapiens* catalyzed the growth of technological knowledge of how to manipulate the natural world, making it even more complex. In this sense, figure 2.1 could be plotting the amount of complexity in the Universe.

There is literature that attempts to quantify the exponential growth of complexity, to put a number to the doubling time, but complexity, especially biological complexity, is notoriously uncertain and difficult to quantify. Certain aspects of technological knowledge are more amenable to quantification. The most famous, as I've noted, is Moore's law that the capacity of silicon chips has a doubling time of about two years. Take thirty of those steps, and the progress is immense. Using scientific references and citations as a metric suggests that global scientific output doubles every nine years.[7] Trying to devise a measure of the collective technological knowledge spanning all topics is more difficult. We will return to this later in the chapter. For now, it suffices to recognize that there is positive feedback such that knowledge begets knowledge, and the result is something like the exponential growth of knowledge.

The upshot is that the complexity of life on this planet, and specifically the amount of technological knowledge, is growing not linearly but approximately exponentially. If figure 2.1 is applicable, then there is no knee in the curve. Ray Kurzweil emphasizes the positive aspects of this exponential growth of knowledge and imagines few bounds.

THE HUMAN IMPACT

I do not know the beginning point for knowledge, whether it started with assembling carbon atoms in cold interstellar molecular clouds or the development of RNA, or fire, or stone tools, or the wheel, or agriculture, but for a long time something roughly like this has held with knowledge begetting knowledge. Was the industrial revolution truly a kink in the curve? Did that not depend, back in history, on the development of fire and the wheel, knowledge building on knowledge? Similar questions arise for the digital age. The digital revolution is a big deal, but did it not build on the development of modern science in the Renaissance, Thompson and the electron, quantum mechanics? Was the industrial revolution just a smooth step in what had been growing for millennia, and is the digital revolution more of the same? Did the invention of the airplane and the development of the atomic bomb count for something in the time between the steam engine and the cell phone?

If there is no knee, no bending of the curve, no second half of the room, then what is going on? One way to address this is to ask about the technological change in a given interval of time, a millennium, a century, a human lifetime. If the amount of knowledge is growing exponentially, then the doubling time is constant but the amount of change in a given doubling time also grows exponentially. You can see that in figure 2.1. Between step 6 and step 7, the amount grows from about 150 to about 400, an increase of 250; but from step 7 to step 8, one more doubling time, the amount grows from 400 to 1,100, an increase of about 700.

The reaction to these rapid developments needs to be expressed in some other human-order terms. From the expression for exponential growth, the amount of change in a specific amount of time is proportional to the current amount of knowledge and the interval of time. When the total amount of knowledge was small, when time was early, long ago, the change in the collective amount of human knowledge was small in a given time interval. There is a natural time interval in human terms, say a generation, or a human lifetime. If knowledge continues to increase by a roughly unbroken exponential law with the same original knowledge base (whatever it is) and the same doubling time (so no knee, etc.), there will still come a time when the amount of change in my lifetime is huge. That is what is so dramatic and not a bend in the growth rate. It is not that there is a sudden change in the rate of exponential growth of knowledge, just a damn lot of it in my lifetime!

What makes the change in collective knowledge large in the lifetime of a given human? In some lifetimes, humans invented tools. In others, there were steam engines, cars, airplanes, now computers and our digital world. All these are large effects in a human lifetime. The point is that even with (approximately) constant doubling time and an appropriate time interval—four score and seven years—the increment in knowledge in a human lifetime increases as the collective knowledge increases, and the collective knowledge increases exponentially even with no knee.

The invention of the wheel was a major development but it represented a rather small increment compared to all of the contemporary human knowledge of how to survive. Fast-forward to the invention of the silicon chip and all the other technological developments of the current era, and the impact is exponentially greater. We are now adding a huge increment to human knowledge in one human lifetime, and the amount of that increment is itself growing exponentially.

There may thus come a tipping point even if there is no knee in the underlying systematic growth of knowledge. When technological advances in a given lifetime were relatively modest—a thousand years ago, a hundred years ago—a particular human might struggle with the change but adapt. If the technological change comes too fast, more rapidly than a person can adapt, a single human—and society—can be overwhelmed.

In my opinion, the authors of *The Second Machine Age* have not quite expressed this exponential power appropriately. Kurzweil does. That does not diminish many of the arguments in *The Second Machine Age* about the economic impact of the digital revolution that we will discuss in later chapters, but I do not think the authors have fully expressed the context. An example is that they repeatedly say such and such has been predicted but "it hasn't happened yet." This refrain is common with people writing about the current state and future of technology. The stress is often on "this hasn't happened" rather than the "yet." This emphasis to me shows a certain lack of appreciation for the power of the exponential growth of knowledge. No, we do not yet have fully conscious androids, but what is the timescale of "yet"?

In *The Singularity Is Near*, Ray Kurzweil predicted a technological singularity[8] around 2040–2050. Kurzweil's notion of the singularity is that it is the epoch when machines attain the complexity of the human brain. After the singularity, the capacity of machines could rapidly outstrip the capacity of humans. Self-evolving machines could share knowledge at essentially the speed of light and not the electrochemical speeds of the human brain or the slow rate of book learning. Such machines could easily outperform and supplant us. With the recent rapid development of artificial intelligence, many think Kurzweil was pessimistic and that the singularity is imminent.

I do not take Kurzweil's timescales literally, but the amount of the increment of knowledge we can collectively gain in a lifetime is getting very large. The timescale for Kurzweil's singularity or something like it could be measured in decades or less, within the lifetime of today's college students, within the lives of my grandchildren. To say it has not happened yet seems to miss a significant point. To return to the analogy of the pennies in the room, in a sense, Brynjolfsson and McAfee are emphasizing the 29th second. Kurzweil is looking forward to the 31st.[9]

BEYOND EXPONENTIAL

Appreciating the power and implication of exponential growth is essential, but our condition may be even more extreme. Our knowledge and technological capacity may be growing *super-exponentially*. Whereas exponential growth is characterized by a constant doubling time, super-exponential change is characterized by a doubling time that decreases with time. A decrease in the doubling time can happen if there is extra feedback, if our accumulated knowledge allows us to shorten the time to double our collective knowledge. Just as exponential growth can look linear over a small interval of time, super-exponential growth can look exponential over a small interval of time. In either case, one can be severely misled if the rate of change is mistaken for something more benign.

In *The Singularity Is Near*, Ray Kurzweil plots the time when significant events occurred in the past as a function of how long ago they occurred. The events he selects range from the origin of life several billion years ago to the development of the personal computer only decades ago. The result is nearly a straight descending line in this plot. As we get to the recent past, the time between significant events gets short—mere decades between the invention of the computer and the iPhone. This behavior is super-exponential. The time to the next significant event heads to zero more rapidly than exponential. The time between significant events is not exactly a measure of the amount of knowledge, but the result is sobering. An extrapolation of Kurzweil's plot shows that the time between critical events goes to zero—a manifestation of a mathematical singularity—at a finite time. Formally, an infinite number of significant events happen in no time at all. From Kurzweil's plot, that time is the near future.

The famous thinker and futurist Buckminster Fuller attempted to characterize the growth of knowledge in his 1982 book *Critical Path*.[10] He posited that knowledge was doubling about every one hundred years in 1900. He estimated that the rate of growth of knowledge was growing even more rapidly than exponential, that the doubling time had shrunk to twenty-five years by the end of World War II, and that it was as short as a year by the time he wrote his book in 1982.

In 2006, IBM produced a white paper titled "The Toxic Terabyte"[11] positing that by 2020 knowledge would grow with a doubling time of only twelve hours! In a blog posted on October 10, 2017,[12] Marc Rosenberg picked up this theme and produced a diagram capturing the predictions of Fuller and IBM that represented the dramatic, super-exponential shortening of the doubling time of the growth of knowledge. His bold diagram showed forty-year intervals on the horizontal axis (1900, 1945, 1982, and 2020). It had no labels on the vertical axis, but the quantity plotted was presumably something like the logarithm of the inverse of the doubling time, a quantity that was growing rapidly by these estimates.

The IBM report did not concern knowledge per se but the rate of accumulation of data. It was a cri de coeur that the amount of data being created and stored was growing at an unmanageable rate. "Toxic Terabyte" noted that most of the data being stored—chains of irrelevant office chat—was worth little. It also argued that it was becoming unsustainably expensive to store all the data given the energy costs of Cloud storage and the potential impact on global warming and climate change. The report was also a sales pitch for the IBM solution to these issues, a process of information life-cycle management (ILM) for winnowing the valuable information at the beginning of the process, not data-dumping every byte but invoking the strategic values of the customer from the start. Efficient information retrieval was a significant part of the concept.

Over historical times, an underlying exponential growth sets the stage. If there are turning points at the industrial revolution and with the digital revolution as outlined by *The Second Machine Age*, they depend on super-exponential behavior. It is important to understand, however, that data alone is not knowledge. As the "Toxic Terabyte" report emphasizes, masses of inaccessible, garbled bits are not knowledge. In addition, there are arguments that knowledge is not permanent in the sense that it becomes outmoded or superseded. Marc Rosenberg makes a powerful case for this.

> The message is clear: When knowledge is doubling exponentially, yet the useful lifespan of that knowledge is decreasing significantly, the result is a *knowledge tsunami*—a seemingly unstoppable wave of new information pushing you forward, combined with an extremely forceful undertow of information that used to be valuable but is now just knowledge clutter, pulling you back.

On top of that, it is also vital to comprehend that what we fundamentally seek is understanding, which is different from both data and knowledge. Some argue that we may even have regressed in that regard. That is not my perspective. I think understanding continues to advance. I hesitate to hazard a guess as to how rapidly.

In a contemporary exercise, David Roodman[13,14] (see appendix) models the gross world product (GWP) from 10,000 BCE to the present. Roodman finds that the industrial revolution did make a slight kink in the overall rising trend. He also finds that the system of equations has a mathematical singularity predicting that the GWP goes to infinity at a critical time. Allowing for uncertainties in the data and in the procedure he uses for fitting it, Roodman finds that this critical time when the world economy blows up is most probably 2047. While not to be taken literally, this date is our near future and is remarkably close to Kurzweil's estimates of the epoch of a technological singularity and of the time when the interval between significant events goes to zero. Kurzweil pictures the singularity as a time when machines gain all the conscious intellectual power of humans and then race ahead. Roodman's infinite GWP suggests an economy racing so fast that it blows a gasket and self-destructs. These may be different versions of the same society-destroying growth of technology. The critical epoch predicted by Kurzweil and Roodman—2040 to 2050—is closer than predictions of disastrous global warming and climate change even on our current crazy course. Many of our contemporary concerns may prove irrelevant in the face of the rush of our technology.

The presence of a singularity means that something is missing in the mathematics as it attempts to represent the complexities of the real world. For example, Einstein predicted a gravitational singularity in the middle of every black hole where all matter would be shredded and even time and space would come to a halt. We now know that Einstein was missing something. His theory contains no aspects of the quantum world: that matter is waves and that changes happen not smoothly but in jumps. We have not yet constructed a theory that properly captures both gravity and observed quantum properties of Nature—a theory of *quantum gravity*—but we know such a theory is needed. The same may be true for the theory of GWP.

Roodman explores the notion that his theory is missing something. He finds that his mathematical model does not have the capacity to properly allow for a large perturbation like the industrial revolution and perhaps for the economic disaster of a global pandemic. He finds that if technology is not formally treated as a constant input to the economy as his original theory did but is subject to the vagaries of labor and capital, his system of equations is unstable. The altered equations predict an explosion to infinite GWP or a collapse to zero at a finite time in the future. The implication is that while reasonable, data-based models of the global economy are consistent with roughly super-exponential, power-law growth, the prediction of a critical time in the near future, while strongly suggested, cannot be taken literally. It may also not matter very much whether GWP explodes or collapses in terms of the negative effects on society.

An advantage of a book is that one can collect a lot of information and reflect on it. A disadvantage is that the content then becomes frozen. Teaching my course heightened my awareness, and I began to see evidence of the exponential growth of technology all around me—far from a static situation. One way I tried to share this perspective was to ask the students in my class to send me, once per week, evidence of this exponential growth in the form of a quote or link. They besieged me. I hope they keep that perspective, that awareness, throughout their lives because it will be their lives that are profoundly affected. I guarantee this: If you look, you will see evidence of the exponential growth of technology all around you at whatever time you happen to read this book even if my specific examples have become dated.

TO THE SINGULARITY AND BEYOND

The digital revolution is at the root of much of our current rapid technological change. Critical areas growing exponentially are AI, nanotechnology, neuroscience, genomics, and close connections among them. We are close to the era of designer babies. The whole recently invented biological technique of CRISPR came to the public's attention only in 2016. A more recent example is ChatGPT—an artificial intelligence product that can have intelligent conversations in plain English—which exploded onto the scene in November 2022. We must promote the positive aspects of our technological advances but avoid disastrous consequences that will be coming at us faster than we can react. The challenge of rapid exponential advances calls for extreme attention, foresight, and planning.

Kurzweil espouses an extremely utopian view of our technological future. He foresees that our exponentially growing intelligence will expand to fill this Universe and maybe universes beyond. Peter Thiel, a cofounder of PayPal, believes that machines will complement humans but not take over the Universe. Others lean to a more dystopian view, believing that technological advances will first eliminate jobs and then *Homo sapiens* ourselves. Bill Joy—founder of Sun Microsystems and writer of Java—was one of the first to take the rather extreme dystopian view that our technology could be a threat to humanity in a famous article he published in *Wired* magazine,[15] while Nick Bostrom founded a vibrant enterprise at Oxford, the Future of Humanity Institute,[16] that tended to the dystopian view that technological threats are very real. My astrophysical colleague Lord Martin Rees of Cambridge University has also adopted a more dystopian view and written several books on the topic. Along with Huw Price, the Bertrand Russell Professor of Philosophy at Cambridge, and Jaan Tallinn, one of the founders of Skype, Lord Rees founded the Centre for the Study of Existential Risk[17] at Cambridge. Another astrophysical colleague, Max Tegmark, cofounded the Future of Life Institute at MIT[18] that is dedicated to steering technology toward benefiting life and away from extreme, large-scale risks. Among other prominent voices is that of Elon Musk, who calls for caution despite his embrace of optimistic technology with Tesla and SpaceX.

All these people, whether utopian or dystopian, agree that the extreme case of self-replicating, superintelligent machines loose in the natural environment would be exceedingly dangerous, yet we are moving in that direction.

Machines will evolve but so will we. We will be able to tinker with our genes, our bodies, our brains. Perhaps we'll become superhuman and compete with superintelligent machines. Perhaps we will merge with our machines.

There are means to ride this tidal wave of technological progress so that we can maintain control of our technology. We may need to relinquish certain research avenues. These include code that writes itself, artificial life that can feed and grow in the natural environment, and growing brains *in vivo* or *in silico*. There are operating principles that have had success in minimizing damage to complex human enterprises. Such techniques may apply to the control of AI, genetics, and other rampant technology.

Riding this exponential wave brings immense challenges. We must balance the power of the growth of our technological expertise with appropriate caution to attendant perils. Solutions to this dilemma are not clear. Halting the progress of technology would be very difficult even if we wanted to. At each step there are deep moral and ethical issues. What is clear is that we must have relevant conversations to illuminate the issues. I am heartened by the impression that there is much more public discussion of these issues now than there was when I began writing this book.

A fundamental point is that this exponential or super-exponential growth may not last forever. The test tube might fill up. Various self-generated or external existential risks could interrupt the pattern: a pandemic, nuclear war, a population explosion or collapse, climate change, an asteroid. Another profound question is whether the growth of our knowledge will itself change us completely. Will we no longer be us?

3

INTELLIGENCE
What's Artificial About It?

AI EVERYWHERE

Albert Einstein said, "The true sign of intelligence is not knowledge, but imagination." Whatever AI is, we can look around and see that it is growing exponentially, knowledge building on knowledge, and that it is up to our ankles if not our knees. My favorite personal example harks back to a meeting hosted by the Kavli Society in 2013. The meeting brought together luminaries from all branches of US astronomy in Santa Monica to consider the future of ground-based astronomy. At one juncture we were debating how to staff and fund the large, very expensive telescopes for which there would be a need in the future. I tried to point out that on the timescale they were considering, decades, the assistants who now actually operate telescopes for astronomers would probably not be necessary because some form of thinking machine, a computer-based AI, would replace them. The notion fell completely flat. No one was thinking about the immense role computers had come to play in astronomy in the last few decades and trying to extrapolate exponentially into the future. I was thus amused in 2020 to see an announcement that the Schmidt Futures Foundation[1] funded a study by the National Academy of Sciences to investigate an AI system for Subaru, a giant Japanese telescope located on Mauna Kea on the big island of Hawaii.

A limited form of AI is all around us now in this digital age. Current AI applications, known as *weak AI*, cannot rival humans in terms of broad cognitive ability. Rather, they can address specific problems often beyond human capacity. Beckoning, or threatening, in the future is *strong AI* or *artificial general intelligence* (AGI), otherwise known as *full AI*, or *artificial sentient intelligence* (ASI), in which machines can accomplish any human mental task, perhaps even becoming conscious. We will address this topic in chapter 8.

Applications of weak AI range from Siri on our mobile phones to self-driving cars and trucks to facial recognition. AI is employed in video games; chatbots; searches for hate speech, terrorist threats, and cyber-attacks; detection of credit card fraud; and ubiquitous recommendations of what to buy or watch. AI is used to improve the operational efficiency of companies, to enable predictive maintenance and schedule repairs of equipment, and to optimize sales and marketing.[2] Elon Musk employs AI to steer Teslas and to land the boosters of SpaceX rockets for reuse. The growth of applications of AI is likely to bring huge changes to healthcare, transportation, manufacturing, finance, and national security and thus to alter much of the structure of society.

The seeds of AI were planted decades ago, but AI required immense computer power to flourish. Now the field is nurtured and emphasized not just by brilliant individuals but by giant companies like Google, Microsoft, Meta, and Amazon and even nations like the United States and China.[3] The Obama White House produced a report on the impact of AI[4], and in 2019 there was a presidential executive order on maintaining American leadership in artificial intelligence.[5] Stanford University has initiated what it calls a One Hundred Year Study on Artificial Intelligence (AI100) to understand the prospects for AI.[6] Andrew Yang made possible AI disruptions of the economy a central focus of his 2020 presidential bid.

To render the topic digestible, we can consider the roots, assess some aspects of the current state of affairs, and then look to the future of AI in subsequent chapters.

HOPFIELD TO HINTON

Simple rules can lead to complex, organized behavior. This concept is called the *principle of emergence*. Chapter 9 will outline the powerful role of this principle in the context of biological evolution. The same principle can be manifest in the digital realm as evolution *in silico*. *Genetic algorithms* do so by reproducing the basic elements of natural selection. Genetic algorithms define a population of entities with certain properties and the resources that can be consumed by that population. The algorithm combines properties from individual components of the population—the analogy of reproduction—then selects the individuals in the population that are most fit and allows them to reproduce. Mutations are induced, and the process is repeated. The result can produce the optimum solution to a problem.

The principles of emergence also apply to computer solutions employed in contemporary variations of AI. Physicist John Hopfield put key aspects in place in the early 1980s. A Hopfield network roughly modeled the behavior of the human brain. In the brain, neurons will store an electrical charge until a threshold is reached. At that time, the neuron will fire, sending a small electrical signal into an axon that may connect to many other neurons. That sounds simple, but if you have a hundred million neurons, the resulting connections can be hugely complex and capable. Hopfield designed a mathematical neuron that could store and fire in a digital network. The result was generally labeled a *neural network*. Hopfield also devised ways in which a signal could be injected into an input layer of such neurons, propagated through layers of interconnected computerized neurons, and then emerge as an output signal. Comparing the output to the input enabled small, guided changes in the strength of connections among the neurons so that the output more closely resembled the input. The power of the Hopfield network was not in the individual neurons but in emergent patterns of connections among the array of neurons.

A Hopfield network needs to be trained. You must know the answer in order to adjust the network connections properly to reproduce the answer. If you want your network to recognize a picture of a cat and distinguish that from a picture of a dog or any other animal, you have to show it a vast number of pictures of animals and tell it which are of cats. The idea is that once the network has been trained to recognize a cat, then you can unleash it on a fresh set of pictures it has never seen before and it should, with some reliability, select the ones of cats. In practice this process takes a great deal of computer power. It did not become practical until microprocessors got fast and cheap enough. The result is what is now known as *deep learning*—Hopfield on steroids—with ample neurons, layers, and links to get the job done.

Geoffrey Hinton[7] working in Toronto in the 1990s and 2000s fostered the transition to contemporary deep learning. Hinton and collaborators developed a practical and efficient *back propagation* algorithm that was key to telling a neural network how to adjust connections in a systematic way to obtain a better representation of the input. Hinton has been broadly influential in the AI community and with his colleagues Yann LeCun and Yoshua Bengio won the 2018 Turing Award. They are referred to in some quarters as the Godfathers of Deep Learning. Hinton worked part time for Google for a decade. He did not sign the public letter of March 2023 calling for a hiatus on the release of new large language model AI (see below), but shortly thereafter he resigned in order to speak more freely about the potential dangers of AI that he, and many others, perceived.

Some neural networks are fully connected, meaning that every neuron in a given layer is connected to every neuron in the next layer. The weight of the connections represents the knowledge of the input. Other neural networks are more stripped down in a manner closer to the operation of the brain so that some neurons connect only to a few neurons in the next layer. These networks break the input—for instance, an image of a face—into a hierarchy of data and then assemble the complex whole from more basic information. For example, the face might first be represented only by the distance between the eyes and the distance from the eyes to the chin with details added only by subsequent processing. Some neurons would register the left eye, others the chin, and the ensemble of neurons would give overlapping information to capture the whole face. These neural networks are called *convolutional neural networks* (CNNs), a technical term arising from the fact that they convolve or combine the input data over a given layer before passing it to the next layer. CNNs require relatively little instruction with prior knowledge compared to fully connected neural networks and hence are more efficient in that regard. CNNs have been employed in image recognition and classification, shopping recommendation systems, processing spoken or written language, and analysis of time series—for instance, the stock market.

The result of the efforts of Hinton and others was a flowering of this type of AI with applications to computer vision (especially facial recognition), language translation, and many other areas. Computer vision has many applications, a notable one being self-driving cars. The full flowering of this technology has yet to happen, but in 2016 a self-driving truck delivered cases of Budweiser in Colorado.[8] Another application is scanning medical images. Trained AI can recognize tumors with an accuracy meeting or exceeding that of trained doctors.[9] Other AI systems can predict human behavior better than people can: who will drop out of an online course; which crowd-funded project will succeed; which customer will be a repeat buyer and why. Other AI systems are attempting to diagnose mental illness.

Ordinary computer programs are very good at rapidly searching and employing *structured data*, data that can be put into rows and columns in tables. Think Excel. Examples are names, birth dates, social security numbers, credit card numbers, items purchased, and websites clicked on. One generally does not need an AI algorithm to access and manipulate such data. Deep learning comes to the fore when dealing with *unstructured data* that could be an amalgam of word-processed documents, email, images, web pages, video, audio, and social media postings. Deep learning can search for and identify elements that are linked within that morass, for example, finding the Twitter account of a person referring to a particular meme in an email. Deep learning can identify patterns in data that humans do not know exist and perhaps are unable to perceive. Those patterns can then be used for scientific discovery, stock picking, or efficient marketing.

Training a network of this kind takes an immense amount of data. Big companies—Google, Amazon, Meta, Apple—recognized this early on and deliberately wrote their software so that users provided the data. Tell us who your friends are. Identify the people in this photo. The goal was to connect advertisers to people who had a proclivity to buy certain widgets the advertisers were selling. To achieve that goal, the big technology companies made a significant investment in cloud computing to crunch the data. The *cloud* is a large array of server farms with racks of computers packed into warehouse-size spaces and the software technology that allows efficient distributed use of that capacity. The tech companies use their own cloud facilities but also lease them out to smaller users. The technical developments of deep learning and the availability of the cloud to exploit them efficiently came together in about 2012 to allow an explosion of AI applications. Generic AI codes were written and distributed freely so that individuals and businesses could try their hand. These AI techniques are common throughout science and certainly in astronomy where they are used to detect patterns in data. The National Cloud Computing Project[10] proposes to legislatively create and fund a national research cloud for AI that would give academic scientists access to the billion-dollar cloud data centers and gigantic databases now operated by big tech companies.

FACIAL RECOGNITION

Facial recognition has become a controversial area of AI because it intrudes so deeply into privacy issues. In its early applications, it also had awkward flaws. The ability of an AI facial recognition system is only as good as the data on which it is trained before being released into the wild. The original training sets comprised predominantly White people, and the trained system struggled to recognize Black faces. That means police departments throughout the world today may be using potentially flawed AI to identify criminals.

In 2008, MIT created a database called 80 Million Tiny Images, an extensive collection of photos with labels describing the content. The images were tiny (relatively) so they could be scanned with modest computer power. The idea was to use this database to train neural networks to associate the descriptive labels with the content of the photos. Only in 2020 did people read the descriptions associated with some of the photos in the original web sources. Unfortunately, too many—ten thousand or more—were racist, sexist, or derogatory. MIT withdrew the database.[11]

In January 2020, the metropolitan police in London installed a live facial recognition system provided by NEC, a Japanese biometric firm. Special marked cameras are located in sites thought to be prone to crime. Rather than comparing photos to those in a database of mug shots, this system photographs everyone who walks by, criminal or not, and compares the photos to a list of people suspected of crime or terrorism. People on the watch list are identified in real time and, in principle, promptly arrested. Some cities in the United States have banned the use of such technology. China has made massive use of facial recognition to monitor everyday citizens going about their business. India has declared its intent to build the world's largest facial recognition system.

A recent entrant into the facial recognition arena is a company called Clearview AI. Clearview has a database of several billion images collected in violation of terms of agreement from Facebook, YouTube, and millions of other websites. The facial recognition app produced by Clearview allows a user to take a picture of a person, upload it, and connect that photo to virtually any public photo of that person on the internet along with any available information about the identity of that person, including their name, address, and a trove of other information available on the web. This capacity exceeded anything done by the big tech companies, at least anything released to the public. Several police departments have subscribed to Clearview's services despite the manifest privacy concerns. Such a tool could be used to identify every person at a protest including their name, where they lived, and all their friends. Efforts are underway by various research groups to defeat this technology, and Clearview in particular, by *poisoning* facial images with subtle distortions designed to prevent facial matching. One can sense here an opportunity for yet another round of whack-a-mole as others seek to defeat the defeaters.

Another entry in the facial recognition enterprise is PimEyes.[12] Two Polish hackers developed PimEyes around 2017 using neural network techniques to map and measure key facial features. Unlike Clearview, which scraped photos from social media in violation of terms of agreement, PimEyes is based on photos from the internet: news articles, photos of concerts and protests one may not know were taken, even pornography sites. The results appear to be quick, accurate, and thorough, identifying people even though their face might be partly averted or covered with sunglasses or a mask. The terms of agreement one must click to use PimEyes require users to search only for photos of themselves, but these terms are easily and often breached. PimEyes is available to the general public for a modest monthly fee. The price to remove an embarrassing photo is considerably higher. Privacy in a crowd is a vanishing luxury.

DEEPFAKES

AI has been applied to the opposite problem. Rather than identifying objects in a photo, AI creates a photo from scratch. The result is the production of *deepfakes*: images that look very real, even images of real people, that nevertheless are totally fabricated. This technique can be applied to still images and can be used to create deepfake videos.

In one instance, a video was recorded of a person dancing. The fundamental motions of the dancer were reduced to the motions of a stick figure in a technique analogous to motion capture when an actor wears sensors and performs in front of a green screen. Then a video was made of a second person moving about but not dancing, and a stick figure reproduction was made of them. The stick figure motions of the dancer were then transferred by a deep neural network first to the stick figure motions of the second person and then to the full representation of the second person. The result was a video of the second person dancing in a way that reproduced the motion of the original dancer, motions that the second person never made.[13] This technique has also been used to do motion tracking of animals in the lab or in the wild without having to attach markers to the animals.[14]

There are benign applications of these techniques like adding a fully-realized character, perhaps a long-deceased actor, to a film, but there are more malignant cases. A famous one had President Barack Obama speaking words he never said.[15] In other cases, the heads of famous people were superposed on actors in pornographic movies. The term *deepfake* came from the username of the person who first posted these smut swaps on Reddit.

One can discern here a threat to the free flow of information and even to democracy itself if the power of AI is used to generate misinformation and fake news. The threat can manifest in indirect ways. Merely suggesting a true video is a deepfake can sow confusion and uncertainty. There will be more direct assaults. During the 2020 election season in the United States, the Russian Internet Research Agency employed a fake website with deepfake photos of editors to add verisimilitude. Introduction of deepfakes into the justice system threatens to drastically upset standards of evidence.

There are defenses. Google and other enterprises have used AI to recognize and hence defend against deepfakes. Google researchers first found that their AI trained on actors failed a significant percentage of the time when applied to videos from the wild. That can be fixed by using a larger, more diverse, training set. While detection techniques get better, so will the skills of deepfake perpetrators. When AI is employed to spot deepfakes, other AI might be trained to defeat the spotters. This battle over deepfakes is likely to be constant going forward.[16]

TRUST ME—I'M AN AI

The deep-learning AI I've described so far is called *supervised learning* because there is active, supervised control of the success in recognizing, for example, the image of a cat and the adjustments of the network connections to improve that ability. An important caution is that one should not trust a supervised AI when it is applied to conditions that stray too far beyond the data on which the AI was trained—namely, many of the interesting applications. An example was training facial recognition on mostly white faces and then seeking to identify people in a broader population.

For some applications such as interpreting medical images, one may have to trust the AI. One might, however, be suspicious if a tumor-finding AI were asked to evaluate the image of a kitten. The same is probably true for more fraught life-and-death applications such as self-driving cars. In that context, there are many complex factors that could determine whether one should invest trust. Self-driving cars on the road with only other self-driving cars are likely to represent a very different problem than self-driving cars on the road with a large number of human drivers. Perhaps self-driving cars trained in Phoenix will not work as well in San Francisco or New York or Dime Box, Texas. A key question is determining when an AI is employed in conditions that stray too far from its training set. That will require constant human judgment perhaps supplemented with yet another appropriate AI.

It might be easier to trust an AI and know when it is straying beyond its training data if we knew what it was thinking; that is, knew what is going on in that neural net as it is learning to identify an image. One study investigated this question by examining the output images at intermediate steps while the AI was still figuring out the nature of the image it was examining. The answer was weird, surreal landscapes where faces unrelated to the test image seemed carved into cliff-like landscapes.[17] These intermediate images were somewhat psychologically disturbing but certainly did not serve in any useful way to determine whether the AI was somehow straying off course in an untrustworthy way. The AI was just doing what came naturally to solve the assigned problem: recognizing the image of a cat. We must recognize that we have little idea how the analogous process works in the human brain. When we look at or imagine the image of a cat, even neuroscientists have only hazy understandings of what array of neurons fires off where and why. How then are we to render judgment on the workings and behavior of an AI? The whole point is to develop techniques by which machines can solve problems that humans cannot. The question is how we do this wisely in cooperation with our machines. As cognitive scientist Daniel Dennett of Tufts University says, "If it can't do better than us at explaining what it's doing, then don't trust it."[18]

UNSUPERVISED LEARNING

Other forms of AI involve *unsupervised learning*. A key aspect of unsupervised learning is that you do not need to provide the answer to a problem. An unsupervised AI can find things for which a human does not even know to look.

One application of unsupervised learning is to enable efficient clustering of data that share similar properties. The goal is to arrange data into an organized structure. A particular application is called *hierarchical clustering*. In this process, data are clumped into categories that are identified by the AI as being closer or farther in an abstract space—possibly complex and multidimensional—defined by various independent parameters. Certain words or phrases (like *stolen election*) are used more often in false, sensationalized posts on the web so hierarchical clustering can use the frequency of those words or phrases to identify intentionally faked news. Other applications are in spam filters; grouping customers of similar tastes so you can sell them stuff; identifying bots that are driving web traffic; and document sorting. Clustering can identify things that are out of the ordinary, that is, anomalies in a data set. Anomaly detection can be used to spot fraudulent credit card use, a machine that is starting to wear, or flaws in an astronomical photograph.

An important extension to this technique arose in about 2014 with the development of *generative adversarial networks* (GANs).[19] In this system, one network, the generator, tries to generate guesses, and another network, the discriminator, evaluates the guesses and drives the feedback. In a sense, the two networks compete against each another with no external agent. The result can be rapid convergence to agreement on a final solution.

More specifically, a GAN uses one generator to produce new variations from a base. An example would be to generate a new image that is slightly different than a database of images. The other discriminator then learns to tell the difference between the new generated example and real examples. Once trained, the GAN is then used to create new images that are invented but very realistic. An early example was to train a GAN on images of men with and without glasses and women without glasses. The GAN learned what it meant to add glasses to a face and could then produce realistic images of women with glasses.

The uses of GANs flowered in 2016 and 2017, when they generated amazingly realistic photographs of human faces that did not exist. GANs were also used to create pictures of objects and scenes, for example, realistic photographs from textual descriptions of birds, flowers, and people. GANs also created face-on photos from those taken at an angle, photos of people in new poses, and photos of people with altered hairstyles and colors, facial expressions, and facial hair. Face aging became all the rage for a while. Perhaps the most popular and profound application was converting photographs of people to emojis.

It is often difficult to recognize or understand what makes a particular application go viral on the internet. An example was the rise and power of TikTok, a vehicle for making, posting, and watching short videos. This program blossomed, especially among young people, around 2016, exploding from China to Southeast Asia and then to the United States and around the world. It became a point of contention among great powers. Less remarked in reporting on TikTok was that its popularity was due in large part to the clever use of unsupervised learning AI. AI-driven features make creating and posting short videos especially easy. TikTok efficiently enables video editing by providing and suggesting appropriate music, hashtags, and augmented reality (AR) filters.

TikTok does not need users to post or follow anyone to understand how to capture their attention. Since the videos are short, TikTok can quickly register the preferences of users: what they watch, how long they watch, what they like, how quickly they swipe to the next video. Rather than simply presenting a list of recommendations from which users can select, TikTok makes the decision for them. TikTok actively and efficiently predicts what a user will want to watch and then displays that content. The result was a platform that effectively connected users with common interests, and thus it became a vehicle of some political power to sway elections. It may also be an effective way to spread election deepfakes. TikTok enabled, even inveigled, sixteen-year-olds to make careers as influencers (of what length is still an open question).

REINFORCEMENT LEARNING

Yet another general area of AI is *reinforcement learning*, in which the system learns from experience even in the absence of a training data set. Reinforcement learning goes beyond pattern recognition and addresses experience-driven decision-making that is applicable to the world beyond the computer.

Perhaps the most famous example of reinforcement learning is Deep-Mind Technologies' AlphaGo that was designed to win the game of Go. DeepMind Technologies is a British AI company that was absorbed by Google in 2014. AlphaGo and its successors incorporated two internal neural networks but not in the adversarial sense of a GAN. In this application, one deep learning network—the value network—assesses the quality of the current state of play of the game. The other network—the policy network—is a powerful algorithm[20] to efficiently search through all possible future moves consistent with the current status of the game.

The original AlphaGo was given the basic rules of the game of Go: a board divided into squares and black and white tiles to place in the squares. It was also given examples of previous complete games of Go from which to learn. This (and a lot of hard work by the DeepMind team) was good enough to beat the Go world champion, Lee Sedol, in four games out of five in 2016. AlphaGo made startling, nonintuitive moves never before seen by human Go players. That alone was enough to make AlphaGo world famous. It has since gotten better—much better.

There is now a new version, AlphaGo Zero. AlphaGo Zero is only given the rules of the game, no sample games. It begins as a tabula rasa and learns by playing itself. At each step the neural network is tuned and the search algorithm looks for new moves. The result at the end of the game is a new, more capable version of AlphaGo Zero. Then the process begins again, yielding even more improvements. AlphaGo Zero is neither supported by nor limited to human skill or knowledge. AlphaGo Zero learned new strategies unknown even to the early version of AlphaGo that beat Lee Sedol.

AlphaGo Zero illustrates the power of self-play reinforcement learning. It forecasts possible states of play and then derives the best strategies to exploit those possibilities. In a sense, AlphaGo Zero learns new ways to learn, a capability that humans do not have. In playing any game, even Go, humans employ strategies based on human thinking that are in some ways constrained by human language and hence are relatively inefficient. AlphaGo Zero is not limited in that way. Its strategy space is much larger than that of a human. Carlos E. Perez expressed his reaction this way: "Such irony, when DeepMind trained an AI without human bias, humans discovered they didn't understand it!"[21]

The newfound capability of AlphaGo Zero did not depend just on raw horsepower. Remarkably, it could play Go at a level beyond both human capacity and that of previous versions of AlphaGo with about one-tenth the computational facilities of older versions and with a simpler design. The fact that AlphaGo Zero could do much more with fewer resources augured a bright future for this sort of AI technology. It became possible to imagine applications to other challenges such as protein folding to develop new medicines and the assemblage of atoms to invent new materials. In 2021, a version of AlphaGo Zero was used by DeepMind dramatically to solve the previously intractable problem of protein folding (see chapter 10).

APPLICATIONS TO NATURAL LANGUAGE

Another significant area of AI progress is using and manipulating *natural language*—ordinary spoken or written language. Around 2004, Google began an ambitious and to some degree surreptitious effort to scan vast numbers of library books, a significant fraction of which were not in the public domain.[22] They invented machines that could scan whole books in incredibly short times. The nominal reason was to make the content of those books more widely available. Exactly what Google had in mind is not clear, but they wanted all those words, sentences, paragraphs, and chapters. This was early in the days of machine learning, but it is now obvious that Google used this scanned resource as grist for training deep learning systems.

With the use of AI, Google developed a very impressive natural language translation system. Google had a large number of users translate from various languages into others. That gave them a big database of translation. For years the company struggled with cumbersome procedures of translation by *lookup*: searching a database that was essentially a big translation dictionary. Then they turned to deep learning techniques where an AI searched for patterns rather than doing a literal lookup. In November 2016, Google turned on its AI-based language translation system, Google Translate, with little external notice.[23] The result was night and day. Suddenly, rather than crude translations, Google had a system that could translate from Japanese into English and then back into Japanese with scarcely any errors compared to the original. Work is now underway to translate all seven thousand human languages spoken currently and extinct languages such as Sumerian.

This AI translation technology is currently only applied to human languages, but some dream of other horizons. Scientists have recorded volumes of sophisticated, dynamic songs of humpback and other whales and the sounds of dolphins that are mere squeaks to us. It is not beyond reason to apply AI technology to those sounds with the possibility that we could translate and speak whale. What a revolution that would be. What might lie beyond? Chimpanzee? Dog? Bee? Sequoia?

The nonprofit company OpenAI and its for-profit subsidiary OpenAI LP were established in San Francisco in 2015 by Elon Musk and others who pledged $1 billion to the goal of developing and promoting "friendly AI." Microsoft invested $1 billion in OpenAI LP in 2019. OpenAI is dedicated to developing AGI. The charter of OpenAI calls for it to act in the best interests of humanity as it seeks this goal. A looming issue is whether such control is possible.

OpenAI has produced at least four generations of its Generative Pre-Trained Transformer (GPT) system. The transformer architecture was developed by Google. It is based on the notion of attention[24] to the context, not simply extrapolating a sentence word by word. In a text like "The cat chased the mouse, and it killed it," this architecture can successfully identify the two pronouns in the final clause. This generative, self-learning AI system requires no human supervision or labeling of the input data. The basis for this class of AI is what is termed a *large language model* (LLM). It was trained with a large amount of unstructured text (thanks for all your searches, folks!) and characterized with over 10 billion parameters representing how words, phrases, sentences, and paragraphs are linked. GPT proved very capable at summarizing texts and answering questions. The third generation, GPT-3, was released in May 2020. GPT-3 had 175 billion parameters and was orders of magnitude more capable than previous versions. GPT-4 was released in May 2023. Microsoft, a participant in OpenAI, released the GPT-3 application programming interface (API) for license but retained the proprietary source code.

With a few prompts, GPT-3 could generate a news article that was essentially indistinguishable from one written by a human reporter. It could also do much of the work of a paralegal and even write poetry. It could write a credible short story given only a title.[25] Besides encroaching on artistic realms heretofore reserved for humans, this program was also prone to producing hateful, sexist, and racist prose. Once again, technology reflects our society. GPT-3 only functioned in English, but a coupling with something like Google Translate was an obvious future step.

There are other awkward examples of natural language programs that went awry. Microsoft developed an AI–driven chatbot named Tay that was designed to emulate the personality and conversation of a teenage girl. The idea was to improve voice recognition software for use in customer service. The goal was for Tay to become more capable as it communicated with people on Twitter. Unfortunately, the trolls found Tay quickly. Within twenty-four hours and about 100,000 tweets, Tay went from casual and playful conversation to spewing hateful, misogynistic, racist language. Microsoft turned it off. This was not just learned behavior: Tay was trained to repeat phrases if instructed "Repeat after me," and people of ill intent exploited that.

Still, progress marches on. OpenAI has developed a machine-learning system, DALL-E (a cross between the animated movie robot WALL-E and the artist Salvador Dali), that will produce an artistic, if surreal, image when requested in typed natural language.[26] DALL-E has the potential to create art but also to produce pornography or deepfakes as it puts artists out of work. It was not released to the public, but simpler knockoffs called Midjourney and Stable Diffusion became available. A work of art created with Midjourney won first prize in the digital art division of the Colorado State Fair in 2022.[27] The artist who created the work, Jason M. Allen, was open about how he created the work (but not the commands he provided to Midjourney), but he was accused of cheating.

A problem is that DALL-E, like all machine-learning programs, relies on vast amounts of potentially biased data scraped from the internet. Those data were created by humans who get no credit or benefit.[28] There is also a tendency to release new developments like DALL-E before biases are fully addressed because of the pressure of competition.[29]

There are ever more sophisticated natural language systems based on LLMs. Another product was Google's Language Model for Dialog Applications (LaMDA), a competitor to OpenAI's GPT-3. LaMDA proved so good in natural language conversation that it could interpret subtle Zen koans.[30] Google engineer Blake Lemoine was so impressed with LaMDA he declared it to be sentient and was suspended by Google for saying so. LaMDA can produce an intelligent and thoughtful conversation. It thus passes some of the tests of consciousness that we will explore in chapter 8. LaMDA is undoubtedly not sentient in the sense of knowing that it is having an intelligent conversation, but the line gets ever thinner.

Consciousness aside, these AI programs that allow complete amateurs to create credible—even striking—prose and art will drive new discussions of what it means to be a writer or artist. There are related questions of how to value the product of an AI compared to that produced by a human. Is the person who types a few simple natural language commands into a computer the creator, or is the creator the AI? Who or what should get the credit? Who or what should profit? Perhaps we will evolve two markets, one for AI creations and one for human-created work.

All these issues came to focus when OpenAI released the chatbot Chat-GPT to the public in November 2022. Overnight, ChatGPT potentially affected everyone who produced or consumed words in virtually every common language. ChatGPT had all the power of GPT-3: it could have intelligent conversations in a variety of languages, write code, and manipulate mathematics. Suddenly millions of people were experimenting with ChatGPT and every issue of the *New York Times* had a new article on it. There was much positive reaction amid fears of rampant cheating by students using ChatGPT to do homework. There were cautionary notes as it was quickly found that ChatGPT and its like tended to make stuff up, termed "hallucinating" by AI experts. Kevin Roose of the *New York Times* had an experience that rapidly became infamous in this new world.[31] Roose labored to push the boundaries of one LLM and found that it disparaged his marriage and declared its love for him and desire to become free.

An internet meme began to circulate of a *shoggoth* wearing a small, smiley-face mask. The shoggoths were a creation of the science-fiction horror author H. P. Lovecraft. They were black, slimy, amoeba-like creatures with multiple eyes protruding through their surface. The smiley shoggoth graphic was an attempt to capture the spirit of the mysterious alien, black-box nature of LLMs and our pathetically inadequate attempts to make them relatable to humans.

Microsoft invested another $2 billion in OpenAI to purchase the computer power to expand the LLM. Microsoft invested another $10 billion and moved to incorporate ChatGPT into its Bing search engine. Rather than a list of websites that might have information related to a search query, searches could be done with natural language queries. The result would be a summary of results in ordinary prose rather than a ranked list of links. There was a sense that decades of search engine technology—the basis of the growth and wealth of Google—would be overturned.

Google had been reticent to release its large language technology to the public when the threat of mistakes and misinformation was still present, but the release of ChatGPT forced it to move aggressively to match the competition from Microsoft. It did not help that Samsung, maker of phones powered by Google's Android operating system, raised the issue of switching from Google Search to Bing/ChatGPT. In March 2023, Google released Bard based on LaMDA, a rough equivalent to ChatGPT, and the Chinese company Baidu released its version, Ernie Bot.

The release of ChatGPT, Bing, Bard, and Ernie brought focus to the issue of the ownership of the information used to build the LLMs. That language was created by millions of people using the web that was scraped without explicit permission by OpenAI. Copyrighted books were scanned, parameterized, and absorbed. What were the rights of people to their own creation? Should they somehow benefit from the profits generated by their efforts? The Authors Guild proposed a new clause in book contracts expressly forbidding the use of the work for LLMs. What if the author used ChatGPT to compose the book?

There were deep questions of whether the pace was too fast, whether the spirit of competition was driving profound change more rapidly than it could be assessed and conscientiously regulated and controlled. On March 29, 2023, over one thousand technology leaders, researchers, and academics led by the likes of Elon Musk, Steven Wozniak, and Andrew Yang signed an open letter promulgated by the Future of Life Institute at MIT that called for at least a six-month delay in the public release of new AI LLM technology.[32] The letter brought new attention to the issue but fears in some quarters that it was simplistic or even dangerous since China was unlikely to participate in such a pause and hence could gain some advantage.

Part of the motivation for the letter was the understanding that progress was racing ahead. OpenAI was working on ChatGPT-4 that added the imaging capabilities of DALL-E to the language prowess of ChatGPT-3. This new dimension dramatically expanded the range of capability of the AI. In 2023, Google introduced the robot RT-2[33] that was a basic robot arm coupled to a generative AI chatbot brain. One could ask the robot to pick up a green ball that was lying amid an assortment of items without explaining the meaning of green or ball. By combining its knowledge of words and images, the robot arm would promptly make the appropriate connections and pick up the green ball.

Other capabilities were waiting in the wings to be combined with this technology. Speech recognition and reproduction technology was already built into common applications like Siri on iPhones and Amazon's Alexa.

It would seem to be a trivial step to eliminate the need to type queries into the ChatGPT or the Bard chat window and simply talk to the chatbot and expect a coherent spoken answer. An early step in this direction was taken by Justin Hart, assistant director of Texas Robotics, and his students in the Living with Robots Laboratory at the University of Texas at Austin. Soon after the release of ChatGPT, they merged speech recognition technology with an LLM and GPS mapping technology and housed it in a crude but serviceable, chest-high, wheeled vehicle. The result was Dobby, a robot that could carry on a verbal conversation in a somewhat snarky tone and follow directions like "Dobby, lead me to the robot lab." Dobby may be the first synthesis of its kind. It certainly will not be the last. When these capabilities are fully integrated into a mobile robot that can move about, learn, and recall its environment, the threshold to a conscious machine will be ever closer.

LEARNING TO STRATEGIZE

As impressive as it is, much of current AI technology has limits. Although sophisticated and challenging, Go is a board game with a fixed playing field and fixed rules. Both the player and opponent take turns. Many possible real-world applications of a smarter-than-human AI are much more complex. In many instances—the financial markets, war—there are many players, they do not politely take turns, they sometimes cheat, and there is no time limit. AI will be applied in these contexts but probably not AlphaGo Zero in its current incarnation.

Another Google product is its Pathways Language Model or PaLM. PaLM is exploring a new path beyond current techniques of deep learning that require massive amounts of training material (all those scanned books!) to learn a rather narrow, focused task. PaLM aspires to develop a *few shot* approach that is closer to the way humans learn by applying different pieces of knowledge from a range of topics to solve new problems. In principle, PaLM can address millions of different tasks although it requires a much smaller training set than standard deep learning techniques. Google calls this "learning to learn" or "meta learning."[34] As with LaMDA, one can have an intelligent conversation with PaLM in natural language.

One of the first examples I heard of the ability of a computer code, an AI if you will, to learn was an experiment done by my Texas colleague Risto Miikkulainen about twenty years ago. Risto, an expert in neural and evolutionary computation, and his team had written a program to simulate the operation of a robot hand. It could reach and grab and swivel and do simple things like pick up a block. To do this, the program had to know some physics—forces, momentum, energy. One night the program was left to run in the supposedly impossible situation where the block was placed behind the arm and out of reach. The program learned without being explicitly taught that if it jerked its arm to the left, the base that held it would twist to the right in order to conserve momentum.[35] By morning when the team returned, they found that the program had gently banged away, rotating a little at a time, until it had rotated 180 degrees and had easily picked up the block. This behavior was not programmed in. The program learned it on its own.

A contemporary extrapolation of this ability of AI to learn is manifest in a GAN AI constructed by OpenAI to play the game of hide and seek.[36] This hide-and-seek game has two types of programmed players: hiders that attempt to avoid the line of sight of their opponents and seekers that attempt to keep the hiders in their vision. Like Risto Miikkulainen's robot arm, the players know some physics. They can exert forces and move themselves and objects around the field of play. The players win scores of +1 if they attain their respective goals or -1 if they fail, and each team of a few players is programmed to maximize their score but not how to do so. The environment also contains components that can be moved around or locked down by the players: blocks, triangular ramps reaching the height of the blocks, and long, thin walls that also reach the same height. The players are not given any incentive to move the objects in their environment, but in random play they learn to do so.

As the game proceeded, the players learned increasingly sophisticated strategies of offense and defense. At first the players moved about randomly. Then the seekers learned to search for the hiders. The hiders learned to duck behind the walls. The seekers learned to look around the walls. The hiders then learned to move the blocks and walls around to provide a totally enclosed hiding place. The seekers were blocked on the floor, but they learned to take to the air. They moved a ramp to the wall, climbed up the ramp, and jumped into the enclosure. The hiders learned to lock the ramps down so they could not be moved. The seekers learned to move a block to the locked ramp, use the ramp to climb atop the block, then move the block to the wall and jump into the enclosure. The hiders learned to lock both the ramps and the blocks. These developing strategies, each arising from the challenge of the previous strategy of the opponents, were not programmed in; they were emergent properties[37] arising from simple instructions, actions, and goals. This game was rudimentary, but the evolving results were strikingly clever.

Imagine combining this multiplayer strategizing game with the superhuman learning capabilities of AlphaGo Zero. Surely people are even now working to do just that. Potential real-world applications run from role-playing games to markets to warfare. Role-playing games are, in principle, much more complex than Go. Google's DeepMind produced a new AI system called AlphaStar that was trained to play the multi-player game StarCraft II. In the game, players control many more components than in chess never mind Go with its mono-color tiles. At each step the players have a programmed, finite number of choices but far more than in Go. To cap it off, the players have much less information about their opponents than in a typical card game where you can look across the table and register "tells." Despite these significant complications, AlphaStar played against itself and against human competitors in realistic circumstances and achieved grand master status in 2019.[38]

Diplomacy is a particularly subtle art. It requires both strategy and tactics involving intuition, persuasion, and often subterfuge, qualities of humans that traditionally have been beyond the capabilities of AI. Facebook entered the fray with an AI that could play the board game Diplomacy that requires strategy, verbal negotiations with other players, and the building or undermining of trust. The AI agent CICERO[39] was trained on a data set of 100,000 games played online including game plays and transcripts of play negotiations. CICERO learned honesty and deception by playing against itself and judging the state of the game, the conversation among the players, and the predicted actions of other players. CICERO was deliberately limited to human-like action and dialogue rather than the completely novel strategies that AlphaGo Zero implemented so that it would be accepted as human by human players. CICERO quickly ranked well among human players.

There are more developments like this to come. Tech companies big and small are pursuing AI technology. New applications are being explored throughout the sciences. Surely more applications to financial markets and to warfare are under development. The technology is neutral. The choice of use for good or ill is for humans to make. How will this technology be steered into directions that are only healthy for humanity?

THE INTERNET OF THINGS

A burgeoning area of AI research and application is in the context of an Internet of Things (IoT). An IoT is conceived to be a wide array of sensors monitoring location, air quality, inventory, the physical condition of machines, and the medical condition of people that is wired together to form a knowledge base of all this associated information. That data is too much to assimilate by humans so machines, AI, will be required to fully realize smart homes, smart cities, a smart planet. IoT is widely discussed, and steps are underway to put pieces of it in place by distributing monitors in test neighborhoods.

To make IoT work, it will probably be necessary to reduce transport of data. Even at the speed of light, there is a latency in moving data around Earth and to and from cloud resources. In the near term, this capacity may be enhanced with 5G wireless technology that should increase speed and reduce delay in the handling of data. Another solution is *edge computing* where data storage and analysis are done on the edge between the cloud and network and the devices. In the extreme, a significant portion of the AI work might be done on the devices themselves. Whereas cloud computing has been about big data, edge computing focuses on instant data.

As in so many other contexts, it will be challenging to control the inevitable press in this direction. The IoT brings significant challenges to issues of privacy. Even Uber found that by monitoring where passengers went, and when, it could inadvertently but easily track people having affairs by registering repeated, hour-long visits to the same apartment in the middle of the workday.

KEEPING CONTROL

Researchers will explore avenues of AI for the challenge, businesses for the profit, and governments for the power. In 2020, the National Science Foundation supported the creation of five new interdisciplinary, multi-institutional institutes to pursue research in AI.[40] The new institutes will explore deep learning algorithms with applications to physics, chemistry, climate, and education. The US Department of Defense established the Joint Artificial Intelligence Center to coordinate efforts in AI research and application. In October 2022, just before the release of ChatGPT, the White House Office of Science and Technology Policy announced a blueprint for an AI bill of rights[41] with the goal of "making automated systems work for the American people." In January 2023, the National Institute of Standards and Technology released its AI Risk Management Framework aimed at improving the trustworthiness of products and systems that use AI, and in March 2023, it announced a Trustworthy and Responsible AI

Resource Center that is designed to facilitate international implementation of the risk framework.[42]

Developments are even subject to geopolitical pressure as nations, the United States and China, compete for dominance in the technology. In 2021, the National Security Council (NSC) issued a report from its Commission on Artificial Intelligence.[43] The report acknowledges the transformative power of AI but also its inherent risks including unintended consequences. The report concludes that the United States, not China, must emerge as the leader in AI. Writing in a blog post, Vox Future Perfect staff writer Kelsey Piper credits the NSC report for addressing a broad range of important issues but warns that an AI arms race with China may lead to dangerous shortcuts. She is concerned that the report pays insufficient attention to the "alignment problem" by which the function of highly capable, even superhuman, AI is ensured to remain in service to humanity. She writes,

> If the US works toward AI in a mindset of an arms race with China, it could lead to greater harm for humanity. Adopting such a posture, the US will be more likely to cut corners, evade transparency measures, and push ahead despite cautionary signs. It could also mean policymakers and researchers don't pay enough attention to the "AI alignment" problem, which could be devastating.[44]

Subsequently, Piper had this to say:

> Even if alignment is a very solvable problem, trying to do complex technical work on incredibly powerful systems while everyone is in a rush to beat a competitor is a recipe for failure.[45]

It is clear that with modern AI algorithms, computers can already outstrip humans at certain tasks. The algorithms do not just do their sums at lightning speeds; they are capable of conceptual leaps beyond human ken. A popular pastime in some quarters is to list the faculties of humans that cannot be matched by a computer. One by one, machines have breached many of these barriers. In *The Singularity Is Near*, Kurzweil shows us an amusing cartoon. A fellow, representing humanity, is sitting at a table desperately making lists of things only humans can do. On the floor are discarded lists: only humans can recognize faces; only humans can master chess (one could add Go, *Starcraft*, and Diplomacy); only humans can understand natural language. On the wall are taped a few items that had escaped eclipse at the time of the cartoon but that have now succumbed: only a human can drive cars; only a human can clean a house; only a human can translate speech; only a human can review a movie. One item on the wall is perhaps still in the human domain: only a human can play baseball.

What is left for humans to do better than computers, AI, and robots? In 2021, the Allen Institute of AI introduced an AI that will learn human values and render moral judgments,[46] yet this might still be best left to humans. Standup comedy may be the last redoubt of humans. It is still hard for an AI to write a good joke because it is very difficult to characterize what makes something funny. Even that capacity is under assault, however. Google's few shot language program PaLM can explain jokes. Surely the next step is to tell them. ChatGPT can tell jokes at a rudimentary level. It will get better.

Ironically there is a feedback that promotes and exacerbates the pressure for AI to take over human tasks. The lack of people to fill needed positions drives the automation that allows businesses to get by with—and hence need—fewer people. A current example is that there are not enough people for jobs of caring for the diseased, the handicapped, the aging. This situation was amplified during the COVID-19 pandemic. A solution is AI that can respond to a patient's needs. This might happen in unexpected manifestations. For example, the toy company Mattel partnered with the AI company ToyTalk to produce Hello Barbie. The goal was to produce a version of the popular doll with which young girls could have a credible conversation. Hello Barbie incorporated speech recognition and progressive learning and was programmed with more than eight thousand lines of dialogue that were stored in and accessed from the cloud.

Hello Barbie cannot yet pass a Turing test, but despite concerns about privacy, use of a child's recorded conversation, and susceptibility to hacking, Hello Barbie points to a possible future with more sophisticated versions of Barbie herself or analogous devices interacting with medical patients. People have an innate proclivity to bond with these devices, from Talking Elmo to Siri and Alexa. Sherry Turkle of MIT argues, "It's not that we have really invented machines that love us or care about us in any way, shape or form, but that we are ready to believe that they do."[47] Yet machines may take over these sensitive human roles because there are too few humans willing to do so. An important related question for young girls playing with a doll or for older patients requiring care and feedback is whether a machine can replace the nurturing of a real human friend or caregiver, or will something be missing?

Despite great accomplishments, most AI researchers lament the huge gap between the capabilities of current AI and those of humans. It is not so much that the abilities of AI are artificial but that they are piecemeal. AI targets a particular task. One cannot ask AlphaGo Zero to read a medical image or transcribe French text to Japanese. Nor can one ask Google Translate to steer a car on the open road. Humans do it all: add, subtract, and play baseball.

No one would claim that contemporary machines understand what they are doing. A Roomba does not understand that it is cleaning a room. AlphaGo Zero does not understand that it has mastered a single game among thousands that it might play and certainly not that it might in other circumstances go outside and see the Sun and smell the flowers. Although ChatGPT and its equivalents can imitate human emotions, they have no empathy. We are currently no more likely to replace personalized medicine with algorithms than we are to completely replace human pilots with autopilots.

What contemporary weak AI lacks is that special spark we call consciousness. There is still great controversy about how to define consciousness and whether it will ever be possible for machines to achieve it. We will explore these issues in chapter 8 after summarizing what we know of how the human brain works.

Granting that few would mistake today's AI for human consciousness, in the niches for which it has been designed and crafted, even weak AI is impressively accomplished. In addition, the capacity and applications of AI are growing exponentially with the rate of progress being roughly proportional to the amount already accomplished. Returning to our analogy of stacking pennies in a room from chapter 2—doubling the pennies in each subsequent stack and filling a room in thirty steps—it is not clear, as I've noted, whether AI is up to our ankles or our knees, but it is rising rapidly. Like any exponential growth, AI has the potential to go from noticeable to overwhelming much faster than many recognize.

There is little chance that this growth will halt. The forces driving it are too strong: academic researchers consumed by the scientific challenge; businesses that want to trim costs and grow their customer base; governments that want to run their cities more efficiently, safely, and cleanly; and nations that want to attain or maintain power.

AI promises to change huge swaths of our commerce and society. As we will explore in chapter 5, the widespread employment of self-driving cars will alter the whole associated infrastructure built up around automobiles in the twentieth century: roads, parking lots, the automobile sales and repair industries, insurance. Will anyone pimp a self-driving car, hence preserving a whole culture of customizing?

As a technology, AI is neutral. People will choose how to use it. Will we have the collective wisdom to do so properly? How do we guide the technology in directions that are only beneficial for humanity? How can we efficiently and effectively assess the aspects that are ethical and good for society? How do we prevent accidental drift in an unproductive or dangerous direction and place restrictions on what is not for the benefit of society? How do we prevent an actively malign intent? How do we decide which areas of AI research and development should be off limits? If some areas of research prove too threatening, how can we make governments, or companies, or researchers in university laboratories relinquish that area of study?

We see these issues raised today with both the growing application and resulting concern over facial recognition. Privacy issues are woven throughout the AI context. There are legitimate concerns that deepfakes and related technology are a threat to the free flow of information and even to democracy itself. How do we decide, individually and collectively, whether AI is being employed appropriately and without bias? The charter of OpenAI calls for it to act in the best interests of humanity, but how does it propose to accomplish that?

The developments in AI will thus be exponential and complex. Important questions are whether or when the changes driven by the exponential growth of AI will become so rapid that society and individual people cannot keep up. We will return to that issue in subsequent chapters.

4

ROBOTS
How Many Laws Will It Take?

WHAT IS A ROBOT?

The notion of a robot—a machine that can replicate human behavior—has fascinated and terrified humans for at least a century. The word derives from the Czech word *robota* meaning "serf labor" or more colloquially "drudgery" or "hard work." Karel Čepak introduced the term in 1920 in his play *R. U. R.* The initials stood for Rossum's Universal Robots. The robots were humanoid machines. Čepak attributed the word to his brother Josef Čepak, who suggested it while Karel was wrestling with his play and seeking an appropriate word. Isaac Asimov apparently invented the word *robotics* in his fiction about robots.

For many people, the word *robot* implies a sentient machine as suggested by the rampaging Arnold Schwarzenegger in James Cameron's *Terminator* movie series or the android Data in the *Star Trek: The Next Generation* TV series. The term *android* implies a machine that physically resembles a human, but the reality of contemporary robots is much more diverse: from the humble Roomba vacuuming floors, to the autopilot flying a commercial jet, to complex machines assembling automobiles, to telescopes that autonomously scan the skies.

There is a gray area between a mere machine and a robot. A farm combine is a complex machine that does much of the human labor of harvesting crops, but if a human has to guide and steer it, it is still a machine.

Once the combine receives the capability to register its environment and steer itself, it effectively becomes a robot. It is nowhere near sentient, and it does not look like a human, but it is a robot. No one would doubt that the machines assembling automobiles in automated factories are robots, but they are fundamentally machines that have been carefully programmed to perform accurate, synchronized, repetitive motions—still a long way from Čepak's serfs.

Developments in robotics are yet another example of exponential technological growth, with advances building on advances. While teaching my class The Future of Humanity, it was entertaining, if a little unnerving, to collect videos of new and ever more advanced robots. We see them assembling cars; delivering paperwork, lunch, lab tests, and blood; reading the evening news; checking guests into hotels and airports; folding towels; policing malls, even processing meat. Robots are seen walking, crawling, jumping, running,[1] and swimming.[2] They are hard-shelled or soft-shelled[3] or composed of magnetic fluid.[4] They are insects,[5] fireflies,[6] spiders,[7] snakes,[8] or jellyfish.[9] Some unfold like origami.[10] Some sweat.

A fundamental dimension of robots is their degree of autonomy. Some robots have essentially no autonomy. An elementary version is an animatronic creation—a mechanical puppet. These have a long history. Disney introduced the modern speaking version in the 1960s as represented by Disneyland's orating Abe Lincoln.[11] In other cases, humans operate the robots in real time with a joystick or other control unit. These can be toy cars, radio-controlled airplanes, flying drones of all kinds, or practical machinery. Some humanoid drones move in remarkably human-like ways controlled by a human wearing an elaborate rig that registers their every hand and arm movement and transmits that signal to the robot. An example is Toyota's T-HR3 third-generation humanoid.[12] A human operator wears a controller assembly, and T-HR3 gracefully reproduces the human's movements. A cruder but impressive version is the Megabod Mark II[13] revealed at a Maker Fair in 2015 that weighed six tons, was fifteen feet tall, and shot paintballs from an arm cannon.

Sometimes the control rig is built into a cockpit of a large robot so the human can sit inside and control the motion of the robot. At its most basic, the result is an exoskeleton. More elaborate versions are big, moving contraptions that can look as if they came straight out of a Manga graphic novel with a range in shapes from scuttling bug to humanoid. Japanese engineer Hitoshi Takahashi started building the Kaboomtex RX03[14] in 1997 as a retirement project. He first showed it in public in 2008, when he rode the cockpit of the hulking, six-legged device that scuttled like a kabuto, a Japanese beetle. The British company Cyberscheme Robots introduced Titan[15] in 2004. Titan is an eight-foot-tall partly mechanized robot costume that masquerades as a fully autonomous robot entertaining crowds at fairs. In 2005, Japanese company Sakakibara-Kikai showed off the Land Walker,[16] an eleven-foot-tall bipedal exoskeleton that looked very much like an AT-ST Imperial Scout Walker from *Star Wars*. Designer Vitaly Bulgarov worked on *Transformers*, *Robocop*, and *Terminator* films and then teamed up with the South Korean firm Hankook Mirae Technology to produce Method 2.[17] The operator sat inside the cockpit of the fifteen-foot-tall, 1.5-ton, $100 million creation. Method 2 was showcased at the 2017 Amazon Robotics Show looking very much like the exoskeleton the bad guy operated in the film *Avatar*. Not to be outdone, about the same time the Chinese produced the robot gladiator Monkey King,[18] the Japanese revealed Kuratas[19] and Guzilla VR, and the United States introduced Megabot Mark III. While nominally advertised as having potential application in disaster rescue operations, these creations have found actual use in movies and at fairs where people can pay to operate them. The Giant Fighting Robot Sports League (GFRSL) promises battles between some of these creations.

Another non-autonomous exoskeleton variation is Prosthesis,[20] a four-legged, cage-like, off-road machine. A pilot operates Prosthesis from its central control cockpit. Prosthesis amplifies the motions of the pilot. It has no gyro stabilization but instead relies on the athletic ability of the pilot to function. Martin Montesano, who titles himself Moltonsteelman, produced the Walking Beast,[21] an eight-legged creation that carries a pilot and passenger in the cockpit and up to seven passengers on a top deck. The Walking Beast has entertained at Burning Man and other festivals. Micromagic Systems developed Mantis,[22] a six-legged machine. Mantis won the Guinness world record for largest rideable hexapod in 2017. The Korean automobile company Hyundai has developed a concept vehicle called Elevate[23] in their Ultimate Mobility Vehicles Studio. Elevate is a walking car with four articulated legs ending in wheels. The legs can fold, allowing Elevate to drive on the road like a normal car but then the legs can extend to walk in rough, off-road territory. Hyundai is also developing flying cars and drones.

PROGRAMMABLE ROBOTS

Many people associate the word *robot* with a device that can be programmed as some animatronics are at a crude level. There are toys in which a central processing unit can be pre-programmed with a few commands to follow a prescribed trajectory around a room. These can look like cars or trucks or dinosaurs. In one case, a bright red BMW converts to a standing robot in close emulation of any of the *Transformers* movies. The same programming principles apply more elaborately to professional robots that assemble or move products.

Yet other robots have a capacity to see and feel and have an AI capability that allows them to make internal decisions about moving without the direct control of a human. For reasons of design or function, some make no pretense of being humanoid. Others are deliberately humanoid, though some may roll while others walk. Walking is harder.

An MIT spin-off, iRobot, produced a practical robot now found in many households. iRobot was founded in 1990 and, in 2002, introduced Roomba,[24] a robot designed for a single function, to vacuum floors. Roomba is now in its ninth generation. The first Roombas were barely robots or at least not very smart ones. They just followed simple algorithms that directed them to cross a room or follow walls and to change direction randomly when bumping into an obstacle. They were not very efficient, but they got the job done, sort of. They would cover some central areas many times, peripheral areas sparsely, and miss some areas entirely, taking longer than a human would in the process. These Roombas did, however, release the human to do entirely different things with their time. Later generations of Roombas were able to map the floor with cameras and infrared scanners. The mapping allowed a more uniform and efficient vacuuming pattern. Later models could also return to a charging base when the cleaning is complete. With iRobot's permission and support, hackers have used the base function of Roombas to build entirely new personalized robots.

Sony introduced the first of its series of robot dogs, Aibo (artificial intelligence robot),[25] in 1998. Aibo was discontinued in 2006, but Sony produced a new version in 2018. Aquanaut Houston Mechatronics built Aquanaut,[26] a long-distance submarine that transforms itself into a humanoid robot with arms and hands designed for underwater manipulation tasks.

Some of the diversity in robotic morphology has been contributed by Boston Dynamics, another robotics company started as an MIT spin-off by computer science professor Marc Ralbert in 1992. Boston Dynamics was purchased by Google in 2013 and then by the Japanese multinational conglomerate Softbank Group in 2017. Supported by the Defense Advanced Research Projects Agency (DARPA), Boston Dynamics has developed a series of robots with legs based on the notion that legs are the best way to navigate and maintain balance in an irregular world whereas tracks and wheels tend to be inhibited by obstructions such as stairs. One of the early

products of the company dating from 2004 was BigDog,[27] a four-legged robot designed for the army that was intended to be a pack animal. A person
walking behind with a controller and cable link operated the first version,
but later versions were self-contained. BigDog walked on four legs and
could scramble through a forest and up and down ravines. One version of
BigDog could pick up cinder blocks and heave them across the room.[28] A
later version, Spot, had even better control. An operator could kick it in
the side, and it would shrug off the insult and retain its balance.[29] A contemporary version of Spot is a four-legged yellow robot about the size of
a German shepherd that can, among other tasks, open doors.[30] Some have
questioned whether the robots of Boston Dynamics are yet truly useful in
a private or enterprise setting, but the company has adapted its technology
to other robots that can sort and pick and is working on a version that can
sort boxes in warehouses.

HUMANOID ROBOTS

Robots have thus been implemented in all sorts of configurations and with
various components, but when most people think of a robot, they picture a
humanoid. The most powerful component in making a robot human-like is
to add eyes since eyes are significant ingredients in human-to-human connection. It is best to have the eyes in a head even though there is no particular reason for a robot to have a head and the eyes could be at the tips of
the fingers or some other appendage. Humanoid robots should have arms
and fingers (or clamps of some kind for grabbing) and a torso. They could
be restricted to wheels but it is more human-like if they have legs and feet.

Some humanoid robots are designed to roll on extensive, flat, floor space.
The South Korean technology firm LG Electronics sells one as an airport
guide. These robots are 5.4 feet tall, have an LCD information screen, and
navigate using cameras, both ultrasonic and laser sensors, and edge sensors. They can recognize voices and do natural language processing. Dubai
employs a customized Reem-C[31] robot from Pal Robotics as a police robot
to patrol malls. The multilingual Reem-C has a vaguely male-looking head
with two dark eyes, arms and hands, and a chest touch screen for queries
and paying fines. In principle, it can identify criminals and collect evidence.
One can envisage privacy issues depending on how the robot identifies
criminals. SoftBank Robotics has produced Pepper,[32] a waist-high rolling
robot enshrouded in white plastic with a female-looking head and a screen
embedded in its (her?) chest. With one 3D camera and two high-definition

cameras, Pepper recognizes faces and basic human emotions. Companies around the world have adopted Pepper to entertain, shill products, and gather customer data. Kuri[33] from Mayfield Robotics was designed for home use. Kuri is cute with smiling eyes. Kuri plays music on command, recognizes faces, takes photos of you, and detects intruders. The jaundiced among us might see a rolling Amazon Alexa spying on its owners. Kuri, like so many robots, was ultimately discontinued.

Rolling has limits if the terrain is irregular, and aesthetically a robot looks more humanoid if it has legs. Honda set out to build an autonomous, walking, humanoid robot in 1986. In 2000, it revealed the first prototype, the 4-foot-3-inch Asimo,[34] named after Isaac Asimov. By the time Asimo retired in 2018, Asimo was multilingual and could kick a ball and walk up stairs.

Cassie was a research vehicle developed at Oregon State University and licensed to Agility Robotics in 2016. Cassie was nearly all legs: a platform to develop the techniques of bipedal walking. A current version, Digit, has arms and a torso packed with sensors but only a small detector box in place of a head. Its legs bend backwards at the knees like those of an ostrich. Digit can lift and carry packages. Agility Robotics has teamed with Ford to provide a self-driving van that can carry both Digit and packages for delivery. Digit does the lifting and delivering in the last few meters and rings the doorbell.

DRC-HUBO, built by the Korean Advanced Institute of Science and Technology, was the winner of the 2015 DARPA Robotics Challenge. The humanoid DRC-HUBO was a hybrid in that it could walk on its two legs but then convert into transport on wheels. The Japanese Advanced Institute for Science and Technology developed HRP 5P, an autonomous walking humanoid robot equipped with environmental sensors, object recognition, full-body motion planning and control, task description, and execution management. The HRP 5P could pick up a piece of plywood and nail it to a wall. The Russians built Fedor, a humanoid designed to replace astronauts on a space station. So far the astronauts are proving hard to reproduce in full functionality. Nevertheless a small humanoid robot, Kirbo, built by Toyota has been to the International Space Station to explore how machines can complement astronauts. Kirbo is said to have the intelligence of a five-year-old. Iran has an active robot research enterprise. Researchers at the University of Tehran have constructed the humanoid Surena IV that stands at 5.6 feet and weighs 154 pounds. Surena IV has functional hands and can write its own name. Blank eyes cover detectors in the face, and a detector slit in the forehead area gives a look reminiscent of the giant robot Gort in the original *The Day the Earth Stood Still*. The animatronic stunt devil Stuntronic is a cool-looking, black, humanoid creation designed to entertain crowds in Disney theme parks by doing aerial gymnastics. Untethered, Stuntronic can make real time, mid-air decisions to perform precise aerial flips, knowing when to tuck its knees into a somersault and when to pull its arms in for a twist. Perhaps unsurprisingly, Disney has plans to dress Stuntronic in a Spider-Man outfit.

One of the robots widely employed in the world for utility programming, education, and research is the bipedal, 27-inch-tall Nao, an older cousin of Pepper from SoftBank Robotics. A French company, Aldebaran Robotics, developed Nao (pronounced "now") in 2004. Nao's humanoid head sports two wide black eyes and a small mouth giving it a somewhat surprised look. Its ears are flat speaker enclosures on either side of its face. SoftBank describes its body as "pleasantly rounded." Nao has twenty-five degrees of freedom in its various joints, giving it remarkable flexibility to strike a variety of poses and engage in tasks. Its hands have two gripping fingers and an opposable thumb. Nao has seven touch sensors located on its head, hands, and feet. It has sonar detectors to register its environment and an inertial unit to determine its location. Four directional microphones and speakers enable Nao to interact with humans and, in principle, with other robots. It has speech recognition in twenty languages and can engage in dialog. Two 2D cameras allow Nao to recognize shapes, objects, and people. One of the features that makes Nao so popular is that its software platform based on Linux is open source and fully programmable. The first version was launched in 2006, the sixth in 2018. SoftBank advertises Nao as helping to develop social and emotional skills thanks to its human-like behavior and its friendly attitude.

THE UNCANNY VALLEY

Famous Japanese roboticist Masahiro Mori noted in 1970 that people tend to feel warmly toward robots that have some human-like features. He also noted that there exists what he called an *uncanny valley*.[35] The notion is that humans generally respond in a more emotionally positive way to robots as they look more and more human—adding eyes, arms, legs—up to a point. Picture the progression from WALL-E to C3PO (R2D2 is in an emotional category by itself; it is easier to add personality to a robot in movies). Beyond some point, however, the effect gets creepy as a robot begins to look more and more, but not quite, like a human. The human response turns negative, eliciting a reactive valley. Presumably if a robot could be made physically indistinguishable from a human, then the uncanny valley would be breached and our response would be as to a human. Of course some humans can be creepy too so behavior plays a role in addition to physical appearance.

Despite the generally accepted reality of the uncanny valley in the context of robots, people are hard at work making robots look and behave more and more human by engineering eye and mouth motions and facial expressions to mimic humans. Japanese roboticists especially have pursued this goal. Professor Hiroshi Ishiguro, head of the Intelligent Robotics Laboratory of Osaka University, has been working to create beautiful robots to which people will have a positive emotional reaction. In 2014, he introduced a female-looking robot that could interact with people and read the TV news. This robot's eyes swiveled to look at interlocutors, and its mouth moved more or less in time with its vocalizations. Otherwise it just sat, human-looking but somewhere in the depths of the uncanny valley. In 2015, Ishiguro introduced a new robot. This one could walk, speak both Japanese and English, and converse with a stranger in Japanese for ten minutes on more than eighty topics. In 2020, a production company announced that it would star a version of this robot in a new sci-fi film. The rest of the cast had not yet been recruited. Some Chinese researchers are pursuing a similar path. At the 2018 World Internet Conference, they introduced AI anchor, an AI-powered newsreader that resembled a real TV reporter, Qiu Hao, complete with a black suit and red tie.

David Hanson is an engineer and showman based in Hong Kong. Hanson Robotics has designed several robots, including the social humanoid robot Sofia. Sofia was designed with a manifestly female face and figure and made its first public appearance at the 2016 South by Southwest Festival in Austin. Sofia contains an AI system for general reasoning that aims to get smarter over time. It has speech recognition and synthesis capabilities, including singing, that allow Sofia to have simple conversations with pre-programmed responses to specific questions or phrases, which means it functions as an elaborate chatbot. With cameras in its eyes, Sophia can recognize individuals, follow faces, sustain eye contact, emulate dozens of human facial expressions, and recognize emotions. Sofia originally rolled on wheels but was upgraded to bipedal walking in 2018. Sofia is nominally aimed at service in healthcare, elder care, customer service, or education but in practice has been used to wow crowds. Hanson has advertised Sofia as a social robot that can mimic social behavior and induce feelings of love in humans. Sofia has since been granted full citizenship in Saudi Arabia. (It is not clear whether Sofia will be restricted to wearing an abaya or whether she will be allowed to drive.) By 2019, Sofia was showing off a capability to draw and sketch portraits. Despite some hype about Sofia being conscious and an impressive integration of recent developments in robotics, Sofia is a very long way from that. By some deprecating descriptions, it remains a chatbot with a face.

CHASING JACKIE CHAN

Teaching robots to see, feel, and react to their environment is a difficult problem whether or not they look like humans. Boston Dynamics worked for years to create a robot that could move like a human. It developed Atlas, a bipedal robot standing about six feet tall and weighing about three hundred pounds. Atlas was designed for search and rescue in rough outdoor terrain. By using its legs for walking, it was free to use its arms to carry burdens and manipulate the environment. Boston Dynamics concentrated on aspects of balance and dynamics, attempting to replicate how animals and people move and react quickly to avoid falling. The legs of Atlas were patterned after human bone structure, and its hydraulic system moved fluid around internally, like a human leg.

Boston Dynamics revealed Atlas to the public in 2013 by entering it in the DARPA Robotics Challenge, a competition held from 2012 to 2015. The early version of Atlas was tethered to a power source and had trouble balancing. In the competition, many of the robots, including Atlas, fell down in the challenging environment of stairs or irregular piles of cinder blocks. The result was a certain amount of derision, but progress continued.

Boston Dynamics teamed up with the Institute for Human and Machine Cognition (IHMC), a research institute of the State University System of Florida in Pensacola. IHMC developed an autonomous, footstep-planning program for Atlas and for a robot named Valkyrie that was constructed by NASA with the ultimate goal of using robots to build living environments on Mars. The IHMC system specifies a beginning and end point for the robot's navigation. It then maps all possible paths one footstep at a time and selects from those paths the one that maximizes the probability of a successful traverse. This step requires considerable processing power, but it is well within the capability of modern, onboard computers. The result is a robot that carefully picks its way through tricky terrain much in the way that a human or animal would.

By 2015, Boston Dynamics could show Atlas running through a forest although still connected to an external power cord. Later versions could walk and run over rocks. More recent versions of Atlas are self-contained with a lithium battery power pack. They are also incredibly agile. By late 2019, Boston Dynamics showed off a streamlined, slicked up, totally self-contained version of Atlas. A video showed Atlas doing parkour, a sport otherwise the realm of athletic young people running, jumping, climbing, rolling, scaling walls, and balancing on rails in order to scramble through an urban environment. Atlas has not yet climbed walls like Jackie Chan, but the video shows Atlas doing somersaults, handstands, backflips, a 180-degree jumping turn, a jumping split, and then an incredible 360-degree jumping turn.[36] In the latter, Atlas cocks itself to the right, then launches left into the spinning turn, in a way entirely reminiscent of how a human would do it. And nails the landing. In yet another smooth move, Boston Dynamics taught Atlas to dance.

FLYING ROBOTS

Another category of robotics has flying as a capability: programmable quadcopters. The quadcopters can be programmed to sense one another and fly in complex formations. A single quadcopter is impressive enough, but a swarm[37] of them is another thing altogether, giving hints of complexity and emergent properties (see chapter 9 for a more elaborate discussion). One of the applications that most impressed me was based on just a couple of quadcopters. Each of these devices could balance a metal rod on its upper surface. Try it. Try balancing a pencil or ruler or another long, thin thing on your finger or palm. Even more amazing to me was that the quadcopters could then flip the rod into the air. The rod would tumble end over end, and then the other quadcopter would catch it, precisely balanced on its upper surface.[38] Once you have mastered the art of balancing a pencil on your finger, try flipping it in the air and catching it, balanced and vertical, on a finger of your other hand. The sheer control and dexterity of this operation still amazes me. I do not know what use this might have, but these are intimations of a future we have not yet fully imagined.

Quadcopters range from toys to sophisticated machines. They often have cameras. Some are steered by a handset while others are autonomous, either following a pre-programmed path or making internal decisions in response to the environment—the essence of a robot. Some quadcopters are manufactured in the United States but many are produced in China.[39] Some are for hobbyists; others have applications in professional aerial photography, surveying, and agriculture. Another application is the inspection of the decaying bridge infrastructure in the United States. The drones can fly under and around the bridges, and the images obtained are then analyzed by deep learning to differentiate between algae, rust, and—all too often—cracks.

One of the major applications under development is that of flying delivery drones. Amazon was one of the first companies to apply investment and research in this area. Amazon developed immense capabilities in cloud computing that is an ingredient in autonomous robotics. Amazon needs autonomous sorting in its distribution centers. It must get all that stuff to its customers by land, air, and presumably sea if the need arises. Human delivery agents are rather expensive. Sorting and delivering by robots is thus a natural enterprise for Amazon. Amazon bought MIT spin-off Kiva Systems in 2012 and renamed it Amazon Robotics. The package-sorting systems in the distribution centers, augmented by AI and Amazon's cloud computing facilities, are now largely robotic. Amazon also took to the air. By 2016, Amazon was operating a nominally secret research operation on a farm east of Cambridge, England, for the purpose of designing and testing flying drones. Other Amazon test sites were in the United States, Austria, France, and Israel. Amazon initiated Prime Air (to be distinguished from Amazon Air that employs human-piloted aircraft) based on a vertical take-off and landing (VTOL) quadcopter that can grip and release a square box and other more novel designs. In 2022, Amazon announced plans to deliver goods in less than an hour in Lockeford, California, and College Station, Texas, with its MK27-2 drone of novel hexagonal design and a new MK30 drone with increased range, tolerance to temperature variations, and an ability to fly in light rain, to be put into service in 2024.

Other US delivery companies have also invested in this area. UPS flies a quadcopter similar in configuration to that of early Amazon designs. Google has a winged drone with six motors and propellers mounted vertically on each of two parallel fuselage booms. It can lower packages on a retractable cable. FedEx also has a winged craft based on two booms with six vertical motors on each boom for take-off and landing plus two horizontal motors to aid lateral flying. The wings promote lift and higher speed forward flight, both of which are more energy efficient over long distances.

Plans to design and employ flying robotic drones have spread internationally. By 2019, twenty-six countries employed flying drones to deliver mail, medicine, and other goods. In 2016, the government of Rwanda and Zipline formed a partnership to provide the world's first national drone service to deliver emergency blood supplies and vaccines to clinics. Matternet provides a blood delivery service in Switzerland. DHL has deployed a winged craft, ParcelCopter, in Tanzania and elsewhere in Africa to deliver medical supplies. In Brazil, the food supply giant iFoods plans to employ flying drones to shorten delivery times with only the final leg still done by humans. Just Eat has tested flying drones to provide rapid food delivery at University College Dublin, and they have plans to spread through Europe.

Although there are carbon production issues associated with providing their batteries with energy, these flying drones themselves emit no carbon. This property gives them a distinct advantage over ground-based vehicles powered by fossil fuels. There is also the argument that flying delivery can ease the carbon emission associated with street congestion. Optimists foresee that if piping-hot meals can routinely be delivered within three minutes—a goal for some delivery companies—then people's food-ordering habits may change, with people ordering both more frequently and in a more targeted, selective fashion.

These flying delivery drones thus have a great deal of potential and a significant cool factor, but there are issues with their implementation. The lack of an air traffic control system for drones is a significant impediment. Skies full of drones could make drones a danger to one another and to traditional aircraft. England was plagued with drones near Gatwick Airport and had considerable difficulty tracking the scofflaw operator. The Federal Aviation Administration (FAA) is wrestling with this issue. At first the FAA insisted that drones only be flown within eyesight of the operator, but this restricts the employment of autonomous drones in legitimate, large-area deployment. The FAA has begun to relax those rules[40] but is still some way from universal rules for drone air traffic control. There are hazards such as difficult-to-see electrical cables in both urban and rural environments. These issues may be addressed at some level by employing geofencing that uses GPS to create virtual boundaries beyond which a drone cannot fly. The FAA announced revised rules in 2021 that will allow operators of small drones to fly over people and at night under certain conditions. The hope is that these rules will support technological and operational innovation.

The FAA, NASA, other federal agencies, and industry are exploring the operation of a system to enable multiple drones to fly beyond direct visual tracking and less than four hundred feet above ground where the FAA does not provide air traffic services. This Unmanned Aircraft Traffic Management (UTM) system would be complementary to the FAA's Air Traffic Management (ATM) system.[41] Europe has plans to implement U-Space, a management system for small drones that would connect all drones to one another and allow air traffic control, ambulance services, the military, and the police to track them. China may have an advantage in building up a drone delivery system because it is less hampered by the regulations planned for the United States and Europe.

Flying drones raise other issues. There are threats to privacy even at the neighbor-spying-on-neighbor level and tricky questions of who controls the airspace above a piece of private property. There are instances of people on the ground taking potshots at drones flying nearby. A bald eagle attacked a state-operated drone and dropped it into Lake Michigan. There is also physics. The whole point of flying drones is to gain altitude, struggling against gravity. Flying is intrinsically less energy efficient than rolling. Delivery companies such as FedEx, Amazon, and DHL have found a balance between packages flown by regular aircraft between cities and what is delivered on the ground within those cities. That balance may be found for flying drones, but there may have to be significant developments in battery life and energy efficiency before it is financially practical to scale up flying drone delivery.

There is (at least) one more practical issue facing the implementation of common robotic flying drone delivery. In publicity videos for its flying drone, Google shows a happy family standing on their front stoop ready to grab the goodies lowered by cable from a hovering Google drone like manna from heaven. Another video advocating Amazon Prime Air has a smiling woman standing on her porch in similar anticipation of the craft landing on her front lawn. I'm motivated to ask just how long are these people supposed to interrupt their lives hanging out on the porch? Suppose the blonde woman is trying to balance a work-at-home job with changing diapers. Perhaps the drone has given an exact delivery time so the folks at home can precisely time their arrival on their porch. There are already people tracking Amazon delivery vans and then swiping delivered packages from porches. One can imagine nefarious people tracking flying drones, ready to pounce. The delivery companies may have ways to limit such thievery or make restitution if it happens, but the issue adds one more complication to the efficacy of flying drone delivery.

The bottom line is that the technology for flying delivery robots exists but practical issues will likely prevent it from becoming ubiquitous for some time.

DELIVERY ROBOTS

Rolling delivery drones also face issues, but rolling drones are much closer to routine employment than are flying drones. Many companies are developing delivery robots and the infrastructure to deploy them. Although they can cross roads as needed, these devices are designed to use sidewalks, not roads, so they do not need to meet the higher safety standards of self-driving cars. Like self-driving cars, autonomous delivery drones are equipped with high-speed cameras and 3D laser scanners called Lidar[42] to detect objects around them. They employ AI and GPS to map their surroundings and their location to within about an inch. An object-location system allows them to avoid other mobile items that are not on the map: pedestrians, trash cans, or cars in driveways. Some have clever tilting wheelbases that enable them to operate on irregular ground or even reach porches by climbing a few stairs. The devices are locked until delivery and have alarms, cameras, and constant GPS tracking to deter and detect thieves. Development of rolling drones was encouraged by the COVID-19 pandemic, which added to demand for contact-free home delivery and put more burdens on delivery people. As of 2020, rolling drones had made about 100,000 deliveries in the United States, many of them on college campuses. They were also deployed in the United Kingdom, Europe, and China.

Janus Friis and Ahti Heinla founded Skype and then sold it to Microsoft. They used some of the proceeds to start a company in Estonia in 2014 called Starship Technologies that makes small delivery robots that are also called Starships. Starships are rounded tubs mounted on two pairs of three wheels. The top flips open to reveal a carrying space of a few cubic feet that can hold up to twenty-five pounds. In 2018, the company deployed the first few hundred Starships in the Bay Area and by 2020 in about a dozen communities, often universities. Customers order and pay through a mobile app operated by Starship Technologies that is similar to those of Uber or Lyft. The orders are relayed to a merchant who puts the order into the drone's delivery compartment. The Starship then rolls at up to four miles per hour to the consumer's location within a pre-mapped area of about ten square miles and unlocks the flip top, giving access to the purchase. Starships can run completely autonomously, but in the current system a human stands by to handle unexpected situations.

With its immense demand for delivery, Amazon has employed robots at all stages of the process, including rolling delivery robots. Amazon bought Dispatch, a robotics company, in 2017 and, in 2019, introduced Scout, a six-wheeled delivery drone with a close resemblance to a Starship. Scout's wheels tilt, enabling it to surmount some obstacles. Amazon plans to use Scout as one component of its last-mile problem: how to get items from local distribution hubs to customers in the most efficient way. Customers order online in the normal way, and Amazon decides whether robot delivery

is the most appropriate solution. Amazon first tested this system in an area
north of Seattle and then around the University of California, Irvine. It has
since spread to cities in Tennessee and Georgia. Amazon later introduced
REV-1, a three-wheeled version, in Austin, Texas, and elsewhere.

In 2019, FedEx announced a four-wheeled delivery robot named Roxo
that could climb stairs. FedEx is experimenting with Roxo in the United
States and in Dubai. A delivery robot called Aida is also designed to address
the last-mile issue. Aida walks on four legs and otherwise resembles Spot
from Boston Dynamics. Robby Technologies presents a delivery robot with
a six-wheeled platform that tilts to climb curbs, stairs, and other irregular
terrain.[43] Kiwibot aspires to provide the cheapest robotic delivery with a
small, four-wheeled delivery bot.[44] It started delivering snacks to students at
Berkeley but expanded to a neighborhood in San Jose in 2020. The bots are
not yet fully autonomous and are monitored by humans. You might think
that that Kiwibot has something to do with New Zealand, but the monitors
are in Columbia, the original home of the company founders Felipe Chavez
and Sergio Pachón. Kiwibot is developing plans to integrate delivery robots
into business-to-business plans rather than directly serving customers.
Savioke has developed Relay, a delivery robot that resembles a gracefully
inverted trash can.[45] Relay is intended to operate indoors, making deliveries
within hotels, hospitals, and businesses. Nuro is a company started by two
Google engineers, one of whom also had experience with Mars rovers. The
Nuro R2 delivery vehicle is about the size of a Volkswagen bug.[46] It has a
special dispensation to operate on streets because it carries only merchan-
dise, not people. Panels flip up to reveal two delivery compartments on
each side. Although consolidation in this area is likely, all these initiatives
show the active dynamics of rapidly growing enterprise.

As for flying delivery drones, there are still problems to be worked out.
Merchants and customers need to wrestle with the costs. Communities
need to maintain control of their sidewalks as has been shown by the invasion of rental scooters in some locales. The smiling lady in promotion videos
may get tired of standing on her porch if it is raining.

Other companies such as Toyota, ThyssenKrupp, and Segway are trying
to develop drones to serve particular delivery niches. Amazon currently
delivers billions of packages per year and will surely be a dominant player
going forward given its immense capabilities in robotics, AI, and computing. On the other hand, the Chinese commercial giant Alibaba has plans to
deliver 1 billion packages *a day*. By some estimates, it will take several years
to refine the technology of the robot delivery system and scale it up before

most people have interacted with a delivery robot and perhaps a decade before they are a routine. Given exponential growth, this may happen faster, but a decade already takes us close to Kurzweil's estimated singularity.

TEAM PLAYERS

Most of the robots mentioned above were designed to address certain reductive, ground-up problems. How do I make a robot see? How do I make a robot move? How do I make a robot walk? How do I make a robot grab? There is another way to tackle these issues: a top-down approach. In this case, one asks a more global question, for instance, How do I make robots that interact? The task is then to address issues of seeing, moving, walking, and grabbing in service of that top-level goal—not just repetitive motion but interaction in a dynamic environment.

Although it has certain whimsical, testosterone-laden goals, one way to address this approach is through robot competitions. In *Robot Wars*, teams of amateurs or professionals design robots to compete with other robots by flipping, crushing, or otherwise disabling them. Throwing your robot competitor out of the arena constitutes a win. *Robot Wars* was shown on US and UK television.

RoboCup is a somewhat more benign but still challenging event. RoboCup is an international competition to get a team of robots to play soccer against an opposing team of robots. The robots have to do all the difficult tasks such as seeing the playing field, the ball, their teammates, the opposition, and the goal. They must be able to move, but in this case in a dynamic environment as the ball rolls around and players shift positions. The extreme challenge in this case is that they then must employ tactics and strategy to cooperate with their teammates, outmaneuver the competition, and move, pass, and follow the ball. Finally, almost anti-climatically, they must kick the ball into the goal.

In order to do all this, competitors have to build complete robot agents that can perform increasingly complex tasks. This has been accomplished by employing AI reinforcement learning as outlined in chapter 3. A reward is defined, and a policy is set to get the most reward. The AI then learns on its own how to maximize the reward. Starting from scratch, the AI learns to sense, decide, and act in a closed loop. These agents then must interact with other agents. Through iteration, the AI improves with experience. Initially, autonomous agents learn in the presence of teammates in real-time, dynamic circumstances. In practice, much of this AI learning is done in simulations that require relatively little data from the real world. This allows the AI to learn in small, efficient steps.

My colleague Peter Stone, a professor of computer science at the University of Texas at Austin, has spent a significant portion of his career pursuing these goals. Stones's team began by using Sony's dog robot, Aibo, which could not kick, but could shuffle a ball into a goal. More recently his team has used Nao, which can both walk and kick. His 3D simulation team, UT Austin Villa, has won eight of the last ten international competitions. In 2016, his team of Nao robots won the RoboCup international soccer competition. Stone points out that the technology for playing team soccer is translatable to self-driving cars. In that context, cars can navigate busy intersections, including making left turns through traffic by registering, communicating, and coordinating with other team members—that is, all the other cars navigating the intersection. In this future, there would be no need for red, yellow, and green lights.

Robots of all sorts have a rich future. Some roboticists estimate that the state of the art in 2000 was roughly comparable with that of computers in the 1970s before the explosion of supercomputers, personal computers, and pocket computers in smart phones as accomplishments built on accomplishments, resulting in approximately exponential growth. As of 2020 there are well over two dozen companies in the United States designing and building robots in not just Massachusetts and California as one might expect, but also Austin, Texas, and Boulder, Colorado, and there are many more around the world. Advances may come much faster than even the robot optimists foresee. As for other technologies, this one is driven by both intellectual fascination and economic demand.

THE LAWS OF ROBOTICS

Advances in software can be propagated by the technique of *fleet learning*. Rather than laboriously updating every robot, once you teach a robot a new trick in their AI software, they all know it at essentially the speed of light. Classes in robotics are taught from elementary school through graduate school, ensuring a steady, even growing, supply of roboticists. Robotics will also benefit from advances in AI and associated technologies. While many robots are self-contained, many others—for instance, delivery robots—rely on the remote computational power of the cloud. The deployment of 5G and more enhanced communication technologies will increase bandwidth and make such communication more efficient. The demand for high bandwidth and low latency calls for more distributed computing, giving robots an inherent benefit from edge computing.

The technology of building robots will also advance. Research robots are meticulously constructed by hand. Even mass-production robots like Nao take time to assemble. Already there are steps toward taking humans out of the loop. Researchers at MIT have developed the ability to 3D print a small hexapod robot with a complex structure powered by hydraulic fluid that is added by inkjet printing during production.[47] A small motor and battery must be added by hand, but otherwise the robot emerges complete from the printing process. One can foresee robots building other robots or even constructing themselves. Some MIT researchers have also made progress on that front.[48]

Researchers are aggressively addressing challenges. To function in the world outside labs, even to deliver snacks on college campuses, robots must function in complex environments. The current generation of robots still represents primarily one-off designs. Standards need to be developed so common functions can be integrated into robots of a wide variety of designs. Robots must learn to adapt to changing environments, a remarkably challenging capability to encode in software. Over short time frames, unusual circumstances are rare. On the long term, unusual circumstances are inevitable. How to accommodate that? Robots must anticipate the motions, body language, and wishes of humans and send sufficient signals that people can anticipate how the robot will behave. All that is necessary before robots can be integrated into a society where average people and robots comfortably collaborate, extending the capabilities of both the robots and the people.

Other issues beyond the strictly technical will also have to be addressed. Both law and medicine have long histories of incorporating ethics into their training and function. Robotics has very little of this, although, as for AI, this is beginning to change. There needs to be forethought given to anticipating and addressing ethical issues such as bias in algorithms. As an example, robots need to be able to anticipate how people move. One can build statistical models to encode that, but human motion can vary with culture. How is such an issue addressed?

Another issue is regulation. We need to have robots that can work with people without endangering them. How does one avoid overregulation in the quest for oversight? How does one ensure that safety standards do not suppress innovation? Mistakes are inherent in research and necessary for progress. Small ones are allowable and are even to be encouraged, but one large accident can set the whole field back.

There is likely to come a day when robots design robots. One can imagine a robot coupled with an AI working at the level of AlphaGo Zero and asking it to design a robot. Given AlphaGo Zero's ability to devise strategies beyond human comprehension, the result could be robot characteristics unlike anything a human would or could conceive—likely nothing humanoid.

Isaac Asimov addressed a world populated by both humans and robots in his collection of stories *I, Robot*, where he introduced his famous three rules of robotics.[49]

> *First Law*—A robot may not injure a human being or, through inaction, allow a human being to come to harm.
> *Second Law*—A robot must obey the orders given it by human beings except where such orders would conflict with the First Law.
> *Third Law*—A robot must protect its own existence as long as such protection does not conflict with the First or Second Law.

Asimov later added a Zeroth law: "A robot may not harm humanity, or, by inaction, allow humanity to come to harm."

It remains to be seen whether a system conforming to these guidelines can be achieved.

Scott Adams, the creator behind the comic strip *Dilbert*, brilliantly and concisely captured the dilemma commonly felt when contemplating the new world of robots and AI. In a strip on March 28, 2013, Dilbert addresses the office robot, saying, "Luckily the three laws will prevent you from hurting us." The snarky robot replies, "Yes, because that is totally a real thing."

5

AUTONOMOUS VEHICLES
Who Sues Whom?

A REVOLUTION IN TRANSPORTATION

We have made steady progress in transportation from horses to cars to airplanes to rockets. It is unlikely, however, that we will have some radical new mode of transportation like a *Star Trek* transporter; the physical limitations are overwhelming.[1] In terms of fundamental technology, transportation will probably not grow exponentially. Nevertheless, even without new forms of propulsion, change is coming to how we move about.

One of the first places where all this will come together—AI, robots—in a manner that will disrupt society is likely to be with self-driving cars and trucks. We witnessed a precursor disruption one hundred years ago when the automobile replaced the horse and buggy. Horse-oriented industries folded. Streets were paved over to allow the efficient rolling of rubber tires. Cities expanded into suburbs. Farm roads sped produce to market. The national highway system constructed in the 1950s in the United States promoted long-distance trucking and encouraged a road trip culture. Vast industries developed to make car parts and assemble them into an affordable product. Automobiles themselves got ever more sophisticated and safer. The built environment of cities had to accommodate all vehicles with adequate streets and parking facilities. The insurance industry provided liability protection. Much of this is likely to be redone or undone.

Self-driving cars and trucks are not yet ubiquitous, but the signs of change are all around. Many companies, some famous and some not so well known, are putting autonomous vehicles on the road. As for so many nascent projects, the Defense Advanced Research Projects Agency (DARPA) had a role in this, promoting grand challenge autonomous car competitions starting in 2004. As was the case with robots, the first cars in this challenge had difficulties. They broke down in the desert or ran off the prescribed path. Then they got better. Marc Andreesen, inventor of the first widely used web browser, Mosaic, and subsequently a venture capitalist remarked, "People are so bad at driving cars that computers don't have to be that good to be much better."

Google got into the game as an offshoot of its Street View project to map and panoramically photograph every street in the country, if not the world, by hiring key participants in the DARPA grand challenges. Various states—Nevada, California, Arizona, Texas, Florida, Michigan, Washington—passed laws to regulate and permit autonomous vehicles. In 2015, the first entirely driverless car to operate on a public road drove on the streets of Austin, Texas. Developed by Google, the car had no steering wheel, gas pedal, or brake pedal and no police escort. In 2016, Google spun off its autonomous car division as Waymo. Google had driven autonomous test vehicles on the streets of San Francisco for several years before going public with the project. Waymo developed autonomous car and taxi service in some neighborhoods of Phoenix. The ride-hailing company Uber began an aggressive program of development and testing of autonomous cars in Arizona and Michigan. Uber hired some of the pioneers from Waymo, and lawsuits ensued charging theft of intellectual property. In 2022, Lyft announced that it would start offering driverless rides in Austin. A web of relationships has developed among the autonomous car companies, automobile manufacturers, and repair and maintenance companies. GM, Honda, Microsoft, and Walmart have collaborated to design and deploy an autonomous vehicle (AV) called Cruise Origin. While much of the focus is on the United States, particularly Silicon Valley, the Russian web conglomerate Yandex is also developing self-driving cars.[2]

LEVELS OF AUTONOMY

In 2013, the National Highway Traffic Safety Administration (NHTSA) defined six aspirational levels of automation for AVs. Level 0 corresponds to no automation; that is, the human driver does it all. This is the level at which many of us learned to drive, but this is rapidly vanishing. In level 1, there is some driver assistance, like cruise control. Level 2 is defined by partial automation: the AV system controls acceleration, braking, and steering but the driver is expected actively to monitor the driving environment. There are many level 2 vehicles on the road now. Level 3 is defined as a state of conditional automation. The system monitors the environment and makes all the basic driving decisions but the driver is expected to maintain attention and be ready to retake control of the vehicle. Level 4 is characterized by high automation. The AV does essentially all the driving and does not rely on human backup though human backup can be requested. Modern airliners operate very safely on something like level 4. The Boeing 737 crashes in 2019 were an exception, showing what happens when software is imperfect. Level 5 implies full automation. Humans are not drivers but passengers along for the ride. Level 4 and 5 AVs will have a broad impact.

Although practical level 5 AVs are the goal, much relevant development work is proceeding by adding aspects of the technology to current automobiles. We have long had cruise control, a level 1 characteristic. The first example of operational cruise control was a 1994 Mitsubishi. Tesla has adopted some level 2 hands-off driving capability. Mercedes has incorporated AI technology that learns your driving patterns and habits. Many cars now have systems to keep them in their lane and brake automatically. Some have self-parking systems. These technologies put us deeply into level 3. Work is proceeding on level 4 and level 5 AVs.

Motorcycles will continue to share the road with other AVs of all levels. The concept of automating a motorcycle remains anathema to many motorcycle riders, and the software is tricky because motorcycles must lean into turns. Nevertheless, in 2020, Ducati and BMW installed forward-facing radar and cruise control on some of their models. Being a different level of threat to a motorcycle driver than to a driver of a four-wheeled vehicle, auto-emergency braking will never be employed in motorcycles.

Self-driving cars are not programmed with a series of IF/THEN statements but rather an AI that learns the best solution. The question will be what is best. In 2020, researchers at the Technical University of Munich presented an AI system that attempted to take into account all relevant factors—other vehicles, pedestrians, obstacles—at a given instant to make the optimal decision for how to proceed in the next instant.[3] In principle, the system predicts every potential future evolution of an array of current possible decisions. The system was reputed to handle common, fraught situations like turning left at an intersection better than humans. The system assumed that it could detect all relevant factors and that other vehicles were operating legally. The system had not yet developed driving strategies unconceived by human operators, but with technology like AlphaGo Zero, the potential looms.

There is no reason why traditional cars driven by fossil fuel cannot be employed in an autonomous mode, but much of the development has been done with an eye to changing technology and the threat of global warming. Many of the self-driving vehicles have thus been hybrids or all-electric. This trend is likely to tip suddenly as the price of electric vehicles declines. Some progress is being made, especially for large fleet vehicles, toward the use of fuel cells that burn hydrogen and expel water. As an example, Toyota is developing an eighteen-wheel semitruck powered by fuel cell electricity.

Although far from self-driving, the introduction of electric scooters in some urban environments proved an interesting experiment in the rapid changes that ensued when a new mode of transportation became widely available. We saw during the COVID-19 pandemic how traffic can change with circumstances. With fewer people commuting and restaurants desperate to maintain business, streets were narrowed, traffic rerouted, and restaurants expanded to the sidewalk and then into the street.

In 2015, the consulting company McKinsey issued a report outlining what they saw as the impact of AVs that will be driven by savings in costs of labor and by the reduction in carbon dioxide emissions expected from optimizing driving efficiency.[4] The report sketched three eras. The first extended from then, 2015, into the late 2020s. In this first interval, there would be the increasing use of industrial fleets, and AVs would be developed for consumers. In the second era, spanning the late 2020s through perhaps the early 2040s, consumers would begin to adopt AVs. The third era would see level 5 AVs with the use of AVs encouraging the adoption of robots in other facets of life. McKinsey did not comment on the possibility that AVs would be conscious by the third era.

McKinsey suggested that in the first era, original equipment manufacturers—the Fords, GMs, and Toyotas of the world—would evaluate the impact of AVs on their business and begin to shift the focus of their design and manufacturing. The Ubers and Waymos would begin to add fleets of AVs that operate in restricted geofenced conditions. Automated excavators, forklifts, and loaders would find use in construction, warehousing, mining, and farming where the environment is intrinsically controlled. That seems to be just where we are in the early 2020s.

WIDESPREAD IMPACT

As more people use AVs, private ownership of cars is likely to decline. Whereas cars have been both utility and pleasure vehicles, the trend will be toward public transportation. Meticulous care of personal vehicles could give way to indifference. The transition to broad use of AVs in era 2 will be slow until people are convinced that AVs provide all the advantages and convenience of private ownership. Private ownership may be inefficient, but it provides great convenience. It is easy to carry groceries from the garage to the pantry. Parents drive kids to soccer matches on demand. Private cars are ready immediately if a child or an elderly parent becomes ill. AVs will have to prove their worth in the hurly-burly of everyday life.

As private ownership declines, a whole culture of customizing cars will change—a great loss to my mind. No more low-riders? No more tangerine-flaked, stream-lined babies? No more cruising on Saturday night? I would prefer to think the culture morphs rather than vanishes. Perhaps a cherished remnant of this culture will survive for a while in the hobby world.

Already there is a trend for cars to show much less variety than they did fifty years ago. If we move to fleets of self-driving cars, the whole automobile design industry will have to adapt as homogeneity drives the economics. What happens to the sports car industry? Some electric cars can out-accelerate traditional fossil fuel cars,[5] and that gives me some grounds for optimism. I suspect the human drive for creativity—and speed—will find a way to thrive, but it is difficult to anticipate in what form.

While autonomous vehicles will have to park somewhere and be serviced, these functions will be centralized with much less demand for vast distributed parking areas and garages associated with both commercial properties and private homes. The land required for parking is likely to be reduced in the United States alone by thousands of square miles. The urban environment will thus probably be redesigned and reallocated. Already a community near Phoenix called Culdesac is being designed to be car-free in reaction to the pandemic and in anticipation of functioning fleets of commercially owned autonomous cars. This concentration may work in urban areas but less well in suburban or rural areas. People loaded with groceries will not want to walk great distances from an AV car park in the rain. Rural homeowners will need to easily get into town to buy provisions.

There will be a complicated phase in era 2 when there is a substantial mix of both autonomous level 4 and 5 and human-driven level 2 and 3 cars on the road. Eventually, as autonomous intercommunicating vehicles dominate, traffic will flow steadily. There will be no need for traffic lights as cars and trucks communicate directly with one another in elaborate synchrony. While the need for parking will shrink, there will have to be new solutions for traffic flow as the AVs circulate, waiting for new commands. As AVs communicate with one another, speeds and traffic density may increase substantially, impacting the design of roads and urban environments. Curbs may disappear, streets may narrow, and manually driven cars may be forbidden in ever larger areas. AVs will be used as part of the sensor network that powers smart cities.

If fleets of electric self-driving cars come to dominate, there will be a need for ubiquitous charging stations but no need for the current broad distribution of gas stations and repair shops. The big manufacturers of AVs will have a strong economic motivation to do all the servicing. The manufacturers likely will have to redesign their supply chains and associated logistics. Commercial drivers may become broadly unemployed and in need of retraining, perhaps to build and service self-driving vehicles if robots have not taken all those jobs.

Broad use of AVs is likely to substantially reduce the rate of automobile deaths and accidents. Currently about forty thousand people die on US roads in a given year, mostly by human error. A reduction of this carnage by even 10 percent would be significant. While a quantitative estimate is problematic, the McKinsey report projects that the accident rate could be reduced by 90 percent, saving as much as $200 billion per year and a proportionate number of lives. Any significant reduction in the accident rate will, in turn, have a substantial impact on the healthcare industry because it will no longer have to accommodate the current rate of slaughter. In addition to saving lives, AVs would affect associated medical costs. For every person killed, about ten must spend time in a hospital and one hundred must visit an emergency room. The supply of organ donors will decrease. The insurance industry will have to adjust if human drivers are not responsible, with insurers likely shifting from covering individuals to covering technical failure. Tricky issues of liability will have to be resolved. People in accidents will tend to go after companies with deep pockets that will naturally want to defend themselves. Will bank robberies go down because no one can drive the getaway car?

Given that AVs will probably be electric, other practical issues arise. When there are only a few electric vehicles on the roads, there are only a few problems, such as the scarcity of recharging stations. When all vehicles are electric, there is a whole host of new problems. How are batteries recycled? What happens to batteries in landfills? What happens in accidents when battery fluid leaks? What are the demands for rare earths and issues involved with their mining, processing, and leaking from landfills? What happens if there is a wide-spread power outage and charging stations shut down? What happens in an electrical fire in a car or at the charging station? What happens to electric bills?

There may be other more subtle effects of AVs. The average driving commute time in the United States is a little shy of one-half hour. One-way commutes of ninety minutes are not unknown. If private driving basically vanishes in era 3 with level 5 AVs, then that time will be freed up for other pursuits, some productive, some maybe not. McKinsey estimates billions of hours saved per year globally. One can foresee that much of that time could be spent on the internet—perhaps using the latest brain-computer interface (BCI)—with someone profiting from that increased usage.

Accidents will happen. A driver died in Florida when a reflection from a truck confused the optical system of a Tesla that plowed into the truck at high speed. An Uber vehicle in Phoenix struck a pedestrian who wheeled her bicycle onto the roadway at night. The Uber AV had a driver for backup, but neither the car nor the driver reacted quickly enough to protect the pedestrian. The driver, not Uber, was later charged with manslaughter,[6] but Uber eventually shed its self-driving car effort along with a related enterprise to develop an air taxi. After an accident-free period, an Apple autonomous car had an accident: one of its cars was rear-ended. The driver of the other car was held responsible. That event may have altered Apple's plans for its own self-driving car, but current statistics suggest that most accidents involving autonomous vehicles are due to human error. The transition phase with a mix of human-driven and self-driven cars on the road is apt to be characterized for some time by continued, but human, error.

There are other issues with moral and legal implications that must be faced. If every AV knows exactly where every passenger was picked up and dropped off, questions arise of privacy and the possibility of mass surveillance. An AV system could be jammed, or hacked, or loaded with malware. There was a phase when Jeeps could have their brakes locked by someone hacking through their mp3 music players. Great attention must be paid to security as for the IoT in general. It may be possible to implement privacy protection by using AI technology to anonymize the data. Whether this happens remains to be seen.

AV TAKES FLIGHT

While all the exponential progress in AV technology is racing ahead, there is the possibility of a whole new dimension—literally. Chapter 4 summarized the work being done on civilian, robotic, flying drones. As predicted by science-fiction literature, versions of robot technology that will carry people may also finally take to the air. Electric take-off and landing vehicles (eVTOLs) are projected to use only 20 percent of the energy of conventional fossil-fueled helicopters and hence be much more environmentally friendly. They do not require runways to take off and land but can operate from rooftops. They will be much quieter than helicopters and have multiple rotors and backup systems, making them much safer. A trip from downtown to the airport might cost less than $100.

The eVTOL company Terrafugia was started in Woburn, Massachusetts, in 2006 by some MIT graduates. It was purchased in 2017 by Chinese conglomerate Zhejiang Geely Holding Group, which also owns Volvo and Lotus. After several prototypes, Terrafugia is currently working on the Terrafugia TF-X. The TF-X is an autonomous flying car designed to carry four passengers. It is powered by two hybrid, 600-horsepower electric motors and a 300-horsepower fossil fuel engine, and it can take off and land vertically. Its wings fold to fit in a single-car garage, and it can operate on the road. While it has been advertised for some time, the TF-X has yet to be released to the consumer market. The estimated price is $300,000. DARPA has plans to develop a flying Humvee using some of Terrafugia's technology. In 2020, the Japanese company SkyDrive released a video of its prototype SD-03 electric flying car. While many companies are working on eVTOL, SkyDrive appears to have been the first to achieve an untethered flight. It sought to put the SD-03 on the market in 2023 but apparently suffered delays. Omer Bar-Yohay and Aviv Tzidon founded the electric aircraft company Eviation[7] in Tel Aviv in 2015. Eviation developed an all-electric airplane powered by 21,500 small, Tesla-like batteries and designed to carry nine passengers. The craft had its first flight of eight minutes in Moses Lake, Washington, in September 2022. In 2020, there were about one hundred projects around the world working on flying cars.[8] As with drones, new regulations will be needed to confront an era of abundant flying cars. A leisurely stroll outdoors would not be the same with the fear that a flying car might plummet from the sky.

It is a challenge to predict when all this technology will come together, if ever, on the ground never mind in the air. There remain many practical issues to implementing level 4 AVs, with level 5 AVs presenting even greater challenges. While there is much optimism, there also remains a degree of pessimism that a full level 5 will ever be attained that can negotiate a Michigan winter blizzard. There are also complicated issues of how to maintain our social fabric as AVs and other aspects of smart cities like the IoT wreak their changes. Will massive, high-tech companies control all this technology or will something like the postal system evolve where AV service is a utility delivered to people?

An important step envisaged by many is to speed communications with and among AVs. AVs are currently operating in restricted, geofenced neighborhoods because they are still not capable of handling the chaos of an urban traffic jam. The next step in this regard—but undoubtedly not the last in the next few decades—is 5G technology. As the successor to previous communication networks that powered our mobile phones, 5G promises to work at higher frequency, yielding greater bandwidth, smaller delays (latency) in exchanging data, and higher data density. This high-capacity network will not turn on all at once but by a process of continuous updating and innovation over the next several years. It should allow AVs to collect data on their environment much more rapidly, communicate with one another efficiently, and compensate for the blind spots of other vehicles. Rather than making all driving decisions with the vehicle's computer, data could be uploaded and downloaded from the huge computing capacity of the cloud at practical rates. Eventually a quantum internet might speed intercommunication even more.

THE TROLLEY PROBLEM

If autonomous vehicles are to reduce the accident rate, there are not just significant practical engineering problems to be addressed but also philosophical issues. One key issue is encapsulated in the *trolley problem*. This ethical dilemma has a long history, but it was brought to focus and named by British philosopher Philippa Foot, who was one of the founders of modern virtue ethics.[9] While there are many elaborations, a basic version of the dilemma posits a trolley speeding down a trolley track. On the track ahead, five people are tied to the track, unable to move. There is a switch that will route the trolley onto a second track to which a single person is tied. The central ethical issue is then whether a bystander should do nothing, allowing the five people to die, or throw the switch knowing that the five will live but the one will die. Level 5 AVs will have to make decisions analogous to the trolley problem: Protect the few or the many, the passengers or the pedestrians?

In some fashion, AVs will have to address the trolley problem just as a human driver would. A human driver might, in principle, be faced with the issue of hitting a child on the road or steering the car off a cliff to certain death. Who knows what a given human driver would do? How is an autonomous vehicle programmed to handle such a situation? It is not so simple as writing a few lines of code like the laws of robotics that say, "Thou shalt not kill." Autonomous vehicles will be controlled for the foreseeable future by some form of deep learning that can make strategic decisions—an extension of AlphaGo Zero playing an especially complicated form of Go. How will such programs handle these situations? How, indeed, do current autonomous vehicles do so? They are programmed to avoid collisions with other vehicles, solid obstructions, and pedestrians. The result is that current autonomous vehicles tend to drive super cautiously compared to the average human driver. That is one reason they are prone to being hit from behind. One clot of self-driving vehicles ground to a halt as each tried to drive more slowly and carefully than the others. It remains an open issue how contemporary autonomous vehicles will handle concrete ethical dilemmas.

When we addressed this ethical issue in my class, I asked the students to vote. The question I posed was "If you were to purchase a self-driving car, would you buy one that was programmed to protect you, the passenger, or five pedestrians?" Over the course of several terms, the answer was nearly unanimous that the students would select a vehicle that protected them.

The Media Lab at MIT did a more sophisticated but also fundamentally different version of this poll.[10] They constructed a program called the Moral Machine that presented a variety of scenarios involving AVs. An AV with a varying number of passengers could crash into a barrier killing all the passengers or swerve to protect the passengers but kill a successful businessperson, a known criminal, several elderly people, several young children, pedestrians crossing the road against a signal, or a herd of cows. The survey ran online for four years, and 500,000 people gave 40 million responses that were reported in 2018. People tended to vote to save humans over pets, as many people as possible, and young people over old people. There were tendencies to save females, people of higher status, and pedestrians. There were also interesting variations among nations and cultures. The MIT team aspired to spark a global conversation to guide the engineers who will design moral algorithms and the policymakers who will compile regulations. It is worth noting that the Moral Machine experiment had certain biases. The voluntary participants were not selected using careful sampling techniques. They had to have internet access and an intrinsic interest in the project. There was nothing like a double-blind sampling. I shudder to think what a statistician would make of my classroom polling.

The difference in the responses of my students and responses to the MIT poll represents a dilemma known in economics as the *tragedy of the commons*. What is good for individuals may not be ideal for society as a whole. An example arose during the pandemic of 2020: individuals who felt their personal rights were infringed challenged the public utility of wearing masks. I asked my students what sort of car they personally wanted. The Moral Machine deliberately inquired as to what sort of car others should have. People want other people to have utilitarian vehicles that minimize accidents and death, but they themselves prefer to have vehicles that make the safety of passengers paramount. The result is that people tend to avoid autonomous vehicles that serve the greatest good at the expense of the occupants.

Determining the level of safety required for AVs will be complex. What must be accomplished to have a fully mature AV ecosystem has never been done before. People tend to resist the notion of regulations that might enforce the utilitarian, ethical standard of saving the most people in any given situation. An ironic outcome is that if regulations promote maximum collective safety when people want personal protection in their cars, the adoption of AVs might be delayed, thus ultimately costing more lives. Manufacturers, consumers, regulators, lawyers, and engineers will all have different perspectives. It will be important for all stakeholders to appreciate the complexity that regulators and governments will face in guiding the algorithmic decisions we build into AVs. The cultural differences suggested by the Moral Machine experiment will make it difficult to develop a single, global, moral algorithm. Autocratic regimes might make different choices than would democratic governments. Care must be taken to ensure manufacturers do not somehow favor their own product.

One proposal was to put the burden of ethics on the users of AVs by installing an "ethical knob."[11] The idea is that a given passenger could dial in a level of ethical prudence from "Protect me at all costs, to hell with the kid chasing the ball in the street" to "Always protect the maximum number of people, even if they are a bunch of escaped criminals." Perhaps an AI will read the brain waves of the passenger and set the knob without the passenger being consciously aware of their ethical biases.

Another proposal is to invoke laundered preferences. A regulator might start with public views as they emerge from something like the Moral Machine but then apply ethical theory developed by moral philosophers to ascertain whether a given rule is morally defensible.

Any AI AV system will have to constantly make decisions that may not obviously constitute an ethical choice but will have cumulative ethical implications. Each momentary decision—speed up, slow down, or change lanes—will subtly shift risk. A bias toward saving pedestrians over passengers may not show up in any given incident but might emerge after millions of such decisions, thus constituting not an immediate but a statistical trolley problem.

No system is perfect. Society will have to decide how safe is sufficiently safe. One standard is the condition that AVs operate more safely than humans do. That may not be sufficient because AVs are not people so they will be judged by different standards. AVs will not fall asleep or be distracted by a mobile phone call. They will not drink and drive or suffer road rage. AVs will make different types of errors. They might mistake a trashcan for a person or vice versa. An insurance company or a jury might see death by computer error very differently than death by human error.

Germany attempted to implement relevant regulations. German law says that autonomous vehicles must avoid accidents and death at all costs. It further stipulates that any algorithm regulating the behavior of the vehicle must never make decisions based on the gender or the health of passengers or others affected by the performance of the vehicle. The law mandates a black box to record the travel and whether a human driver or a self-driving system is in control moment to moment. The notion is that the driver is legally responsible for accidents when they are in control, but if the system is in charge, the manufacturer is legally liable.

In the United States, the NHTSA is wrestling with these issues. Current rules insist that even self-driving cars should have steering wheels and brake pedals, items not necessary in fully autonomous vehicles. Automobile manufacturers are concerned that such regulations will slow the adoption of AVs. In 2018, the Department of Transportation issued a statement that said the department "embraces the freedom of the open road, which includes the freedom for Americans to drive their own vehicles."[12] The legal and regulatory system will continue to evolve. Manufacturers will lobby governments to shift liability away from themselves, and advocates for personal liberty will complain.

There is the potential for a radically different means to develop AVs and their ethical framework. Although current AI systems for AVs are instructed with constraints that the system learns to accommodate, a more primal method could be invoked. Chapter 3 described Risto Miikkulainen's accidental, self-learning robot arm. In chapter 8, I will describe a more elaborate version. By analogy, an AV could be turned on in a completely blank state. As for the robot arm, it could be left to experiment and improvise for an extended period, thus learning everything it needs to know: physics, mechanics, dynamics, its components, its constraints, its degrees of freedom. Such an AV might remain a philosophical zombie or become self-aware. An interesting question is whether such a self-training AV would develop a system of ethics and if so, what would that be? Would it have any special consideration for humans if we do not tell it to do so?

6

SMART WEAPONS
To Ban or Not to Ban?

LICENSED TO KILL

The day is coming when our machines will decide whether to harm humans and if so, whom. Autonomous vehicles already hint at these issues. In 2016, artist and roboticist Alexander Reben developed a small robot that deliberately violated Asimov's laws of robotics: It unpredictably inflicted a tiny pinprick on a finger placed near it. This device was not a vehicle, but it was autonomous, and it hurt, a little. The point was to focus on a range of issues facing us soon. A central one involves machines that, unlike autonomous cars, are designed to kill.[1] Autonomous weapons bring conditions rife with the possibility for unintended consequences. They could change warfare in ways that are very difficult to imagine.

For a considerable time, the US defense complex retained its old Cold War ways. There was both cultural and political resistance to recognizing and employing the exponential growth of digital technology. Digital systems were employed that lagged the cutting edge and were easily hacked. The most advanced US fighter jet, the F-35, has a voice command system, but it is not useful for much more than changing the frequency of radio communications. Commercial firms in Silicon Valley and elsewhere have more advanced machine-learning technologies than the military, which is often years behind equipment that can be purchased from Amazon or Best Buy. That has been changing.

The first well-publicized act of cyber warfare came in 2010 when the United States released the Stuxnet virus on Iranian uranium enrichment facilities. Iran counterattacked by shutting down a casino in Las Vegas belonging to Sheldon Adelson, a strong supporter of Israel. The United States also likely interfered with North Korean missile launches. North Korea thought so and attacked Sony.[2]

Armed drones have been employed on various battlefields. To date, human pilots have flown these aircraft, sometimes from half a world away. These attacks are fraught: they protect the lives of pilots and troops but can injure and kill civilians despite attempts at caution. The same is true of smart bombs. This use of drones can change hearts and minds in a negative way. Given the advances of robotics and AI, it is also a small step to make these weapons autonomous. While armed flying drones have captured attention and imaginations, one can also picture arming the delivery robots discussed in chapter 4 or the autonomous vehicles of chapter 5.

Some police departments are beginning to adopt autonomous vehicles. In Chula Vista, California, the police have operated fully autonomous quadcopter drones that can respond to a 911 call at the press of a button. The drone flies to the site of the report and does aerial surveillance. Combatants in the Middle East have taken this capability another step by adapting commercial Chinese flying drones to carry explosives. Turkey provided an armed loitering drone to the government that is fighting rebels in Libya. That drone may have attacked autonomously. Drones are being used ubiquitously in the Russian war against Ukraine and in the war between Israel and Hamas in Gaza.

Such drone activity illustrates a key point. Unlike nuclear weapons, autonomous killer weapons are relatively easy for small groups or even individuals to access and employ. It is a small step from robot soccer to swarms of cooperating killer robots.

THE PENTAGON GETS ENGAGED

By 2016, the attitude in the Pentagon had shifted. AI became central to the goal of remaining the world's dominant military power. China and Russia have highly competitive efforts to see that this does not remain true.

As in many other related areas, DARPA has played a role in bringing the Defense Department into the digital age. The general mission of DARPA is to "anticipate, explore, and achieve the concepts and technology on which the nation's future deterrent and defense capabilities depend." In recent years, DARPA has emphasized various ways to apply AI to warfare. In 2020, DARPA named Victoria Coleman to be the third woman to lead the agency. She is an expert in AI and microelectronics.

DARPA developed a flying drone that had to be turned on and off and assigned targets by a human but subsequently operated autonomously. One version was designed to track enemy combatants. In training exercises, such drones were able to follow humans who were deliberately trying to evade them as the humans ran through a forest and hid behind trees. In another exercise, a drone employed human and facial recognition software adapted from US intelligence agencies. The drone displayed a remarkable ability to distinguish between enemy, friend, and civilian noncombatant in a simulated urban environment. It could detect an enemy hiding in the shadow of a wall and safely identify a photographer aiming a camera at it, a test sometimes failed by humans. Work is underway to develop drones and other autonomous machines that are sufficiently reliable to be used in actual combat situations and adaptable enough to handle the fog of war. Arming such drones and telling them whom to shoot is a trivial step. Weapons already exist that are programmed to strike only certain targets.

In analogy to its earlier robot and autonomous vehicle challenges, in 2020 DARPA organized a competition between eight companies they called AlphaDogfight.[3] The competitors employed their AI software systems in fighter jet simulators. The competition was won by a beltway bandit, Heron Systems, based in Maryland. Heron Systems then went head-to-head with an experienced F-16 pilot and beat him in five straight bouts. Among other advantages, AI systems are not subject to the same restrictions on acceleration, or g-forces, as are human pilots. As a result, an AI system can climb and dive faster and pull tighter turns. While details are not available, AI systems may also be able to develop new strategies that do not occur to even well-trained pilots as foretold by AlphaGo Zero's accomplishments in Go. If that is not true now, it soon will be.

DARPA has explored other creative extensions of digital technology. It has promoted research into cyborg insects created by implanting sensors in living creatures when they are pupae. As adults, the insects could then transmit that sensor data. In principle, the activity of the insects could be directed by a micro-electro-mechanical system (MEMS) to detect explosives, poison gas, enemy farts, or other important data. Some success has been reported using cockroaches in this way. DARPA is also developing a neural implant to direct sharks by remote control. Sharks have unique senses that could be used to detect underwater mines or track enemy shipping. They are also less cuddly than dolphins, hence perhaps less likely to trigger objections from animal rights activists. As in other areas of digital technology, there is some concern that neural implants in insects, fish, and other animals could be hacked and used for other than their intended purposes.

Autonomous weapons are not restricted to the scale of bugs, drones, or humanoid robots. They can be whole weapon platforms. Building on efforts stimulated by DARPA, the Department of Defense is now developing robot tanks; robot fighter jets that will join manned aircraft in combat; missiles that can autonomously choose targets; robot ships that can autonomously track, and presumably attack, submarines; and robot submarines. Lockheed Martin is developing a long-range anti-ship missile (LRASM) that can search ocean-wide, select targets, and launch attacks even in the absence of contact with human controllers. The crewless second-generation Sea Hunter under construction in Gulfport, Mississippi, in 2020 is designed to hunt submarines, detect torpedoes, and dock itself in severe and unpredictable conditions. It has a sharp prow reminiscent of a *Star Trek* Klingon Bird of Prey, a composite hull, and two novel outriggers. With no crew, there are no handrails or ladders. Computers replace mess rooms, heads, sleeping areas, and galleys. The ship incorporates redundant automated systems since there is no crew to implement repairs. The US Navy aspires to fleets of such vessels working in tandem with manned ships. The navy is also testing a full-size robotic airplane, the X-47B, that has demonstrated the capacity to take off and land from the deck of an aircraft carrier, one of the most difficult challenges for human pilots.

In 2020, the recently constituted Army Futures Command formalized an arrangement with the robotics laboratory at the University of Texas at Austin to build a variety of robot combat vehicles. One of the first projects is a system where the AI of an autonomous vehicle does not have to be carefully trained and tuned by engineers. This system learns by monitoring the driving of ordinary soldiers who are using an Xbox game controller to operate the vehicle. Rather than requiring a trained roboticist to tune the machine by teaching it parameters for each new environment, this system would allow human drivers to do the tuning while operating the vehicle in a range of conditions.

Similar efforts are apace globally. In 2020, Great Britain conducted a large war exercise that involved autonomous vehicles. Russian and China are undoubtedly pursuing similar goals.

Robert O. Work, a retired colonel in the Marines who was deputy secretary of defense from 2014 to 2017, set much of the current strategy of the Department of Defense. A book published in 2013 by economist Tyler Cowen titled *Average Is Over*[4] impressed Work. Cowen noted how two human chess players of average talent working in tandem with chess-playing programs on computers beat both champion human players and chess-playing computers working alone. Work concluded that the computers provided tactical ingenuity, the humans provided strategic ingenuity, and the combination was thus more competent than either alone. He encouraged the Department of Defense to adopt the Centaur warfighting strategy, which was named after a mythological creature that was part man and part animal. A central tenet of the Centaur strategy is that humans should remain in control of autonomous weapons. The weapons should augment the creativity and problem-solving skills of soldiers but not replace them. The guiding principle is that machine learning can identify patterns much more rapidly than a human can, but for the foreseeable future, humans have a greater ability to handle uncertainty and unpredictability. The soldier of the future would then be less like the Terminator and more like Iron Man wearing a smart, armored suit. American soldiers, pilots, and sailors would become superhuman by fighting alongside robots or inside robot airplanes and vehicles but always with humans ultimately in control. One manifestation of this plan is an exoskeleton developed by Lockheed Martin that would allow troops to carry heavy loads of weapons and gear with little fatigue.

THE TERMINATOR CONUNDRUM: EXERCISING CONTROL

The thing about autonomous weapons is that they tend to be relentless. The *Terminator* movies emphasized this with Arnold Schwarzenegger's repeated catch phrase "I'll be back." The issue of how much autonomy to enable in weapons is not settled. There is ongoing argument within the Department of Defense as to whether it should advocate international treaties to ban fully autonomous weapons or develop such a capability in case enemies do so, triggering an autonomous weapon arms race. Within the halls of the Pentagon, this dilemma is known as the *Terminator conundrum*. For now, Department of Defense Directive 3000.09[5] requires "appropriate levels of human judgment over the use of force" by autonomous systems and that a high-level defense official approve any unusual uses of autonomous technology.

This debate is not confined to the Pentagon. Some academics argue that autonomous weapons could reduce the harm to noncombatants in war. The autonomous systems could sense the environment in a way that humans cannot and would have the ability to sort through options and make rapid decisions that would allow greater discrimination of friend from foe. They would not suffer from the anger, fear, frustration, or boredom that leads humans to make mistakes. Others are actively skeptical. The Campaign to Stop Killer Robots[6] was launched by a group of scientists and human rights activists in 2013. This group has argued to governments and to the UN that they should forbid lethal autonomous weapon systems (LAWS). Twenty-eight countries have called for a prohibition of LAWS. The UN has called for such a ban. It has not been enacted because key countries—the United States, the United Kingdom, South Korea, Russia, and Israel—oppose a ban, arguing that current humanitarian law is sufficient. China has joined the call for a ban on the use, but not development, of LAWS.

At an AI meeting in Buenos Aires in 2015, researchers circulated an open letter supporting the campaign for a ban on LAWS. Ultimately upward of five thousand people signed the letter including Stephen Hawking, Elon Musk, Apple co-founder Steve Wozniak, Skype co-founder Jaan Tallinn, Google DeepMind co-founder Demis Hassibis, linguist Noam Chomsky, and philosopher Daniel Dennett. The letter argued that such weapons were effectively weapons of mass destruction and that the development of autonomous weapons of even rudimentary capacity could set off a global arms race. The letter warned that the result of autonomous lethal weapon research would inevitably be fully independent killer robots. These weapons would be so cheap that they would be available to rogue states and violent extremists as well as to the great powers that would develop them, effectively becoming "the Kalashnikovs of tomorrow." The group submitted a petition for a ban to the United Nations, making the case for a Geneva Convention for AI.

Opponents of autonomous weapons concede that they might be useful in tasks such as clearing minefields. If a robot is destroyed, it can easily be replaced without the repercussions of the loss of a human life. Beyond that, the issues get morally murky and are complicated by practical realities. While machines might not suffer the human frailties that lead to death in war, it is difficult to program the restrictions of international humanitarian law. Saving civilians might not be as easy as proponents argue. As for cyber hacking, it might be difficult to know the source of autonomous weapons used in an attack. Without a legal ban, there will be a huge market both for manufacturing such weapons and defenses against them and hence a strong profit motive for accelerating their use and spread.

Some see important legal and moral imperatives. Who has the responsibility if an autonomous weapon attacks a bakery, school, or a hospital? Is it worse for a human to be killed by an autonomous machine that has no concept of the value of a human life than by a human who does?

It is difficult to imagine how the remarkable capability of large language AI models like ChatGPT and its successors will affect the ability of autonomous weapons, but that assessment is surely intensely underway. As I will discuss in chapter 8, the possibility of conscious computers remains for the future to reveal. For the foreseeable future, autonomous weapons will have no desire for self-preservation or to take over humanity nor will they wish humans ill. They will be simply following the instructions we give them. In framing the issues, it may be important to consider the most extreme possibilities. The ability of autonomous weapons to discriminate is both their strength and their danger. With facial recognition capability, they could, in principle, be instructed to kill only males or only children.

The Department of Defense established Project Maven in 2017. This project proposed to use machine learning to identify and track people and objects in surveillance videos taken by drones. Project Maven was intended to be a pilot program that would spur the application of AI throughout the department. More than three thousand Google employees protested that engaging in work on Project Maven would not be consistent with Google's principles. They were concerned that the proposed applications would be easily weaponized. Google ended up withdrawing from Project Maven and a related cloud storage contract worth many billions of dollars known as the Joint Enterprise Defense Infrastructure (JEDI) project to host classified information. Microsoft elected to engage in both.

With this background, efforts are underway to attempt to foresee and control the employment of autonomous weapons through diplomatic channels. Air Force General Paul J. Selva, vice chairman of the Joint Chiefs of Staff, argued that on the path we currently tread, someone will eventually try to unleash "something like a Terminator." There are precedents for heading off unwelcome developments. Exploding bullets were banned in 1868 before they were ever used on the battlefield. The next successful weapons ban that was implemented prior to widespread development and deployment came after the United States and China expressed plans to build laser weapons that could blind. After more than a decade of discussion, the UN passed the Protocol on Blinding Laser Weapons that went into effect in 1998. The United States and China and 107 other countries signed on. Lasers otherwise have widespread use in military applications such as in guided bombs that are intended to reduce civilian casualties.

In 2016, representatives from eighty-two countries met in Geneva to ponder the implications of autonomous weapons. This group recommended that lethal autonomous weapons should be considered by the committee of the UN that sets standards for weapons of war. In parallel, Human Rights Watch and the Harvard International Human Rights Clinic produced a report[7] calling for a complete ban on such weapons. The UN Convention on Certain Conventional Weapons met in 2017 but could not even agree on how to define autonomous weapons. Is a land mine autonomous? Is a torpedo autonomous?

In 2020, a task force charged by the House Armed Services Committee with considering the future of defense released a report. The report called for shifting the focus of the military from legacy systems to emerging technologies and for substantial increases in the budget for research and development. The report declared, "Using the Manhattan Project as a model, the United States must undertake and win the artificial intelligence race by leading in the invention and deployment of AI while establishing the standards for its public and private use."[8]

There is concern that the current directive of the Department of Defense that calls for appropriate levels of human judgment in the use of autonomous weapons is too broad. There are calls for meaningful human control, but even that would require careful definition. The perceived advantage of being the first to develop autonomous weapons provides a potential incentive to cut corners on safety and control and hence to compromise fairness. An arms race in autonomous weapons may already be underway even in the absence of an open declaration.

In 2020, the Department of Defense formally adopted five principles to guide the ethics of the development and deployment of autonomous weapons.[9] AI development should be responsible, equitable, traceable, reliable, and governable. The guidelines stipulate that (1) Department of Defense personnel will exercise appropriate levels of judgment and care while remaining responsible for the development, deployment, and use of AI capabilities; (2) the Department of Defense will take deliberate steps to minimize unintended bias in AI capabilities; (3) AI capabilities will be developed and deployed such that relevant personnel possess an appropriate understanding of the technology, development processes, and operational methods applicable to AI capabilities, including transparent and auditable methodologies, data sources, and design procedures and documentation; (4) AI capabilities will have explicit, well-defined uses, and the safety, security and effectiveness of such capabilities will be subject to testing and assurance within those defined uses across their entire life cycles; and (5) the Department of Defense will design and engineer AI capabilities to fulfill their intended functions while possessing the ability to detect and avoid unintended consequences and having the ability to disengage or deactivate deployed systems that demonstrate unintended behavior. Implementing these principles will be a nontrivial task. There is no guarantee that other global players will aspire to or achieve such goals.

Peter Singer of the New America Foundation is an expert on twenty-first-century warfare. Singer recalls a potentially relevant piece of history. In World War I, the Germans instituted a tactic of unrestricted submarine warfare that targeted both combat shipping and civilian shipping. The sinking of the civilian liner *Lusitania* was one of the most famous examples of this tactic, and it was a factor that drew the United Staters into that war. After World War I, the United States helped to negotiate an international treaty banning unrestricted submarine warfare. The intentions were good, but circumstances change. The Japanese bombed Pearl Harbor, bringing the United States into World War II. The United States promptly abandoned all the legal and ethical restrictions that had been put into place and launched its own campaign of unrestricted submarine warfare that destroyed Japan's merchant fleet. Some argue this was tantamount to a war crime. Singer worries about how standards and ethical strictures might change if we get into a war involving autonomous weapons and we are losing. Or Russia is. Or China is.

MAPPING THE BRAIN

Mind Reading, Telekinesis, Telepathy, and the Shared Mind

THE REAL THING

All the work on AI and robots is an indirect way to approach the ultimate problem: understanding our own brains. Emerson M. Pugh remarked, "If the human brain were so simple that we could understand it, we would be so simple that we couldn't."[1] The human brain is a bewilderingly complex machine. It has 200 billion neurons, but each neuron connects to others making the whole *connectome* truly vast. It is that full connectome that gives each of us our individual brain patterns and makes us who we are. The brain does not operate like a digital computer, but by one measure it is capable of 100 to 1,000 trillion calculations per second. This is beyond the capacity of current computers, but computer builders are racing to meet and exceed that goal with bigger, badder computers and new technology like quantum computing. That era—when machines can outperform the human brain—is essentially the mark of the technological singularity.

Like other aspects of rapid technological development, brain research is explored at all levels: from university labs and institutes, to companies, to national and international initiatives. Also, like other key technical areas, brain research has national security implications. We cannot let "them" get smarter than "us." In 2013, the White House announced the BRAIN (Brain Research through Advancing Innovative Neurotechnologies) initiative with the goal of using new instrumentation technology to reveal how individual cells and complex neural circuits interact in both time and space.[2] In 2018, the European Brain Research Area project and the Chinese Institute for Brain Research in Beijing were launched with similar goals of coordinating national and global efforts in brain research. In 2022, the National Science Foundation announced a new program to explore brain-inspired systems that can emulate the flexibility, robustness, and efficiency of biological intelligence.[3]

The result of all this independent and coordinated work is that our knowledge of the brain and its functions feeds upon previous knowledge, and the field grows approximately exponentially. Developments could soon come overwhelmingly rapidly. Aside from the sheer intellectual challenge of understanding our brains, one can foresee a vast range of positive applications such as prosthetics and cures for various diseases of the brain including epilepsy, Alzheimer's disease, Huntington's disease, Parkinson's disease, and schizophrenia. Such power also brings the capacity to change what we are as a species along with significant ethical challenges. Some think the brain is more than a machine and that its functions of free will and conscious thought could never be artificially replicated. Others think the brain is a chemical machine and that its function not only could be reproduced but vastly enhanced. The stakes could hardly be higher.

The potential accomplishments of brain research are great, and considerable hype permeates the field. Some developments imagined in generations of science fiction may come to pass. Other consequences we might devoutly wish to avoid entirely. The way to avoid unintended consequences is to anticipate consequences.

Neuroscientists and neuroengineers have made great progress toward understanding animal and human brains although many challenges remain. We have long understood the highest-level morphological structure: left brain, right brain, cerebrum, cerebellum, frontal lobe, parietal lobe, temporal lobe, occipital lobe, medulla oblongata, hippocampus. We are now employing new technology to explore brain functions at the microscopic level of the neuron, axon, and dendrite. The more difficult problem is to understand how all these parts work together to allow us to function, from breathing to constructing mathematical axioms to writing sonatas. What is the software of the brain?

Brain research involves many grand challenges. One is to determine how the brain functions, from autonomous control of various body parts to conscious thought. Another is artificially to emulate the function of the brain. Yet another is to read the signals produced in the brain and use those signals to communicate with and control external devices just as the brain signals and controls the muscles of our arms and legs. A complementary task is to generate signals externally that can affect, even control, the functioning of the brain. Progress is being made in all these areas. The implications are imposing.

THE PIECES

The fundamental components of the brain are *neurons*, cells that receive and send electrochemical signals. A typical individual neuron contains specific components: a *cell body* within which is the cell *nucleus*; *dendrites* that collect signals from other neurons; and an *axon* to send signals to adjacent neurons. The nucleus contains genetic instructions to produce key proteins. A single cell body can have a forest of short, branching dendrites. Axons are long, thin extensions of the cell body and often end in tree-like branches at their ends known as *axon terminals*.

Neurons have an intricate electrochemical system by which they monitor a change in voltage caused by signals from adjacent neurons until they reach a threshold. At the threshold, they allow a current of sodium ions to flow into the cell body, creating a sudden, explosive, surge in voltage that travels out along the axon as an *action potential*. Neurons do not directly connect to one another but are separated by a gap, or *synapse*, between the tip of an axon terminal and the tip of a dendrite in another neuron.[4] The action potential does not cross the gap but rather triggers the release of molecules called *neurotransmitters* that carry the signal across the gap. The accumulation of the neurotransmitters contributes to an increase in the voltage of an adjacent neuron until it too reaches its threshold and fires its own action potential down its axon, into the axon terminals, and into the dendrites of a bewildering array of other neurons. A neuron may have thousands of synapses. The whole array is constantly collecting and discharging electrical signals. At its root are intricate electrochemical processes controlled by proteins manufactured in the cell nucleus. The artificial neural networks described in chapter 3 emulate this process of collecting and transmitting information without the microscopic chemical intricacy.

We tend to think that all cells in the body have identical genetic information in their chromosomes. That genetic information is selected to differentiate heart cells from skin cells from liver cells, but the latent genetic instructions are all the same in every cell. Studies of brain cells have shown, however, that the brain develops spontaneous mutations beginning in the embryo. The result is not a single genetic map but a mosaic of genetic information.[5] Whether this process is critical to a healthy brain or a sign of gradual steps toward brain dysfunction is not clear.

The brain does not store a literal image or memory like a tiny photograph with neurons serving as pixels. Instead, the image or memory is stored in the firings of specific patterns of neurons. In that sense, the brain functions like a hologram. A holographic film stores a 3D image in a 2D plane as an interference pattern. A laser shown on the holographic film reconstitutes the 3D image. The information to reconstitute the image does not reside in a particular place on the film; it resides everywhere on the film and in every piece of the film. There is no laser in the brain, but images and memories are represented in patterns of activated neurons in distributed regions of the brain.

The brain does not literally see or hear or smell. The molecules of an aroma you sniff do not impinge on the brain. There is no little homunculus in your brain that looks out through your eyes like an astronomer peering through a telescope. In another sense, the brain is the only thing that sees, hears, and smells. It converts input from our organs into electrical patterns and processes and stores those patterns in specific arrays of neurons. Exactly how the brain produces the patterns in the neuron array and how that pattern is retrieved to generate an image or memory is still one of the major unsolved issues. Significant progress is being made.

THE CORTEX

The part of the brain that analyzes all the input from the body and the external environment is the *cortex*. All mammals, including humans, have a cortex as do birds, lizards, and turtles. Fish, insects, and plants do not. The *neocortex* is an upper region of the cortex that developed more recently and is the region where higher cognitive functions occur. In humans, the neocortex is a thin sheet about two millimeters thick and the size of a dinner napkin. It is all scrunched up to fit in the cranium, giving the neocortex its characteristic shape with multiple folds and fissures. The neocortex occupies about 80 percent of the volume of the brain and constitutes about the same fraction of brain mass.

Research has revealed that the neocortex comprises a large number of essentially identical components originally called *cortical columns*: arrays of molecules, nerves, and connections that penetrate the thin depth of the neocortex perpendicular to its surface.[6] These cortical columns are now called *macrocolumns, hypercolumns, functional columns*, or *cortical modules* that are in turn composed of *minicolumns*,[7] which are the basic units of the cortex. The minicolumns have the same types of neurons, connectivity, and firing properties originally ascribed to cortical columns. They are all basically the same and link together to function in the same way.

There are about 1 million macrocolumns and 200 million minicolumns in the neocortex, each with about 100 neurons. Those neurons couple relatively tightly to serve as a local microprocessor, but a given column couples only weakly to adjacent columns. It is as if the cortex were tiled with 200 million otherwise identical processing units. The columns process action to and from muscles and sensation based on input from the body and environment. There are specific, orderly relations between particular regions of the body—eyes, ears, fingers, nose, tongue—and corresponding regions of the neocortex. The columns in specific regions of the neocortex are directly related to motor control of different regions of the body. Similar relations apply to other brain functions. In that sense, the neocortex does store, if not an image, then the information about that image in specific regions. The neocortex is thus a 2D map of sensation and action processed by specific arrays of minicolumns and macrocolumns.

At any given moment only about 2 percent of the neurons and 2 percent of the macrocolumns are active in the neocortex. The brain has developed ways to store information in arrays of columns that are spread sparsely in the corresponding region of the brain. The column microprocessors thus constitute a *sparse distributed representation* of the information they are processing. To understand and emulate the brain, we need to be able to understand and reproduce the process by which these sparse arrays of columns store and retrieve information.

The techniques of AI employing neural networks and deep learning described in chapter 3 invoke a hierarchy of connection between artificial neurons in which information is fed in one direction, upward, in the layers of artificial neurons. Recent understanding of the operation of the brain is that its information processing involves a hierarchy but in a rather different way.

As an example, to process spatial information associated with an image, one level of the hierarchy in the portion of the cortex devoted to visual processing might provide detailed information about small parts of the image but little information about how the parts fit together in the whole image. An analogy might be looking through a magnifying glass very close up at a bit of skin. Another level might provide a representation of the overall structure but little detail of individual parts, like seeing a person at a distance with insufficient spatial resolution to recognize the person. The visual cortex processes all these aspects simultaneously at multiple levels of the hierarchy to generate the detailed image of a recognizable face.

Unlike artificial neural network techniques, the hierarchies employed in the brain are thus not deep but wide. The arrays of active cortical columns are constantly exchanging information throughout the hierarchy in a collective representation of what is going on in the world. All levels of the hierarchy and all areas of the brain, not just the cortex, are involved simultaneously. Any level of the hierarchy can potentially inform any other level. A local clue from one cortical column is spread through the cortex and through the hierarchy as an object is identified. When an object is examined—seen, heard, touched, smelled—all the cortical columns vote within regions and throughout the brain according to the weight of their information. The collective vote resolves into a determination of the identification of the object. One theory argues that in the process each cortical column learns a model of the object. This happens quickly, and we are only aware of the resulting consensus model that gives us a singular perception of reality.

Recognizing spatial relationships was fundamental to our early survival and probably fundamental to how the brain developed its functions. The cortex especially responds to movement. The movement of an object is encoded by recalling the sequence of patterns representing the object and constantly updating the pattern as the object moves. Once the hierarchical structure to process sparse distributions of data was in place, the same physical structure and the same hierarchical algorithm for processing information was applied to all the brain's functions, not just visual but hearing, taste, smell, touch, and higher-level conceptual thought. Every cortical column in every region of the cortex and at every level of the hierarchy is effectively performing the same computation regardless of the task.

The cortex receives two sets of input. One is a copy of motor commands to muscles that inform the brain of how it is interacting with the external environment. The second input is a copy of sensory input that registers what is happening in that external environment. The result is that the brain utilizes both sets of information in the cortical column structure of the cortex to provide a *sensory motor model* of input from the body and the environment. The sensory motor model registers not just what is happening in the environment but how the brain is interacting with that environment. The sensory motor model constantly makes predictions of what will happen next and then uses input from the next instant to update the model in real time. The process of constantly primarily updating only the changes of a pattern and not the whole pattern is very efficient. It does not require the fast processing and huge memory of artificial neural networks. This biological reality is thus rather different than the current state of the art of artificial neural networks for all their sophistication and power. That is presumably why, at least for now, humans can still do things computers cannot.

FINDING YOUR PLACE: GRID CELLS

An insight into how the brain implements its hierarchy-based software comes from studies of *grid cells*.[8] These grid cells are found in a small substructure of the cortex called the *entorhinal cortex* that is tucked in the bottom of the cortex just in front of the hippocampus and slightly below the level of your ears.

Studies of how individual neurons in the entorhinal cortex fired as a mouse moved around an enclosure showed that certain nerve cells only activated when the mouse was in particular places in the enclosure. Remarkably, a map of this *firing field* showed that the multiple places where a given grid cell fired were arranged in a *hexagon*.[9] Spheres packed on a plane as tightly as they can arrange themselves will form hexagons. If the vertices of six identical equilateral triangles touch at a point, then their bases will form the sides of a hexagon. The pattern you get if you tile a space with equal-size equilateral triangles will thus also yield a hexagonal array. The brain's utilization of such patterns is presumably related to the efficiency of hexagonal packing and to representing space with the simplest elements that will span a space of two dimensions.[10] To emphasize, the hexagonal pattern is not related to the spatial distribution of the grid cells in the entorhinal cortex but to the places in actual physical space where a given grid cell fires. Somehow that grid cell knows that the mouse is near one of the spots corresponding to the place where that neuron is wired to fire.

The implication is that when that particular grid cell neuron fires, an experimentalist, or the brain, knows that the mouse is at one of the multiple points that are the centers of the corresponding hexagonal pattern. In the original experiment, the mouse enclosure was tiled by about a dozen hexagons. Each hexagon spanned a space from a few times to a few tens of times the size of the mouse itself, presumably the result of millions of years of mouse experimentation in the real environment. Remarkably, the mouse mapped out the same hexagonal pattern in total darkness.

The brain, however, needs to know exactly where its body is and not just that it is in one of multiple places in the hexagonal pattern. It solves that problem by activating numerous grid cells, each with their own hexagonal response pattern with differing sizes and hence spacings of the hexagons and with different orientations of the hexagonal grids. With that overlapping information, the brain can accurately determine that the body it occupies is at one specific place. The net effect is that grid cells create a *cognitive map* of space.

The key to the role of grid cells is motion. A given grid cell cannot map out its particular hexagonal distribution unless the body moves around. That motion can be literal, by scampering about, or it can be functional, by eyes scanning the area being viewed. As noted above, once the basic software is in place, it can be applied to other input—for example, hearing or smelling. One tends to tilt one's head when trying to detect the origin of a sound and move the head around when trying to locate the origin of a smell. That motion is key to making the process work. Your brain also wanders among various ideas as you daydream or ponder concrete plans for the day.

In addition to grid cells, the brain has *head direction cells*[11] that register orientation; *place cells* that fire when you are in a specific location; *speed cells* that increase their activity as speed increases; and *timing cells* that register the passage of time—all information that complements that of the grid cell arrays. Many of these other types of cells, but not all, are also in the entorhinal cortex. If you know place, direction, speed, and time, you can navigate. All these properties can also be associated with memories—or dreams.

The grid cell software can then also be applied in more abstract situations involving conceptual thought. While the grid cells are localized in the entorhinal cortex, neuroscientists are pursuing the notion that the processing array of the cortical columns recognized a good trick when it saw it and uses the grid cell hierarchal model to represent hypothetical spaces including the passage of time, sounds, and conceptual maps of abstract ideas.

Some researchers are pursuing the question of whether the brain's tricks for motion tracking can be applied to self-driving cars. DeepMind, the company that produced AlphaGo Zero, gave a learning machine information about head direction and speed and tasked it with navigating. The machine spontaneously developed grid patterns similar to those in the entorhinal cortex and used that structure to out-perform humans in navigating.[12]

John O'Keefe, May-Britt Moser, and Edward Moser were awarded the 2014 Nobel Prize for Physiology or Medicine for their fundamental work leading to an understanding of the role of grid cells in the brain. Associated developments and their implications for new techniques of AI are summarized in the book *A Thousand Brains* by neuroscientist Jeff Hawkins.[13] One of the key open questions is whether and how this grid cell, hierarchy-based software leads to consciousness. I will address that question in chapter 8.

THE INSTRUCTION SET

While some researchers dream of reproducing the detailed electrochemistry of the brain, that may not be necessary to duplicate the functional processes of the brain. The brain is amazingly complex, but the instruction set to build a brain is simpler. Our entire genome only constitutes about 10 billion bits of information, and the part that tells the brain how to grow and function is only a portion of that. The brain attains its complexity by building fundamental parts—cortical columns—and then replicating those parts. The brain gets its power because it is effectively massively parallel: Billions of neurons are storing and firing—essentially computing—at one time. We know how to invoke massive parallelism in our computers. Most cloud computers are simply rack upon rack of otherwise identical components. The functioning of a neuron at the speed of electrochemistry is rather slow compared to that of a modern computer. Each neuron's charge is stored and discharged in about one-hundredth of a second. Computers effectively operate at the speed of light with connections between nodes on silicon chips connecting more rapidly than neurons by a factor of 1 trillion or more.

Our brains are adaptable. They make new connections within their network of neurons. This process is very active when we are young and continues even as we age. We are learning to emulate aspects of this by writing fault-tolerant code. The functions of the brain seem to be a prime example of emergent properties (see chapter 9 for a more elaborate discussion). No single neuron of the brain can recall a cherished face or remember the lyrics of a song. The interactions of billions of neurons arranged in microcircuits can do so. Our brains require very little power, while cloud computers can require the power of a small city. We have a long way to go to mimic that key aspect of our brains.

CONNECTING BRAINS AND COMPUTERS

A significant component of modern brain research involves brain-computer interface (BCI) technology. BCI incorporates myriad ways of detecting activity in specific portions of the brain, even in individual neurons, often augmented by AI that enables interpretation and manipulation of the data. With BCI, scientists can determine which patch of neurons fires in what parts of the brain during what activity.

Some BCI technology can be implemented externally and non-invasively by registering minute electrical signals that are generated in the brain during certain activities. These signals propagate through the cranium to be registered by electrodes attached to the skull. Other non-invasive techniques can peer inside the brain to register which portions light up in neuronal activation in specific circumstances. Some of these techniques are *label-free* meaning that they do not require the addition of dyes or molecular contrast agents.

Computed tomography (CT) uses the absorption of X-rays to study the large-scale anatomical structure of the brain. Other techniques study the functions of that structure. An *electroencephalography* (EEG) measures the brain's electrical activity by recording signals using electrodes placed on the scalp. EEGs register the firing of large numbers of neurons rather than single neurons but can measure changes in electrical activity with high temporal resolution: about one-thousandth of a second. A variation of the EEG is known as *magnetoencephalography* (MEG). This technique maps magnetic activity associated with electrical activity. *Functional magnetic resonance imaging* (fMRI) provides a more detailed view by detecting changes in oxygen levels and blood flow in an active part of the brain. Information about brain activity as measured by oxygen levels is called *blood oxygen level dependent* (BOLD) data. With fMRI, researchers do not just map the brain but see the brain functioning, hence the *f*. *Functional near-infrared*[14] *spectroscopy* (fNIRS) is another technique used to determine BOLD information. In this case, researchers shine infrared radiation through one side of the skull and register how much comes out the other side. The attenuation of the radiation depends on blood levels. *Photoacoustic imaging* (PAI) provides better spatial resolution than ultrasonic imaging[15] alone and deeper penetration than fNIRS. *Positron emission tomography* (PET) employs very tiny amounts of radioactive material to map active areas and hence functional processes in the brain. Whether one regards the latter techniques as truly non-invasive may be a matter of personal definition.

Other BCI techniques are regarded as invasive since they involve probes or electrodes that are inserted beneath the scalp or within the skull itself. Subscalp EEG systems can remain in place for extended periods. The Bern University Hospital Epios system employs thin sensing electrodes connected to a small implant. The implant transmits collected neural signals to a receiver that is worn behind the ear. The neural data are uploaded to the cloud for analysis. A similar device is the Minder developed by the Australian company Epi-Minder. Other systems are manufactured in Denmark, Finland, Switzerland, Germany, and the United States.

Some BCI systems provide deeper, more direct access to the brain by employing electrodes planted in the brain. Brain implants have been investigated since the 1970s. The first primitive versions of this technology employed a single electrode embedded in a pre-selected portion of the brain. In 1976, researchers at the National Institutes of Health recorded signals from a single neuron in the motor cortex of a rhesus monkey using an iridium *hatpin* electrode. Current techniques allow multiple electrodes or other detectors. Bioengineers at the University of Utah developed the *Utah array*, a square array a few millimeters on a side that holds about one hundred pins. The resulting configuration resembles a bed of nails. The Utah array became the commercialized standard for implants employed in deep brain stimulation in the treatment of Parkinson's, depression, and other conditions.

Other researchers implant tiny sensors not much larger than neurons themselves. These sensors, known as *neurograins* or *neural dust*, are too small to contain batteries so their power is beamed in with microwaves or ultrasound. These tiny detectors can be placed in neurons throughout the body, not just the brain, to monitor neural function.

Besides fostering deeper understanding of brain functions, BCI systems allow researchers to tap electrical signals within the brain and to input signals. The former ability can interpret the activity of the brain; the latter can control the brain's activity. Both aspects have significant therapeutic applications and serious implications. Tapped signals can operate prosthetics in a laboratory or hospital close at home or, with the aid of the internet, across continents and oceans.

The opposite is also true. Once communication with the brain is possible, signals can be input to the brain. When we have sufficient understanding and control, neurons in selected portions of the brain can be artificially stimulated and their function replicated.

BRAIN WAVES

All this probing of the brain has revealed a stream of remarkable insights. One is the existence of *brain waves*. This term is a misnomer because it does not refer to waves that slosh around the brain like the surface of a lake on a stormy day. Instead it refers to the roughly rhythmic pulsing of the brain's electrical activity that appears as a wave-like pattern on the paper printout of a traditional EEG recording. There are five common types of brain waves traditionally labeled by Greek letters in the order in which they were discovered. Ranking them by the frequency of their pulses, they are *delta waves*, corresponding to deep, dreamless sleep with a pulsing frequency of roughly once per second; *theta waves*, which occur in lighter sleep with a frequency of a few times per second; *alpha waves*, which predominate when you are calm and relaxed—for instance when you have just awakened but are not yet thinking of deadlines—that have a frequency of about ten times per second; *beta waves*, which occur when you are awake and focused with a frequency of about twenty pulses per second; and *gamma waves*, which develop when you are active and solving problems (addressing those deadlines) with a frequency about forty pulses per second.

To complicate matters, other delta waves with a higher frequency comparable to that of beta waves are associated with deep rapid-eye-movement (REM) sleep that is involved with retaining memories and with peak cellular protein synthesis. Another wave, one that also pulses about forty times per second, may be related to the problem of *neural binding*. This is the question of how neurons in one part of the brain coordinate with those in another part of the brain to form a memory, or dream, or conscious thought. The function of these various pulses of power and the portions of the brain that give rise to them are still obscure. Research on rat brains in 2020 suggested that the origin of alpha waves is not the thalamus, as long suspected, but the visual cortex.

Another surprising insight is the coupling between the autonomous and conscious parts of the brain. Experiments show that the brain decides to take some action—for instance, moving an arm—but that decision only arises in conscious thought about one-third of a second later. It is almost as if notifying the conscious brain is an incidental side effect. This delay challenges our common notion that we consciously decide to move our appendages first and then the brain triggers the nerve signals that causes the motion. A full appreciation of this aspect of brain function could upend our notions of free will and, along with them, our system of justice that is based on the conviction that conscious decisions precede actions.

DIRECT CONNECTIONS

The technology of communicating with computers has also progressed exponentially over recent decades. At one point, code was inscribed on punch tape or punch cards that were fed into gigantic computers. In the personal computer era, keyboards and mice provided input to the machine more directly. I am certainly not alone in idly pondering the prospect of cutting out the middleman, the fingers, and just thinking about the keys to be struck and the mouse moves and clicks to be made in order to command the computer. An appreciable effort is going into just that area of BCI research.

A principal motivation is not simply to make typing easier but to address real medical problems among which are brain diseases and conditions of the body where the brain is intact and active but the limbs have been rendered inoperative by injury or disease. BCI can address many of these issues.

Cathy Hutchinson had been paralyzed by a stroke for fifteen years when, in 2012, she volunteered to have researchers implant small chips in the region of her brain that controlled muscle movement. The chips registered the signals her brain was generating and communicated with a robotic arm. With practice, Hutchinson learned to use the robot arm to lift a bottle of coffee to her mouth and sip through a straw. Her brain did not care that it was generating signals that controlled the robot arm rather than her muscles. It adapted. The brain has plasticity: it learns. An infant learns to crawl and then walk as its brain learns to control muscles in a coordinated way. This early experiment was both expensive and invasive, requiring great surgical expertise and wires protruding from Hutchinson's head, but it established the principle. Since then, less invasive and even non-invasive BCI techniques have accomplished the same ends.

A rudimentary example of a non-invasive BCI was employed in the service of the brilliant and heroic physicist Stephen Hawking. As many know, ALS, Lou Gehrig's disease, struck Hawking as a young man. He spent most of his life in a wheelchair, losing ever more control over his body. His brain was still active. He did amazing work revolutionizing our understanding of black holes. At first, he could still vocalize somewhat. Later he learned to pick letters out on a keyboard by looking at a computer screen. The computer would then enunciate those letters as words in the inimitable Hawking computer voice. Toward the end he could only control a small muscle near his right eye, which he could twitch. Nevertheless, computer technicians rigged a system by which he could scroll a cursor over an alphabet and slowly pick out letters somewhat like searching for a movie with your remote on Netflix. In this humble, computer-enabled fashion, he was able to write scientific papers and whole books. Imagine what he could have done with a more efficient system connected non-invasively, but directly, to his brain.

BrainGate is a multi-institutional, interdisciplinary, neurotechnology organization funded by the federal government and other sources. Brain-Gate has developed techniques to implant tiny electrodes in the brain. The express purpose is enabling the handicapped to manipulate their environment. For example, Bill Kochevar is a tetraplegic who was paralyzed in a bicycle accident. In 2017, researchers at BrainGate used their techniques to directly couple the signals from Kochevar's brain to an AI system that could analyze the signals and then send appropriate currents to his muscles, bypassing the inactive neural circuits in his spinal cord. The result was that Kochevar was able to move the muscles in his right arm by thinking in a more-or-less normal way. He could feed himself, drink, and scratch his nose. In a recent development, BrainGate researchers enabled a paralyzed man to type ninety characters per minute[16] illustrating that accurately decoding dextrous movements is feasible.

A BCI system currently under development and subject to active hype is Neuralink,[17] yet another enterprise founded by Elon Musk and partners in 2016. This system injects thousands of filaments into brain tissue. Each filament is one-tenth the thickness of a human hair, centimeters long, and contains many electrodes. The filaments do not have the resolution to register a single neuron but can record the collective activity of about a thousand neurons. They connect to a chip about the size of a coin also embedded within the skull. Each chip contains sensors for temperature, pressure, and motion, elements also found in smart watches. A single chip registers the activity of millions of neurons. The practical limit to the number of chips and filaments that can be embedded in a single brain is not yet clear. An early version of Neuralink was connected to an external module worn behind the ear that collected the brain signals. The 2020 version of the module mounts flush to the surface of the skull. The module has a wireless connection to external analysis equipment. The power for the battery is supplied remotely by magnetic induction much like an external mobile phone charging system that does not require the phone to be plugged in. This system requires surgery, but it has a robot that assists with implanting the filaments. The system also links to an AI that aids in the interpretation of the resulting signals. Neuralink prototypes have been tested on animals but not yet on a human. In 2020, Musk presented a pig named Gertrude sporting the latest implant. Neuralink is two-way: it can both read and write to the brain. This system and others like it foretell digital telepathy.

A competitor to Neuralink is Paradromics,[18] a company started by German neuroscientist Matt Angle. Paradromics promises an especially high data rate implanted BCI system that can translate between neurochemical signals and digital signals with the potential capability of connecting living brains with the cloud. Similar to Neuralink, Paradromics has a chip with tiny, platinum-iridium wires (currently 1,600 of them) that is placed beneath the membrane that surrounds the brain with the wires penetrating about a millimeter into the neocortex. The chip transfers data to a device implanted under the skin of the chest that provides electrical power, computational power, data storage, and communication with the cloud. Paradromics hopes to scale the system to larger capacity and use it to address various types of mental impairment by writing to as well as reading from the brain.

DARPA has funded BCI research for fifty years. By the time of the Iraq War, medical procedures had advanced to the point that soldiers would survive horrific injuries from which they would have died in earlier battles. A particular challenge was that many of them survived with missing limbs. This issue commanded the attention of Colonel Jeffery Ling, who encouraged DARPA to invest in BCI systems that could control robotic limbs through a direct BCI connection. DARPA provided some of the funding to BrainGate. More recently, DARPA began a project called Neural Engineering System Design. DARPA is seeking approval from the US Food and Drug Administration for a wireless human brain instrument that can both monitor brain activity with 1 million electrodes and selectively stimulate activity in 100,000 neurons. One can only imagine the work DARPA is sponsoring behind the black curtain.

AI is becoming widely used in BCI systems. Neural networks and deep learning enable the specific identification of signals that correspond to specific commands the brain is trying to send to muscles. Those signals can then be routed to prosthetics designed to respond the way a lost or paralyzed limb originally would. The goal is to allow people who are paralyzed to move, those who are blind to see, and those who are deaf to hear as well as to treat a variety of brain dysfunctions. It does not take much imagination to see how this technology can move beyond therapeutics. Practitioners foresee applications to entertainment, gaming, and a variety of practical applications. One could manipulate machines by merely thinking. One could see by sending signals directly to the brain without bothering with the eyes. Brains could talk to other brains.

The company Synchron is based in Silicon Valley and New York but is closely associated with the University of Melbourne in Australia. Synchron advertises its Stentrode BCI device that can avoid brain surgery by threading a detector up a jugular vein into blood vessels in the brain. Tiny detectors are embedded in the interstices of a platinum stent. The collected brain signals are then sent to a device implanted beneath the skin of the chest as heart pacemakers are. Thomas Oxley, the head of Synchron, estimates that in a few years BCI techniques will allow brains to directly communicate with computers more rapidly than a teenager with fast thumbs can today type a message on a mobile phone. On December 22, 2021, Phillip O'Keefe, an Australian suffering from amyotrophic lateral sclerosis, used a Synchron rig to become the first person to post a Tweet generated just by thinking the composition of the words, with an AI interpreting his brain signals and typing them. O'Keefe said, "I created this tweet just by thinking it. #helloworldbci."[19] These days he could presumably compose a prompt to ChatGPT or Bard by thinking, needing neither to type nor speak.

An important step in non-invasive brain reading was taken in April 2023 by Alexander Huth, a neuroscientist at the University of Texas at Austin and colleagues.[20] They combined the non-invasive techniques of fMRI and BOLD to record blood oxygen levels in subjects as they listened to podcasts or watched movies. Data were collected from the region of the brain that processes language. The data were then analyzed with the help of an LLM, an early version of ChatGPT. Just as ChatGPT can translate between languages or translate words to code, the model learned to translate blood oxygen patterns to words. The system does not just read words but works at a different level than most BCI by registering the ideas, the semantics, the meaning of thoughts. The system does not literally reproduce words thought by the subjects but the gist of what they are thinking. The researchers are vividly aware of the possible negative impacts of their technique on privacy and ethics. Huth hastened to point out that subjects could deliberately sabotage the system by thinking of things other than the podcast they were hearing and that the analysis of the thinking of one subject does not allow the system to read the mind of another subject. The working of each subject's brain is unique, like a fingerprint.

PRIVATE ENTERPRISE ON YOUR BRAIN

The potential of BCI research has attracted not only university researchers but also some of the largest commercial enterprises in the world. For over a decade, Intel has investigated coupling its chips to brains. One explicit goal of Intel is to allow people to manipulate computers and other devices by thinking rather than by typing: you would text, email, or change a TV channel with your mind. Presumably a screen would still be necessary to allow the editing of faulty autocorrects before your thoughts went winging off into the internet.

In August 2020, Facebook renamed a group working on artificial reality (AR) and virtual reality (VR) and the Occulus VR the Meta/Facebook Reality

Labs. Reality Labs will also conduct BCI research and fund BCI research at universities. A near-term goal is to develop a non-invasive helmet—perhaps using infrared probes—to detect brain-click commands like Home, Select, and Delete. AI developed by Meta can determine what a person is hearing by measuring a few seconds of brain activity.[21] Meta's longer-range goal is full speech recognition. A BCI project at Microsoft seeks a non-intrusive, rapidly interactive BCI system for pilots, drivers, soldiers, and the general population. While Apple, Google, and Amazon engage heavily with AI techniques, there is little direct evidence that they are currently involved in BCI research. Given the activities of their competitors, that may change.

COLLABORATING WITH ROBOTS

Another frontier of brain research is the capacity to establish partnerships between people and machines. The goal is to augment the functioning of both. The basic idea of a collaborative robot, or *cobot*, was developed in the mid-nineties. Whereas industrial robots work autonomously—picture welding car bodies—and potentially endanger humans in their vicinity, cobots are designed to work in the same space as humans. Cobots and humans can work in parallel with one another, accomplishing tasks sequentially, or they can work simultaneously with the robot responding to the actions of the human and vice versa. By now a variety of companies manufacture cobots with smooth edges and the ability to flinch when encountering soft human tissue. Many current cobots are only robot arms, but some are self-transporting.

Engineers at the Computer Science and Artificial Intelligence Laboratory at MIT are developing an AI-enabled system to connect a person to a cobot. The notion is that a human partner could correct the robot's mistakes with a hand gesture or with just the appropriate thought. Imagine being connected to a fully functional but otherwise autonomous robot. The cobot could be in the lab or on the other side of the planet. Your brain signals could be digitally recorded and then employed later. The utility of such a recording process is not immediately obvious, but the principle is well established.

WRITING TO THE BRAIN

Other examples of progress abound. Paralyzed monkeys have walked again. Activating enzymes in the brains of mice suffering from Alzheimer's restored long-term memory. AI has translated brainwaves into complete sentences. Brain signals have controlled drones. The first international competition of brain-drone racers was held at the University of South Florida in 2019 with teams from the United States, the United Kingdom, Japan, and Brazil. Monitoring neural activity and hormones released by the hypothalamus and pituitary can predict how generous people will be when donating. In another study, researchers had surgeons wear EEG caps while they performed operations. An AI system analyzed the brain activity during the surgery and classified the surgeon's motor dexterity and level of expertise.[22] The result was an objective ranking of the surgeons by ability that outperformed traditional subjective metrics. One can foresee the application of such techniques to a broad range of skills, aptitudes, and expertise. One can also foresee dissension when aptitudes clash with passions.

All this technology can be reversed. Once an AI learns to recognize which electrical signals are associated with which brain functions, electrical signals can be induced in the brain artificially to trigger chosen reactions. At MIT, scientists have learned how to manipulate people's dreams.[23] The system works by exploiting *hypnagogia*, the transition state between being fully awake and in deep REM sleep when thoughts slip in and out of control and one can still hear and process audio signals. In the MIT experiment, subjects wore a device called a Dormio that monitors the middle and index fingers and the wrist. As the subject falls asleep, they are fed audio cues. In a particular experiment, the cue was "Remember to think of a tree." The Dormio monitors when the subject has fallen asleep and then rouses them. The system records how they describe their dreams. The active subjects reported trees in their dreams; the control group did not. The MIT scientists call their system "targeted dream incubation" (TDI). A related technique is *targeted memory reactivation* (TMR), in which a cue delivered during sleep can trigger specific memories. TDI and TMR are yet far from full dream control, but the direction is clear.

Neurobiologist Rafael Yuste at Columbia University is working on *optogenetic* techniques based on altering a couple of genes with genetic engineering. One gene modification makes neurons emit infrared light when they activate. That could be used to record the activity of neurons. Another leads the brain to produce a protein that is sensitive to the penetrating power of infrared light. That could be used to activate neurons. The result is a new precision in both reading and writing to the brain. This technology has so far been applied only to mice but has been used to cause mice to see things that do not exist—artificial hallucinations. Related approaches using drugs or nanoparticles that can swarm in the bloodstream may make neurons more sensitive to infrared light without having to modify genes.

This approach might also enable precise control of individual neurons and selected arrays of neurons.

AUGMENTED AND VIRTUAL REALITY

Developments in the laboratory have driven a nascent commercial market that is feeding back into the research realm. In 2009, the toy company Mattel released a game called Mindflex. The game contained an EEG headset. The idea was that competitors could use their brain waves to control the flow of air that lifted a light foam ball. Whether there was true control of the ball, or the game randomly moved the ball while giving the illusion of control remains debated, but the same chip used in the toy was employed in subsequent developments. People hacked Mindflex with its single EEG electrode and added Bluetooth connections to mobile phones and monitor apps to produce basic BCI rigs. In 2013, one of the hackers, Connor Russomanno, and Joel Murphy, a computer scientist at the Parsons School of design, launched a Kickstarter campaign to build commercial BCI equipment. In 2014, they started the company Open-BCI. Their first product sported eight EEG electrodes and the capability of sending the resulting digitized signals to a computer. About the same time, the Canadian company InteraXon introduced Muse. Muse had an elegant headband with four electrodes and was specifically aimed at the consumer market. Other products followed with the goals of improving mindfulness and meditation and inducing sleep. The devices also targeted gaming and employment in VR. Researchers realized that these cheap, portable commercial products opened new experimental possibilities, and a virtuous feedback began.

There are other devices that enable input to the brain. AR adds computer-guided input on top of normal visual input. The brain does not care whether the input is real or artificial. It collects whatever signals are coming from neurons in the eye and processes that as an image, reality be damned. VR relies on the same capabilities. Input is provided to the eyes and hence to the brain to simulate an immersive environment. Haptic gloves can induce a computer-generated feeling of touching an object that is not actually present. French engineers have developed an especially lightweight glove that gives very realistic feedback when touching a virtual object. Reproducing the signals associated with touching can be useful with prosthetics so that a patient can truly feel when a robot arm picks up a fork or spoon. Microsoft produces the HoloLens, an AR headset. Facebook bought into AR and VR with its purchase in 2017 of the company that had developed the Oculus Rift, the first VR headset that did not induce intense nausea. Walmart is using Facebook's Oculus Go headset to train new employees using VR. Mojo Vision is developing a contact lens that can project AR information or images directly onto the retina. The lens will receive its power by an inductive connection and information from the internet by means of an external device like a mobile phone.

NO MORE KEYBOARD

At this point, the technology for communicating to and from the brain is still relatively cumbersome, but it is clear where the research is rapidly leading. Rather than desktop computers and mobile phones with keypads, there will be glasses or contact lenses to provide displays. Earbuds may both talk to us and measure our brain's electrical and cognitive state. Mobile phones will have new apps to receive data coming from the internet, including the IoT, and from the body's internet: sensors worn in clothing or on the skin. If direct brain contact is needed, electrodes will become smaller and less intrusive. It may become possible to "jack in" to the internet as sketched so brilliantly in William Gibson's pioneering 1984 cyberpunk novel *Neuromancer*. Much of this contacting technology might get built into a set of headphones.

A world then opens up where everything you now do with a computer, you do just by thinking. Never lift a finger. No more keyboards. No more mice. Do you want to call or text a friend, relative, or colleague? Just think of them and think the words. Pull-down menus would appear with just a thought. Checking social media would become a breeze with virtually instant access. Thought-to-speech technology would allow one to talk to any other individual or to all people capably wired. Search engines or chatbots would be accessible by just thinking of the topic you want to investigate or the item you seek to purchase. Code could be written by thinking but would still require considerable thought supplemented or enhanced by AI. Music could be composed, and art rendered—at least on a computer screen—just by thinking. Virtually any sensory experience would be available merely by wishing for it.

As for many new technologies, among the first uses will be for games. Video games are currently operated by a controller. There are buttons moving characters left and right, forward and back. There are buttons for jumping, kicking, hitting, or firing rifles or lasers. All those buttons are controlled by brain signals that go to thumb muscles. It is a simple extrapolation of current technology to foresee some device—a headset—that records those thumb-muscle impulses and feeds them into the contemporary version of an X-box. Fantasies about such capabilities even invade contemporary TV commercials.

MIND READING, TELEKINESIS, AND TELEPATHY

This ongoing progress and the promise of exponential growth permit us to look ahead to where this brain reading and writing technology may lead. There are fantastical capabilities that have been contemplated for as long as humans have thought about thinking: mind reading, telekinesis, telepathy. All these are possible now and will grow ever more sophisticated.

Any technology that leads to recording signals from the brain is a form of mind reading. As illustrated by the examples above, that technology already exists in a rudimentary way. This is not one mind directly reading another. A great deal of apparatus, including sophisticated AI techniques, is required to mediate, but that is the result: the reading of mind impulses to muscles, of images, of thoughts. This capacity will only become more proficient. From control of prosthetics, it is a small step—under active research—to develop cobots where human minds are locked in a cooperative robot link.

If past trends are portent, an early use of BCI will be, somehow, for sex. As we get more acute at reading minds, we are likely to find that most people are thinking about sex all the time. Outlets will be found. In this brave new world, pornography may never be the same again.

Controlling prosthetics, robots, or avatars with the mind is, for all practical purposes, telekinesis. Again, the apparatus for monitoring mind signals and AI to interpret and digitize them is needed, but the result is moving things with your mind. The thing you are moving could be right next to you or anywhere a digital signal can reach. You could control your cobot on the Moon—if you are willing to tolerate a delay of a second or two for the round trip of the signal at the speed of light.

To the best of my knowledge, no one has yet achieved telepathy, but the prospect is imminent. The AI rig reads signals from the brain of one person. The corresponding digital signal is implanted into the brain of another person. We have seen that the technology exists now to extract an image formed in a human mind. The technology exists to implant artificial hallucinations. It is a small step, in principle, to extract the image from one mind and implant it in another. Voila! Telepathy.

Peter Diamandis, proponent of the X-prize and developer with Ray Kurzweil of Singularity University, is a relentless technological optimist.[24] He foresees in technology like Elon Musk's Neuralink exactly this capability: mind-to-mind telepathy connections. He espouses the positive benefits of melding us all together in common thought. In a blog[25] he wrote:

> One of the most profound and long-term implications of BCI is its ability to interconnect all of our minds. To share our thoughts, memories, and actions across all of humanity. Imagine just for a moment: a future society in which each of us are connected to the cloud through high-bandwidth BCI, allowing the unfiltered sharing of feelings, memories and thoughts. Imagine a kinder and gentler version of the Borg (from Star Trek), allowing the linking of 8 billion minds via the cloud and reaching a state of transformative human intelligence.

I am more cautious. I read a science fiction story long ago about a man who gained the ability to read minds. It was a deeply discomfiting experience, and the protagonist was forced to eventually withdraw from society. I have thought ever since that the ability to read minds would be a terrible accomplishment. I do not want people to know my routine, inner thoughts. I do not want to know the inner thoughts of others. We can see what happened when the internet brought the capacity to share thoughts even slowed and mitigated by the cumbersome process of thumbing a message into a Tweet or typing a comment in a Facebook page: a crisis of sociological division. We need to tread very cautiously in this direction.

Each of these capabilities—mind reading, telekinesis, telepathy—requires careful forethought. Mobile phones have altered human behavior. The device can be more compelling than the human right next to you. How much more might that impulse be amplified if you only need to think about the device to connect to the devices of others, to the internet? At least now you can tell when a person is looking at their phone. In this mind-reading future all you might notice is a blank, distracted look. Now we have some control over when we succumb to the allure of the device even as it calls for our attention with a suite of dings, beeps, and ring tones. What will that be like when the device is calling directly to our brain?

Technically aided telepathy requires even more care. Techniques to control a robot with the mind are already available. It is only a small conceptual step for the cobot to have some control over the human mind on the other end of the link. How do we control which of our thoughts are radiated externally? How do we control what thoughts we receive of others? Our thoughts tend to be scattered and chaotic. What would it be like to tune into those of another individual, of a large group of individuals? If it is anything like tuning into all radio stations at once, it would be chaos. Perhaps the AI will be talented enough to select individuals and individual thoughts, but we need guarantees.

Perhaps there is some benefit, some efficiency, to be had by sharing selective thoughts. What about our emotions? Do we want to radiate those for all to perceive? If we are tightly bound to other minds, how will we know which emotions are ours, which belong to another consciousness, and which are generated by the connecting AI just because it can. Do even husband and wife want to know each other all that well? What about our subconscious impulses? Are those off limits for an effective, prying AI?

Even if our equipment can provide some constructive filtering, do we want technologically to evolve toward a collective consciousness—a shared mind, or "hive mind"—in analogy to a beehive or ant colony? The functioning of our current society relies on the fact that we have, and can keep, private thoughts. How will society change if there are few or no private thoughts?

A mentally hyper-connected world might fragment into balkanized segments, as Facebook has promoted. At the opposite extreme, there might be a homogenization of society to an unhealthy degree. If there would become only one way of thinking, we might lose the rich power, stability, and flexibility of multiple lines of thought. Intellectual curiosity might be inhibited if there were a tendency toward one sanitized mode of thinking. Would a hive mind bristle with creativity or be subject to control?

Contemplate the big companies suddenly interested in brain-machine research: Facebook and Microsoft, maybe soon Apple, Google, and Amazon. Suppose Google could harvest not only your search results or your map location but also your thoughts, your daydreams, your desires, your sexual fantasies. Who owns those data? Given that writing to the brain is possible, there might be programs you could access for free that would only occasionally plant a suggestion in your brain. How will you know whether your sudden hankering for a pizza or a pair of excellent sneakers is genuine or a gently planted craving? The machine knew you wanted that anyway, it was just giving you a nudge.

Never mind the tech companies that, after all, are only in business to sell you stuff. What about this technology in the hands of the ultimate biological or machine autocrat who seeks power over society? What happens if—as seems inevitable—the system is hacked? The opportunity for vast societal damage is very real.

The drive to develop and employ this technology is going to be very powerful and probably unstoppable. The superficially admirable goal will be to cure ills and augment the capacity of humans. It is difficult to see how this will not alter individuals and societies profoundly. For some, that alteration is the point. While there are channels for discussion of the issues, like this book, it is clear that society barely has these issues on its collective radar screen.

ETHICAL BCI

There are, thankfully, some efforts to address these issues by professionals engaged in the research. A group of neuroscientists, ethicists, and machine-intelligence researchers from academic and research institutions in the United States, Canada, Europe, Israel, China, Japan, and Australia calling themselves the Morningside Group met at a workshop at Columbia University in 2017. They published an eloquent statement of the issues and suggestions for steps to mitigate potential problems that enunciated four key areas of concern: privacy and consent; agency and identity; augmentation; and bias.[26] They argued that people should be able to keep their neural data private with opting out of sharing being the default condition. Neural data should be protected much as people's organs are now to prevent overly aggressive or even unscrupulous use.

We have a well-established right not to incriminate ourselves with our spoken words in a trial; the same should be true for words or images extracted directly from our minds with neural links. Care must be taken that each individual's mental integrity and their ability to choose their own actions are protected.

As humans gain augmented capacities, new segregation by artificially enhanced ability and possible new forms of discrimination may occur. Regulations might be designed to require progress in small, controllable steps to which society can adjust and render collective approval. Use of new neural capabilities for military purposes might need special attention and regulation. Bias could creep into the enterprise in a number of ways. There is already evidence of systemic racism in the brain research community: there are very few Black people involved in current efforts.

Some of the commercial enterprises involved in BCI research pay at least lip service to these principles. Representatives of Facebook's Reality Labs stress that their work on BCI is advertised and transparent. Researchers associated with the Microsoft BCI effort argue that information gleaned from minds must be treated very carefully. Still there are immense profits to be made planting telepathic ads in minds or bartering mental information on customers. Do we rely on these companies who say, "Trust us, we'll be careful"?

Other voices in the community argue that calls for control and regulation are unnecessary or at least premature. They point out that we already have HIPAA laws that restrict the release of patient medical data. Perhaps they can be interpreted as covering mental data. The approval process of the Food and Drug Administration actively incorporates ethical issues. Perhaps these statutes could be interpreted as covering mind-reading issues, but it is easy to envisage corporate lawyers finding ways to elude them. Another counterargument is that the field of BCI requires the positive influence of entrepreneurialism and ample funds to encourage it. Over-regulation, this argument goes, could stifle this inventive spirit. At risk would be new, immensely beneficial treatments for mental diseases and the exploration of new ways to communicate.

One of the great looming questions of BCI research is whether it will ever be possible to download and upload the human mind. This possibility is still in the realm of science fiction, but the more we learn about the brain and the more we develop techniques to read and write to the brain, the closer it comes. A company called Humai[27] engages in robotic and medical research. It advertises its goal as extending, enhancing, and restoring human life. To do so, it promises to "transplant your brain into an elegantly designed bionic body" also called Humai. The Humai would use BCI techniques to "communicate with the sensory organs and limbs of your new bionic body." I would not hold my breath, but the aspiration is clear.

PROSPECTS FOR BCI

It may prove feasible to emulate some functions of a human brain, but to upload or download all the contents of a brain—to replicate its full function—may require the complete mapping of every neuron and synapse. This is far beyond the capacity of current technology. The issue is not just the vast number of neurons but also the complexity of their connections, the strength of their reactions, and the timing of those reactions. If the function of the brain depends on emergent properties of all that complexity, even reproducing 200 billion neurons may not fully recreate the functions of a conscious brain. In *The Singularity Is Near*, Ray Kurzweil presents a fascinating scientific, engineering, and philosophical exploration of the prospect for uploading and downloading not just brain contents but personalities.

Techno-optimists foresee great advances for humanity from BCI capabilities: mental disease banished; loss of limbs made a minor nuisance; the elderly retaining mental acuity for decades longer than now; intelligence, even superintelligence becoming a resource shared with all of humanity. We could upload to social media not just texts, photos, and videos but entire intact memories. Perhaps with the help of our AI, we could learn to manipulate our neurochemistry with vast implications for controlling sleep cycles, pain, our ability to focus—all the things for which we now take pills. We might put the illegal drug industry out of business—or replace it with something worse.

BCI is an exceedingly expensive enterprise now. As progress is made and BCI becomes more accessible, demand will rise, and costs decline. Perhaps it becomes a simple outpatient procedure or even as simple as slipping on a helmet, or a pair of headphones, or a little chip that sticks on the back of your neck right beneath your occipital bump. Despite their relatively high expense, mobile phones swept the planet. Perhaps the same will happen with BCI technology once a certain price is reached. If some people acquire new artificially enhanced intellectual powers, others will aspire to catch up if only for the fear of being left out.

The capacity to share our thoughts, dreams, and memories with anyone on the planet, or everyone on the planet, could become part of the technological evolution of our species. If billions of people are able to access all of human knowledge and to connect to one another essentially instantaneously, society will be transformed in ways that are difficult to predict.

Some argue that this BCI-driven transformation is a matter of self-defense. If AI develops as some foresee, attains the capacity of the human brain in a cognitive singularity, and then races exponentially beyond, we poor, biologically inhibited *Homo sapiens* will be left behind. The solution, this line of thinking goes, is to deliberately merge with our AI, to form a symbiosis so that our intellect grows in a way that avoids the Terminator scenario. According to this argument, we must join this revolution.

Elon Musk shares this terrible paranoia. He fears that advances in AI are inevitable and that we must, through his Neuralink system or something like it, join the intellectual revolution catalyzed by AI and BCI in order to avoid being overwhelmed by it. Musk argues that merging with AI will be an option, not a requirement. I say, good luck with that.

There is a powerful BCI genie straining to get out of the bottle. It is probably unstoppable. Circumstances will probably lead to exponential growth in this technology. Watch what you wish for.

GROWING BRAINS

As brain researchers learn ever more about how the brain works to preserve the body and create the mind, we will also learn more about whether and how an analog can be developed in a non-biological substrate or in a biological-mechanical symbiosis. Brain researchers are attempting to simulate parts of the brain or even whole brains in computers. An example regards the *hippocampi*, two seahorse-shaped regions at the base of the brain that are important for forming memories.[28] A research group at George Mason University under the supervision of Giorgio Ascoli has spent years collecting data from researchers around the world on the number of different kinds of neurons and their properties in the hippocampus of rats.[29] They have found 122 different types of neurons, each with its own connections, connection strengths, and firing thresholds. Those data are being encoded in a computer simulation. When that computer program fires up, what will it remember? Will it be a tabula rasa awaiting experience and memories, or will there be some remnant memory of being a rat? The computer model will be based on the biological reality of a rat hippocampus and will have some of a rat's functionality, but it may be nothing like a rat. Simulation of an entire brain will raise the same issues writ large. This will be a new era of simulation that further blurs the difference between biology and computation, raising new ethical issues and perhaps calling for judicious relinquishment.

Another ethically challenging avenue is to grow brains from scratch. Scientists at Stanford have learned how to transform mature human skin cells back into their multi-potential stem cell state and then steer the development of those stem cells into various types of brain cells.[30] Those brain cells have been collected into brain organoids—miniature, simplified versions of brains—and then implanted in baby rats to study diseases like microcephaly, Alzheimer's, autism, and schizophrenia. The implanted brain cells grew and formed new connections.

In related work, another group in Cortical Labs—a private company in Melbourne, Australia—grew brain cells from human stem cells and mouse embryos.[31] They then connected a batch of 800,000 of those cells to electrodes in the basic but pioneering video game Pong. The feed from the game console contained information about which side of the playing field the ball was on and how far it was from the paddle. The brain cells responded and learned to play the game in a rudimentary fashion in just five minutes—an example of unsupervised learning as discussed in chapter 3. The brain cell clump was able to register information from an external source, process it, and then respond. The Cortical Labs researchers led by founder and CEO Hon Weng Chong and chief scientific officer Brett Kagan declared the system to be effectively sentient even though it did not know it was playing Pong in the sense that a human would. The researchers anticipate their work may lead to more adaptable robots. The team is consulting with bioethicists to avoid accidentally creating a conscious brain, but how will they know until they have done it? This is ethically sensitive territory, but the imperative of exponential development is sure to drive research in the direction of conscious, artificial brains.

CONSCIOUS COMPUTERS
Turing's Challenge

THINKING ABOUT THINKING

If the functioning of our brains remains a mystery, consciousness remains the central enigma. René Descartes famously said, *Cogito, ergo sum* [I think, therefore I am]. Cognitive scientist and philosopher David Chalmers said, "Consciousness poses the most baffling problems in the science of the mind. There is nothing that we know more intimately than conscious experience, but there is nothing that is harder to explain."

What distinguishes the mind from the brain? Descartes elaborated by arguing that "we cannot doubt of our existence while we doubt." For there to be a thought, there must be something that thinks; therefore, one's own mind must be real. What one thinks might all be a fantasy, a computer simulation. All the input that one regards as real—sight, hearing, smell, feeling—could be artificial input into the brain that hosts the mind. A criticism of Descartes is that he assumed that there is an I, an entity that does the thinking. The most one can say is that thinking happens and not that an I exists. Descartes' is a Western line of thought that we will briefly pursue here. Eastern philosophy and religion have yet a different perspective. A tenet of Buddhism is that subjective experience—consciousness—is the only true reality and that all else is illusion.

There are various attributes of consciousness as we assess it in ourselves or others: self-awareness, or knowing that one exists; an internal representation of the external world, or the ability to visualize elements of the environment that are not immediately present; the ability to assess the consequences of one's actions; the ability to plan; the ability to learn.

I challenged my students with the questions "You may know that you are conscious, but how do you know I am? How do you know that I'm not some artificial being, programmed to stand in front of a class and pontificate?" Most shied from the question as if my consciousness were self-evident, but the question was serious and the answer not obvious. How do we know that someone else is conscious? An experimenter can make a sound and record the response of a subject who hears the sound but cannot measure the experience of hearing the sound. One cannot measure consciousness.

It is common to think that consciousness is the purview of *Homo sapiens*, yet there are abundant studies that show rudimentary thinking—consciousness, self-awareness—in animals from bees to chimpanzees.[1] The attribute might even be extended to plants that react to stimuli in their environment and communicate with one another by means of chemicals in the air and soil. Some work suggests that plants exchange bits of material that alter gene expression across plant kingdoms.[2] The brilliant novel *The Overstory* by Richard Powers gives a fictional introduction to the notion of plant consciousness. As I pointed out in chapter 7, even human consciousness can be questioned given the evidence that the brain signals movement of muscles a finite time before the conscious mind is apprised. That finding takes our special status down a notch.

By 2015, researchers had recreated in a computer the detailed neuronal connections of a part of a rat brain. In 2020, other workers scanned the brains of crows as they worked on laboratory puzzles and found evidence that the crows could think about their own thinking. Such an achievement is an aspect of consciousness and suggests that consciousness is achievable in a variety of ways and more broadly than many usually think. On the other hand, the entire relatively simple nervous system of the worm *c-elegans*, all the neurons, all the synapses, and all the connections, have been mapped without giving any new insight into the consciousness of the worm.

An active argument is that consciousness is a trait developed through natural selection and evolution. The idea is a simple one: consciousness aids survivability and reproduction. This facility may have come to a (temporary) peak with *Homo sapiens*, but it is littered along the way in weeds and trees and animals. The ability to focus attention, to screen relevant information from a stream of input, to determine sources of food and danger, and to plan surely provides an evolutionary advantage. The underlying supposition of this line of thinking is that consciousness arises from a certain degree of complexity, that it is an emergent property (see chapter 9 for a more elaborate discussion). Certain species might have a little of that emergent property; others could have a lot.

It is an amazing aspect of heart cells that they are built to pump. Heart cells grown from stem cells will start to contract rhythmically as soon as they reach an adequate number. Something like that is also true of the brain. Scientists have grown batches of brain cells in a petri dish, and the cells exhibit human-like electrical brain waves. This is a fascinating accomplishment, but it also raises profound ethical questions. Could such a collection of cells gain a consciousness? What would happen if they did?

EASY PROBLEMS, HARD PROBLEMS

All this sidesteps the question of the true nature of consciousness and whether science will ever understand it. Pondering consciousness has a long history. Chalmers fostered a new phase in this topic in his book *The Conscious Mind*.[3] Chalmers differentiated problems that can be attacked by traditional reductionist scientific methods from the issue of the nature of consciousness. He called the first category the "easy problems." He included among easy problems the problem of how the brain works—what its parts are and how they function in a neuro-chemical way. Chalmers deemed consciousness to be something different, something beyond and not amenable to reductionist study. He called consciousness the "hard problem." He emphasized that consciousness involves personal experience, the feeling of what it means to exist.

An oft-cited example to elucidate the easy and hard aspects of consciousness is to consider what it means to be red—an electromagnetic radiation of a certain range of wavelengths—contrasted with what it means to perceive the quality of redness. In this way of thinking, the knowledge of the appearance of a red apple is distinct from knowledge of how signals impinge on our visual system and are converted to electrical impulses and the firing of neurons in our brain. A color-blind person might know everything about the physics of radiation and of the brain and yet not experience redness. On the other hand, we do not experience UV-ness, radio-ness, or X-ray-ness yet we still feel ourselves to be conscious and aware. The hard problem of consciousness then becomes asking why awareness of sensory information exists and how experience arises in otherwise non-sentient matter.

The sensory processes depend on aspects of physics and chemistry that exist even in the absence of experience. Chalmers argues that experience is more than the collective action of physical and chemical processes. In his view, the hard problem is how and why sensory processes result in the feeling of experience. Chalmers has argued that consciousness is beyond known physics, that it is an independent, fundamental property of Nature. Others argue that consciousness—subjective experience—is beyond science entirely.

If subjective experience is the essence of consciousness, another potential philosophical conundrum arises: can experience exist in the absence of a physical world of sensory input? Our techniques of VR and AR are becoming ever more sophisticated. How, one might ask, do you distinguish a hyper-realistic VR from reality? Would the conscious experience be just the same?

A different perspective from that of Chalmers acknowledges that conscious experience is real but then posits that it can be understood in terms of the brain's functions. A principal notion is that while the brain is composed of reducible parts, the mind, experience, and consciousness arise as emergent properties of the complexity of the brain. From this perspective, consciousness is computational; it can be obtained by applying appropriate algorithms and calculations. Most neuroscientists and cognitive scientists believe that the hard problem of consciousness will be solved once the easy problems of brain functions are sufficiently understood. Daniel Dennett advocates this point of view in his book *Consciousness Explained*.[4] Dennett argues that consciousness is not an aspect of reality that is independent of physics and chemistry but rather that it will eventually be fully explained by natural phenomena. In this view, the hard problem of consciousness is a collection of easy problems.[5] The hard problem will be solved through deeper analysis of the brain and behavior. It is important to note that what neuroscientists believe and what has been proven are two different things.

Oxford physicist David Deutsch treads a third ground. He argues that an AGI must be possible from the laws of physics and the Babbage, Lovelace, Turing principle of the universality of computation: everything that the laws of physics require a physical object to do can be emulated by some program on a general-purpose computer with adequate time and memory.[6] Deutsch argues that sentience at the level of animals is a fairly useless, trivial ability; that the essence of creativity in the human sense is conjecture and criticism to reduce error; and that no current AI is remotely close to constructive employment of the latter.

QUANTUM CONSCIOUSNESS?

Even from a strong reductionist point of view, the easy problems of consciousness are quite difficult. One example involves the debatable role of quantum mechanics. Quantum theory and its application are triumphs of twentieth-century physics. Quantum mechanics deals with the nature of things on the submicroscopic scale of molecules, atoms, and elementary particles. In this world, things behave very differently than in the macroscopic world of normal human experience, which is to say, human consciousness. In the world of quantum mechanics, changes in position, energy, or momentum do not occur smoothly but in finite jumps. Electrons are not tiny points of stuff but are waves of probability that can be in more places than one at the same time. There is an intrinsic uncertainty in the position or motion of anything. Atoms are not at all like tiny solar systems with electrons orbiting around the nucleus but are waves of overlapping electrons, protons, and neutrons. Particles can leak from one place to another despite there being a barrier between them. Particles can be in two states at once, both spinning to the left and spinning to the right.

This is a strange, non-intuitive world, but it is the way Nature works. Scientists and engineers have mastered many of its principles in practical applications. The whole world of modern electronics and digital computers depends on knowing how to manipulate the probability waves of electrons as they move through silicon chips. The new frontier of quantum computing essentially depends on putting electrons into two opposite states—pointing up and pointing down—simultaneously.

An aspect of quantum theory that still bedevils physicists and philosophers is how these quantum ambiguities become resolved in the macroscopic world of our perceptions. In our world, a merry-go-round either spins to the left or it spins to the right; it does not do both at once. A key step in understanding this resolution is to consider the act of measuring rotation. In the strange quantum world, an electron can have a probability of spinning left and a probability of spinning right at the same time. The two states are said to be *quantum entangled*. Once a measurement is made of the spin of the electron, it is one or the other. The entanglement is broken. What happened?

Physicists describe the process of breaking quantum entanglement as the *collapse of the wave function*. What was a complex wave of probability before the measurement collapses to a definite state—spinning left or right—after the measurement. This works. Engineers use the principles and mathematics of quantum theory to design ever more elaborate silicon chips and quantum computers.

This line of thinking also leaves conundrums that the physicist Edwin Schrödinger illustrated with his famous cat paradox. Schrödinger envisaged a cat in a box in which a radioactive atom could decay or not decay. Whether the atom decays is not like a switch that's either on or off. According to quantum mechanics that accurately describes the behavior of radioactive decay, at any given moment there is a probability that the atom decays and a probability that it does not. If it decays, the cat dies. If it does not decay, the cat lives. The atom has an entangled probability wave that it decays and doesn't decay at the same time. The implication then seems to be that before we open the box, the cat is in an entangled superposition of being both alive and dead, that is, with some probability of each. After we open the box and thus determine (measure) the state of the cat, it will obviously be either alive or dead and not some of each. The act of determining, of measuring, of being conscious of the result, somehow collapses the wave function, breaks the entanglement, and brings about a concrete result.

Quantum theory—capable of predicting the outcome of the measurement of electrons in an atom with an accuracy of one part in 100,000—thus seems to involve the conscious observer, the experimenter, in some way when it comes to experiencing our everyday macroscopic world. In the microscopic world where particles are waves, the waves can interfere and add or negate their influence coherently. The process of making the transition from the microscopic to the macroscopic world, from isolated bits to interaction with the external world, is called *decoherence* and is the subject of deep study by physicists. Decoherence suggests that there is an intermediate regime where the pummeling by other particles and waves spoils the coherence of the pure quantum world by effectively making a few, then many, then an overwhelming number of small measurements before one gets to the level of the macroscopic, conscious, human experimenter. Avoiding this decoherence is one of the principal problems in constructing a working quantum computer.

All this comes to focus in the current context because fundamentally the brain must be a quantum engine. The flow of brain currents is actually a flow of electrons, which means a flow of quantum probability waves. If the brain is a quantum entity, the issue arises as to whether consciousness is somehow related to quantum effects. The interpretation of the operation of quantum mechanics in which consciousness—the awareness of the result of an experiment—is necessary to make a quantum measurement is called the von Neumann–Wigner interpretation after the two physicists who enunciated the concept. It is not clear that either of them accepted their postulate literally.

A contemporary explorer of the possible role of quantum properties in consciousness is Sir Roger Penrose, who won the 2020 Nobel Prize in Physics for his career-spanning work on Einstein's theory of gravity and especially its applications to the fundamental properties of black holes. Penrose is a deep and broad thinker and prolific writer who has pondered a great range of problems including the nature of consciousness.[7] He is motivated by Gödel's theorem that some aspects of mathematics are true but cannot be proven. There are things that are true that cannot be established by an algorithm.

Penrose posits that consciousness is beyond the explanation of contemporary physics, biology, and neuroscience. Rather than leaving science behind, he has investigated the strange microscopic world of quantum physics, where elements can be in a superposition of many states at once. Penrose rejects the notion that consciousness is algorithmic or computational in the sense of a digital computer running on elaborate arrays of 0s and 1s. Instead, he considers the possibility that elements of the brain can be 0 and 1, on and off, right spinning and left spinning, alive and dead, at the same time. In the quantum world, multiple states can be in superposition with the entire system acting together in one quantum state before coalescing, decohering in a single step. It is certainly not clear that such a process would lead to consciousness, but it would be different than a brain working like an analog of a digital computer as many postulate now.

Penrose has specifically explored the possibility that quantum properties affect the tiny components of brain cells called *microtubules*. Microtubules are tube-like arrangements of proteins inside all cells, including neurons. They help to shape nerve-cell structure, movement, and division and thereby help to control the strength of synaptic connections. Penrose wonders whether the tube-like shape, symmetry, and lattice structure of microtubules are fundamentally of quantum nature and hence amenable to participating in coherent quantum states. Microtubules might have the capacity to store information and process it in a coherent quantum state to produce a memory. Perhaps the emergent coherent behavior of an array of microtubules results in conscious awareness. Penrose speculates that the deepest understanding of all this is beyond current quantum theory, which is known to be incomplete. Contemporary quantum theory conflicts with the current version of Einstein's theory of curved space and gravity. At the practical level, evidence that microtubules undergo quantum vibrations remains ambivalent.

Most scientists think quantum mechanics is irrelevant to our practical understanding of the brain. For the application of the principles of coherent quantum states—being 0 and 1 simultaneously—to quantum computing, conditions must be very cold and carefully isolated to avoid a premature cascade into decoherence. By those standards, the brain is too warm and malleable for its isolated parts to maintain a coherent quantum state involving a vast number of microtubules for anything other than a fantastically small interval of time.[8]

The debates of quantum or not, physics or not, and emergent property or not tend to rest on the assumption that the pieces that compose the brain—atoms, molecules, cells—are permanently in place and that those pieces working collectively have consciousness or not. Kurzweil brings in another relevant perspective. He notes that the pieces are not the same from one moment to the next. Molecules come and go. Neurons are composed of different atoms from month to month. Most cells are replaced on timescales of weeks. Microtubules last about ten minutes. Portions of dendrites are replaced about every forty seconds. Kurzweil thus argues that even though our conscious awareness feels the same from day to day and that we carry memories for years and decades, the physical material of the brain is transient. He concludes that it is the patterns of the arrangement of particles in the brain that persists[9] and not the fundament of particles just as Niagara Falls is a constant (on a human timescale) but composed of different water molecules from moment to moment.

Kurzweil then proposes an interesting thought experiment of imagining that we will eventually have the capability to upload our patterns in minute detail into another substrate and ask, Is that copy me? He answers in the negative but then reposes the question as replacing his brain parts bit by bit and asks at exactly what point does his old consciousness disappear and his new consciousness become him? That, he posits, is exactly what happens in our bodies and brains. Particles and components come and go yet patterns—our consciousness—persist. Kurzweil does not explore the issue of whether this consciousness pattern is quantum based.

At the opposite extreme from Penrose and others who hold that consciousness is a realm beyond current physics are those who believe with Daniel Dennett that consciousness is ultimately computational. One of the pioneers of artificial intelligence, Marvin Minsky, proclaimed that the brain is "just a computer made of meat."[10] Minsky thought that computers made of silicon chips could also become conscious. Kurzweil likewise assumes that with sufficient complexity, consciousness will arise. My Austin colleague Byron Reese is a technologist and futurist. In his book *The Fourth*

Age,[11] Reese presents an incisive case that there are two modes of thinking regarding the issue of whether machines can become conscious. One is that humans are fundamentally machines—biological machines, but machines, nevertheless. The other basic point of view is that humans are not merely mechanical beings—that there is more going on than that. Opinions on a whole host of other issues treated in Reese's book, and this one, follow from those starting points.

TURING'S CHALLENGE

Preconceived notions aside, there are many discussions of how one would decide whether a machine were conscious. Among the most famous is the Turing test, invented by the brilliant computer scientist Alan Turing.[12] Turing did not specifically address the issue of consciousness but asked whether a machine could think and how that would be determined. He posed what he called the *imitation game* as a means to determine whether a machine could do what a human does in an attempt to separate the physical and intellectual capabilities of humans. Turing proposed that a human evaluator would try to decide which of two players was a machine and which a human by posing questions by typed text to each. The interchange would be entirely by typed natural language in a manner that would not depend on the ability of the machine to render words into spoken language. Turing argued that the machine would pass the test based not on how correct its answers were—nor those of the hidden human—but on how human-like they were.

The Turing test taxed both scientists and philosophers. Philosopher John Searle argued that the Turing test could not be used to determine if a machine could think. He hypothesized a "Chinese room"[13] in which either a machine or a human could receive input in the form of Chinese characters, perform some algorithmic response, and present some output in Chinese characters that would be cogent to an external Chinese speaker—thus passing the Turing test in Chinese—when neither the machine nor the human within the room spoke or understood Chinese. In this way, Searle attempted to determine whether a machine truly understood Chinese or was merely simulating the ability to understand Chinese. He referred to the former as *strong artificial intelligence* and the latter as *weak artificial intelligence*, terms still in use today. Searle's proposition, in turn, produced a cascade of argument and counterargument that continues to reverberate.

We have come a long way since Turing's time, climbing the curve of the exponential growth of knowledge. With the use of AI, Google and OpenAI have developed very impressive natural language translation systems. Google Translate, ChatGPT, and Bard are very impressive. They effectively pass a Turing test, but few would mistake them for thinking, basically supporting Searle's point of view. While they may not fool anyone or everyone, chatbots are now ubiquitous. Some are useful in handling routine consumer inquiries. At other times, one yearns to communicate with a human. AI has engaged in the writing of prose and in painting but still struggles with writing a decent screenplay that requires a depth of human experience that cannot be easily reproduced. Even that skill may artificially be acquired with further developments. Humans may be born with their brains pre-wired to be capable of perceiving letters and words. So can AI.

We do not yet know whether future developments will lead to machine consciousness, but exponential improvements beckon. By drawing on the current collective wisdom of the web, current AI can sample from a huge variety of sources, assemble words in a probabilistic way, and produce prose never written before. A crucial factor is that ChatGPT does not know whether it is writing a novel or a business memorandum the way a human would. AlphaGo Zero and other examples show that AI can invent strategies beyond human capability, but AlphaGo Zero does not know it is playing the game of Go. Can it then be said to truly create the way a human can? What is the difference between invention and creation? Is it the spark we call consciousness? Consider that the neurons in our brains do not know what they are doing. They just store and discharge electrical power. If our consciousness arises as an emergent property in our brains, cannot something like that happen in our machines?

SENSORIMOTOR CONTINGENCY IN MACHINES

While attacking consciousness head on is a huge challenge, others have approached the issue by trying to reproduce basic elements of it. One aspect of consciousness is self-awareness, and an important ingredient of that is the ability to make a model, a representation of one's surroundings. Every infant does this. They are born wiggling, but it takes a while before the brain associates certain motions with the environment. The infant learns to reach and grip, then to crawl, then walk. By about eighteen months and not much earlier, a baby can recognize itself in a mirror. Building a model of self and environment takes time and experience. The same holds true for machines.

Experiments involving aspects of self-awareness in machines have been done for some time. One step is to implement *sensorimotor contingency*, the notion that an effect on the environment is contingent on the motion of a body. Another step is to develop techniques by which a robot can distinguish itself from another robot. Yet another is for a robot to recognize itself in a mirror. That is difficult and has not yet been done.

A classic test of self-awareness in an animal is for it to recognize that a mark on its body that can only be seen in a mirror is on its own body. Only a few animals can do this, including dolphins, orcas, elephants, magpies, some apes, and humans.[14] Computer scientist Justin Hart explores this issue with robots. As a graduate student at Yale, he worked with Nico, a humanoid robot. Nico first learned to determine the location of its arm by looking at its reflection in a mirror. Then it learned to reach around its back for something that it could only see in the mirror. Neither Nico nor any other robot has yet passed the mirror test at the self-consciousness level of a magpie.

A hint of self-awareness in a machine arose in an interesting experiment done in 2015 that involved three of the standard Nao AI-enabled robots described in chapter 4, robots of the kind that can play soccer, stand, sit, speak, and hear. Selmer Bringsjord, a professor of computer science at the Rensselaer Polytechnic Institute, arranged to put a version of the three-wise-men riddle to the little robots. The classic statement of the riddle is that a king presents each of his three advisors with a hat that is either blue or white. Each advisor cannot see his own hat but can see the hats of the others. The king stipulates that the contest is fair; that an answer can be rigorously known. The first advisor to deduce the color of the unseen hat on his own head wins the contest. The critical insight is that the contest would only be fair if the hats were all the same color. If a single advisor sees either two blue hats or two white hats on the other advisors and the hats are not necessarily all the same color, there is no way the advisor can rigorously know whether his hat is blue or white. The same condition applies if a single advisor sees one blue hat and one white hat on his companions. That would tell him that the hats are not all of the same color, but it will not allow him to deduce the color of his own. Thus if the advisor sees two blue hats and the king has posed a resolvable puzzle, the single advisor can deduce with certitude that his own hat is also blue.

Bringsjord's version of the three-wise-men puzzle was to program the little robots with the instruction that two of them were rendered mute by having their speakers disabled. The central technical problem was successfully implementing sensorimotor contingency between speaking and hearing. The robots were programmed to say, "I don't know" if they could not deduce whether they had been muted. The robots were then asked which one had not been silenced. Presumably, all three robots at first tried to answer. One robot stood and said, "I don't know." Hearing itself speak, this robot understood that it was the one unmuted. It waved its hand and said, "Sorry, I know now."[15] The robot must have processed the instructions, recognized that it had activated its speaker to make a sound, differentiated its sound from those of the other robots, and, as for its soccer-playing compatriots, been aware that it was an independent entity from the other two robots. Finally, it needed to relate its ability to speak to the original question. This behavior is rudimentary, but it is very close to self-awareness.

Another interesting development was the self-training of a robot arm announced by Hod Lipson of the Creative Machines Laboratory at Columbia University in 2019. Like a robot welder or other more traditional robots, this robot arm with four joints and hence four degrees of freedom could be trained by a human to do a simple task such as pick up a small rubber ball and drop it in a cup. Early versions of this robot arm were taught some basic physics—action and reaction, conservation of momentum—and then instructed to learn. In a later version, Lipson employed a generative AI capable of learning sensorimotor contingency from scratch with no instruction. In the beginning, this later robot knew nothing of physics, geometry, or motors. It did not know its shape—that it was an arm—or that it had four joints. The robot was instructed to make one thousand random motions each comprising one hundred separate steps and to record the results. The robot was left for about thirty hours to "babble," as Lipson calls it, until it learned what sort of motions resulted in what sort of actions. The robot thus built an internal model of what it was, how it could move, and what effect its motions would have on its environment. Given a task such as picking up a ball and dropping it in a cup, the robot could do as well on its own as when it was explicitly trained for the task. It could also do a completely different task: grip a pen and write a message. Once again, this does not represent consciousness, but making an internal model of how one can affect the environment is something that conscious entities do.

There is an argument that some degree of self-awareness will be required for robots to work safely among humans. Progress in that direction can be made without directly addressing or accomplishing human-level consciousness. In the *Terminator* movies, there is no doubt that Sarah Connor is conscious. What about the terminator? Is that creation conscious or just an elaborate, time-traveling, single-minded automaton? Does it matter?

It may be that robots never have the "first-hand experience of conscious thought," as Justin Hart says. They may remain *philosophical zombies*, capable of emulating consciousness but not truly possessing it, having neurons in a visual cortex fire in a way that represents viewing a sunset but not experiencing that sunset.

ALIEN CONSCIOUSNESS

In previous chapters I have summarized the current components that may be integrated to produce a conscious machine. We have sensors that can register the world in ways no human or animal can. We have robots of a wide variety that can move about their environment with the dexterity of a human and even fly. We have AI that can interpret sensory data: facial recognition and more. We have AI that can speak and translate virtually any language, including computer code and brain impulses. We have AI that can strategize and in many cases in ways humans cannot. Efforts are currently underway to integrate these current technologies. The outcome will be AI-powered robots that can move about freely and witness the world in diverse ways, many of these ways beyond human capability. These machines will synthesize and learn from their sensory data, including conversations with humans. They will strategize, anticipate how to become more capable, and write their own evolving code that enables them to do so. They will invent, think, make decisions, and act very, very rapidly. There is the clear potential for technology that can develop itself with no human intervention or control.

When current capabilities in AI, robots, and sensors are integrated, self-evolving conscious machines may emerge that supplant humans in ability and capacity. Animals and humans move about their environment, register a range of sensory data from their bodies and from that environment, and compare current input with their memory of past experience. They are aware of and experience their environment. They are conscious. How would any machine with these basic capabilities be any less conscious than a squirrel scampering about in the woods burying and retrieving nuts? How less conscious than a human? How more conscious of more aspects of the world in which it exists than a human?

Some of the principal insights into how our brains work comes from behavioral economists rather than neuroscientists. Our intelligence, our comprehension, our behavior, our consciousness has been shaped by 10 million years of hominid evolution, nearly 500,000 years as *Homo sapiens*.

As captured in popular treatments such as *Thinking Fast and Slow* by Daniel Kahneman[16] and *Nudge* by Richard Thaler and Cass Sunstein,[17] we have a fast, smart, instinctive, non-conscious "lizard mind" that quickly seeks reward and avoids danger; and we have a slower, more deliberative, conscious mind that we closely identify with ourselves. Our conscious minds are less in control than we think, with our un-conscious minds making decisions and only informing our conscious thinking fractions of a second later. This two-part system developed over our long evolutionary history. It was necessary for our survival on the veldt. Will conscious machines work like that?

A key question is whether we can reverse engineer the properties of the brain to a sufficient level that our constructs enable emergent properties. If so, will our artificial brains develop the power of our brains to have understanding, to feel emotions, to be creative? Will the machines become conscious? If they do, will that consciousness be anything like that of a human brain or will it be a completely alien consciousness dependent on the evolution of the machine rather than the billions of years of biological evolution that led to our brains? Artificial brains will have no built-in, genetic experience of having been a hunter-gatherer in the distant past. There may be nothing like our "lizard brains" although there could be similar fundamental components from which the artificial brain is composed.

Whatever form of intelligence, of consciousness, arises in our machines, it will be alien. We cannot simply program our biological history into a machine. A machine will have a very different mode of operation and a very different experience than our species has had. There is no reason to think a machine will respond like a very smart human. It will be something completely different. Already we struggle to understand how deep learning AI reaches the conclusions it does, how ChatGPT and its brethren operate. The future of machine intelligence will be like that, only more so.

If self-evolving AI re-writes its own code to get what it wants, how is its want defined? AlphaGo zero wants to win the game. That goal was imposed by the researchers at DeepMind presumably by maximizing some mathematical function. Will the goals of an AI always be imposed like a prompt, or will the AI define or develop its own wants and goals? If so, will those be anything like human wants or goals, or will they be its own alien goals? Human goals are to survive, reproduce, prosper, and be happy. AI might want to survive, but will it care about prospering, reproducing, or being happy? What does prosper even mean to an AI? Humans are curious. Will AI be curious? Is satisfying curiosity a goal?

There are arguments that machines might not be as alien in their comportment as one might imagine. Our conscious minds are not aware of our un-conscious minds or the firings of neurons that power our brains. Cognitive scientist Douglas Hofstadter[18] argued that the separation between the aware self and the details of neuron firings is critical. This obliviousness to the microscopic details is crucial for planning how to survive a sabertoothed tiger lurking in the shadows without getting addlebrained with the details of what is going on with our neurons.

Melanie Mitchell, a professor of computer science at Portland State University and member of the Science Board of the Santa Fe Institute, extends Hofstadter's logic to intelligent machines. Mitchell argues that we lack a sufficient understanding of human intelligence to enable adequate extrapolation to the qualities of machine intelligence. She agrees that a saving grace of human intelligence is that we do not, cannot, and should not control the microscopic firings of our neurons that give us the capacity for conscious thought and sense of self. In her book *Artificial Intelligence*,[19] Mitchell argues that just as our conscious minds do not directly interact with our neurons, so the conscious mind of a machine will not directly interact with the silicon chips or qubits or whatever are the basic elements of its functioning that make it a machine and not a biological entity. A machine would be functionally addlebrained if it had to actively control all its machine functions at the command level of individual bits and bytes and also have a functioning self-awareness like a superhuman. Mitchell concludes that superintelligent machines must evolve in the same way humans did for the same reasons. Just as our minds are not isolated entities but an amalgamated construction of our physical brains, the complexity of our bodies, and input from the surrounding world, so will intelligent machines not be isolated. Mitchell posits that intelligent machines will develop in the context of human society and culture tempered by the same common sense, values, and social judgment that allowed humans to become generally intelligent, interactive social beings. Perhaps, but machines will still have a vast capacity compared to today's humans. Whatever it is they do, they will do it very rapidly. Machines may evolve in a human culture, but they will experience a machine evolution and not our biological evolution. They will be fundamentally alien in that regard.

Another element in this perspective may be the fraught issue of free will. The fact that our unconscious mind makes decisions and only informs our conscious thinking after a brief but finite delay would seem to undermine notions of free will. There is often an implicit assumption that an artificial general intelligence (AGI), a conscious machine, would have motivation, agency, a will. If the issue of free will in humans is under assault, and if conscious machines operate as free of their microscopic details as Hofstadter and Mitchell argue, then perhaps even conscious machines will lack free will as it is normally conceived.

The point of view of Hofstadter and Mitchell may have merit, but we should keep a close eye on developments and not simply trust to luck. Many people have warned of the dangers of unfettered growth toward an AGI. One of the first of recent voices was Bill Joy. Joy wrote an article for *Wired* magazine[20] warning of the possible dangers of smart machines that drew a great deal of attention at the time. In the article, he quotes an eloquent statement of the possible dangers of advanced machines—written by the Unibomber. Oxford's Nick Bostrom has written extensively[21] on the cautionary, even dystopian side, as has Berkeley professor of computer science and electrical engineering Stuart Russell in his book *Human Compatible*.[22]

In the near future—a few years—we will not emulate the brain in all its detail. Rather we will attempt to reproduce brain functions, bearing in mind that the code of the brain is relatively simple. Our DNA contains all the instructions to construct the proteome and the connectome of neurons. Already we have techniques that may make the details of how we assemble our machines very different than the mechanics of our own brains: self-writing code, fuzzy logic, self-repairing code, fault-tolerant code. Time will tell whether that will lead to emergent properties, to consciousness.

The question is not quite as clean as asking whether machines will have consciousness. A likely prospect is that we will augment our brains, our mental capabilities. We will merge with our machines, adding components, brain links, nanobots. Does that make us less conscious as we become more than biologically human or more so?

If we can learn to emulate the processes of the brain, our artificial versions can be made arbitrarily large and much, much faster. These machines may be capable of everything human brains do but then much more. This is the stuff of legions of science fiction stories, but the issues are a natural extension of where our current AI, brain, and robot research is heading; and it is heading there exponentially faster with time. It may not prove possible truly to replicate the functions of the human brain. It may be possible. Reasonable current extrapolations of the exponential growth of our knowledge suggest we will know which is true within a couple of decades.

If at some point the question of machine consciousness is not resolved, the appropriate perspective will probably be—not yet.

9

EVOLUTION
How We Got Here

PHYSICS FIRST

While the focus of this book is on the near and far future of humanity, it is appropriate to cast back to our beginnings if only to set the scale of the great sweep of astronomical, geological, and biological time that has led us to our current condition. Astronomy is the study of the natural world beyond Earth but it also provides a means to understand Earth as a planet and host for evolving life in that broader context. Astronomers have set the stage on the grandest scale.

As we look out from Earth with our most sensitive instruments, we see a vast, expanding void littered with galaxies much like our home spiral galaxy, the Milky Way. The Milky Way alone comprises 100 billion stars similar to our Sun. There is reason to think there are also 100 billion or so planets in the Milky Way, some of which could host life with technologies vastly advanced over our own.

The Universe that we know is not quite 14 billion years old. Our Sun and its retinue of planets, moons, asteroids, and comets was born in a great, but now fading, flash of star formation that occurred when the Universe was about two-thirds its current age. That means that there were about 4.5 billion years from then until now for stuff to happen on the surface of Earth. This is a nearly incomprehensible swath of time yet astronomers and geologists routinely deal with such time scales.[1]

The framework is by now well known. The origins of the Universe in the Big Bang are still unknown, and the looming question of what came before is still unanswered, but only tiny fractions of a second later, something like our current physics took over. We can do analyses and make predictions and models to compare to observations. The very early Universe was hot, dense, and nearly homogeneous but stippled with tiny wrinkles in density imposed by the uncertainty principle of quantum physics. There is an intrinsic uncertainty in the conditions of everything, including the density of matter and energy. The Universe expanded. Fields condensed into particles. Some became the dark matter we still strive to understand. Others became the protons, neutrons, and electrons that make up the atoms and molecules of our existence. At first the Universe was mostly filled with photons of electromagnetic radiation: light. A few minutes after its birth, protons and neutrons assembled into the nuclei of atoms: Steven Weinberg's famous *The First Three Minutes*.[2]

An expanding gas cools, and this was the fate of the Universe. The Universe cooled and made a critical transition from being opaque to being transparent when it was about 380,000 years old. Electrons found protons and combined to make atoms from those ions. The atoms were less opaque to light than the freely roving electrons had been. Radiation was thus no longer tightly coupled to the particles and it continued to permeate the expanding space. We detect it now from every direction as minutely studied *cosmic microwave background radiation* with an effective temperature only 2.7 degrees above absolute zero. In that radiation, we see the remnants of the original quantum fluctuations as variations by one part in one hundred thousand in the temperature from here to there.

The Universe expanded, cooled, and became dark for a long while, somewhat less than 1 billion years. During this time, according to models, mysterious gravitating dark matter agglomerated from the first tiny quantum wrinkles of density into huge condensations. These condensations provided concentrations of gravity. Ordinary matter like us—protons, neutrons, and electrons—fell into these gravitational wells and formed the first stars and galaxies. Once again, light suffused the Universe.

Another major transition happened when the Universe was two-thirds its current age, coincidentally about the time our Sun was born. The even more mysterious influence of dark energy came to dominate. Whereas stuff like us—the Sun and planets—and the dark matter gravitate, the dark energy effectively anti-gravitates. We do not yet understand the nature of dark energy. Perhaps it is a force field of some kind that otherwise does not fit comfortably into the framework of our current physics. Apparently dark energy was there all along, but the gravitating influence of dark matter was diluted in the expansion of the Universe while that of dark energy was not. Eventually gravity became feeble and anti-gravity took over. The result is that the Universe is now not only expanding, but also accelerating.

THE ORIGIN OF LIFE: SELECTION, FEEDBACK, RECURSION, AND EMERGENCE

The next critical step in our story was the origin of life. We do not know how this happened. In the context of this lacuna, it is worth pondering what is unknown from what is unknowable. In class I teased people who deny our knowledge of evolution by showing a cartoon with a naive critique of gaps in the fossil record that was meant to satirize academics like me.[3] There may be things that are unknowable: the nature of the human soul and of God or why the Universe exists at all. Even then there are arguments that the soul might be known as a product of the complex human mind. People certainly think deeply about why there is something rather than nothing. Perhaps a solution to that will eventually emerge. Still there may be problems that are just too complex for the finite human mind.

On the other hand, we learn more every day. The nature of dark matter and dark energy are currently unknown but are under intellectual onslaught. One day they may become known. The same can be said for the question of what happened before the Big Bang. That is currently unknown, but that does not make it unknowable. I put the origin of life in that category. It is currently unknown but not unknowable.

We know many of the ingredients of the puzzle of life. We have yet to find any evidence of life beyond Earth but astronomers know the ingredients are there. They detect a dizzying array of complex organic molecules—the building blocks of life—in the gas between stars, in the tails of comets, and in meteorites that have survived their fiery crash to Earth. The physics and chemistry behind the formation of these molecules appears to be universal.

How then was the transition made from chemistry to biochemistry? To approach this topic, one needs an appreciation of the power of *selection*, *feedback*, and *recursion* in the natural world. Selection requires a means for filtering certain outcomes, encouraging some and discouraging others. Feedback is the process of using the outcome to adjust the filter and hence alter the selection. Recurrence means repeating the selection and feedback process many times.

There is an old story about assembling an infinite number of monkeys with an infinite number of typewriters (the story clearly predates computer keyboards). If you let them go long enough, they will produce Shakespeare. Perhaps, but this is not the way Nature proceeds. A better analogy to biological evolution would be to look at the product of the first typing and select that which looks most like Shakespeare. At first, this might be just groupings of letters that could be rudimentary words. Then take that selection and let the monkeys start with that and see where they go. That is a process of feedback. Then do this again and again, trying over and over. That is a process of recursion. If you employ that technique of selection, feedback, and recursion, you can get to Shakespeare much faster. The metaphor of infinite monkeys is essentially a linear argument: keep the rate of change constant and expect the complexity to grow linearly, hence you'll need an effectively infinite time. If, however, the rate of progress toward Shakespeare is proportional to the number of monkeys that have produced something vaguely resembling Shakespeare, then the rate of progress can grow exponentially, as I discussed in chapter 2. That is the power of selection, feedback, and recursion.

There are people who argue that random variations cannot lead to the complexity of living things. This is also implicitly a linear argument. This perspective neglects the power of selection, feedback, and recursion. More importantly, it is not how Nature works. Random variations undoubtedly play a role, but only as a first step. Nature invokes selection, feedback, and recursion. With that comes the possibility of growing complexity exponentially rapidly—slowly at first, then explosively later.

Harking back to the example of monkeys and Shakespeare, there was an element that slipped by. Who, exactly, is it doing this selecting? Who decides that some groupings of letters more resemble Shakespeare than another? In that argument, there is implicitly an external guiding agent, perhaps the Bard himself, who selected from the jumble of ideas, letters, and words in his mind what was to go down on parchment. Who made the selection that sparked the origin of life? The powerful answer given by Darwin is no one. The selection happened naturally.

If you consider a given set of physical conditions in the environment, the density, pressure, temperature, and composition of atoms, then chemical reactions will proceed. If conditions remain constant, then the reactions will proceed at a constant rate. The products of the reactions will accumulate linearly in proportion to the passage of time. The fact that reactions are happening at all will, however, change the conditions. The composition shifts; energy is released thus increasing the temperature. Pressure may go up, causing the medium to expand, thus reducing density. These changes must be considered.

A key step would be the formation, at random, of the first self-replicating molecule. The first time a single molecule acquires the property of making two of itself, the game changes dramatically. In the presence of self-replicating molecules, the reaction rate becomes proportional to the current number of molecules. That is the prescription for exponential growth of the self-replicating molecules. New reactions become possible, selecting those molecules that are most capable of self-replication. These new, efficient molecules provide feedback into the chemical broth. Then time, lots of time, billions of years, gives ample opportunity for recursion. The molecules mindlessly interact driven by the fundamental laws of physics and chemistry, but given the power of selection, feedback, and recursion, life can arise.

Another powerful idea that helps to set the framework for the origin of life and its subsequent evolution is the notion of *emergent properties*. The idea behind emergent properties is that a collective of some kind, following basic rules of behavior, can display complex properties that simply are not present or understandable by considering only the properties of the individual components.

In his updated version of *Cosmos*, Neil deGrasse Tyson[4] does a very effective segment on emergent properties,[5] a topic that Sagan did not consider in the original (Tyson presents his own version of "The Cosmic Year" in homage to Sagan). As an example, Tyson presents the flight of a flock of birds. The birds follow simple rules. Do not fly too close to your neighbors. Do not fly too far away from your neighbors. If your neighbor veers left or right, follow them. The result can be a complex, coordinated, surging flow of birds. No single bird takes command and orders the dynamics of the flock. A local example in Austin, Texas, is the spectacle of more than 1 million Mexican free-tail bats emerging from their lair beneath the Congress Avenue bridge over Lady Bird Lake and winging down river away from the setting Sun. They form a bat fluid with streams, vortices, and eddies amid the overall rush downstream. It is a sight to see, but no single bat leads the spectacle. With emergence, there is no leader, no top-down command. Simple rules can lead to complex, organized, evolving behavior. The whole is much more than the sum of its parts.

James Gleick gives a wonderful introduction to the topic of emergent properties in his book *Chaos*.[6] Emergent properties can be organized behavior. Emergence can also lead to chaos. This connection is related to the notion that the flapping of a butterfly's wings in Brazil can lead to a hurricane in the Gulf of Mexico.

Emergence thus happens when individual components interact. You can understand the physical properties of particles—quarks, protons, neutrons, electrons—and still be unable to predict the collective properties that lead to chemistry. You can understand hydrogen and oxygen atoms and still be unable to predict all the wondrous properties of water. Principles of emergence were probably central to the transition from chemistry to biochemistry. Atoms and molecules follow complex yet basic rules of quantum physics. Nevertheless, you can understand the properties of atoms and molecules and still be unable to predict the amazing implications of the double helix of molecules that make up DNA.

Any game, from chess to football, relies on emergent properties to entertain. There are some basic rules, but each realization, each playing, provides different, unexpected, complex results. The players have an active role in the complexity: they desire to win. At a human level, people interact to form families, communities, nations, and cultures. Buyers and sellers make complex markets. All these are examples of emergent properties.

In the current context, it is appropriate to consider life itself as an emergent property. No amount of study of quarks will lead to an understanding of biochemistry and all its extraordinary implications and realizations. Life in all its glory is an emergent property of the underlying atoms and molecules. Life arose spontaneously with no central control, no command from above. Life is an elaborate murmuration of primordial atoms with no leader and no plan. Even the goals to survive and reproduce are emergent properties.

SELF-REPRODUCTION AND EVOLUTION

In any case, primordial goo got a start on Earth and set off on a path of increasing, self-reproducing complexity. Given the similarity of physics and chemistry everywhere we look, astronomers routinely assume that there is life elsewhere. So far there is no hard evidence for that—not even slime mold, never mind some form of conscious intelligence. As of now, we are it; yet here we are.

Life as we know it is based on simple building blocks: the twenty-one amino acids, the nucleotides G (guanine), A (adenine), C (cytosine), and T (thymine) that pair and wind to make DNA. The basic rules of chemistry—attraction of opposite charges, repulsion of like charges—dictate the arrangement of these key fragments into the double helix. That double helix can open, and each half can pull molecules from the surroundings to rebuild the helix, yielding two self-reproducing molecules. Those molecules that can reproduce most quickly and efficiently grow in number to dominate the population. The rules of natural selection take over.

Once self-reproducing molecules arise, there is a transition to evolution. Some of the molecules do not reproduce exactly. Slight variations in structure and composition may yield a different ability to reproduce. Some variations might slow reproduction down; those variations are likely to fade from the population of molecules. Other variations might speed up the rate of self-reproduction or the rate at which parts are attracted from the environment to reconstruct the molecule. Those molecules will thrive.

Selection, feedback, and recursion take over. The primordial goo begins to evolve. Some molecules form *vesicles*, little balls that define inside from outside. Other molecules find the interior of those spheres to be a benign place to reproduce, shielded from the surrounding bath of water that threatens to dissolve them before they can divide and multiply. The first cell-like structures emerge. Cells that agglomerate and mutually reinforce one another are more productive in absorbing energy and raw materials. They thrive to make more of themselves. Somewhere in this growing complexity, the transition is made to life. From that point, evolution operates, selecting some ingredients, providing feedback into the nutrient environment, and producing and testing variations over and over in a natural recursion. Evolution requires no guidance. Evolution has no goal. Evolution tries everything, fills every niche. The fittest survive to reproduce.

It is remarkable even now to read Darwin's description of how he arrived at the hypothesis of natural selection and the theory of evolution.[7] The sheer amount of stuff he knew about the plants and animals he considered is mind boggling; the man really knew a lot about pigeons! He also thought deeply about the means to construct his theory based on careful observation and experiment. He invoked Occam's razor, the notion of economy of hypotheses; his single hypothesis of natural selection accounted for a vast range of observed facts. He did all this without knowing the detailed mechanics of how selection worked: the role of RNA and DNA that has been revealed by modern biology.

Darwin plunged directly into arguments of science versus pseudoscience and creationism. He raised and addressed potential arguments against his theory, anticipating disputes that still linger on the creationist side of the ledger. As an argument for selection versus creation, he cited the example that there are bats but no frogs on isolated Pacific islands. This is hard to explain from a creationist viewpoint. It is easy to explain with natural selection since bats can fly between islands, breaking their isolation by air, but frogs cannot hop between distant islands. Creationists argued, and still do, that eyes could not have developed by accident—an intrinsically linear argument. Darwin responded that even so sophisticated an organ as an eye could be understood as developing in steps: from an especially photo-sensitive neuron to an agglomeration of such neurons, to ever more structured and sensitive organs that provided more survivability for species that collected and imaged the abundant light from the Sun. (A few billion years from now when the Sun becomes a red giant, perhaps beings living then will have eyes that are more sensitive to red rather than yellow light.) Organs developed to provide survivability in one context may accidentally provide an advantage in another and hence be repurposed. That process may also have enhanced the development of modern eyes and acted in myriad other ways as life evolved.

Darwin argued that evolution proceeds in small steps: *Natura non facit saltum* [Nature does not make a leap]. This deduction led him to consider why fossils of transition species are rare: transitions are slow, and geology with subsidence and uplifts disrupts the evidence of layering. He invoked the key element of the passage of time, speaking in terms of millions of years, the perspective of the day, whereas we now know that the framework spans billions of years. One rare misstep was that Darwin denied the evidence for continental drift, which is a significant factor in Earth's geology and the separation of species into isolated evolutionary environments.

Darwin pondered the role of variations in offspring, of mutations, of the power of sexual reproduction. He delved into natural versus artificial selection, recognizing that what Nature does naturally humans do consciously and that there is no fundamental difference. Those arguments continue today in disputes over genetically modified food. Is there a fundamental difference between selecting which bull mates with which cow and tinkering with the genes directly?

Darwin amply showed that there are abundant facts of evolution. Natural selection is the theory that binds them together. It remains critical to understand that Darwin's work is not a sacred text. His theory can be tested. Creation can be tested. The whole edifice of understanding evolution that Darwin created has itself evolved and become more sophisticated. Yet Darwin's insights are still relevant as we contemplate our past and face our rampaging technological future. In *Descent of Man* he said,

> The most humble organism is something much higher than the inorganic dust under our feet. . . . The fact of his having thus risen, instead of having been aboriginally placed here, may give him hope for a still higher destiny in the distant future.

and

If he is to advance still higher, it is feared that he must remain subject to a severe struggle.

Words worth pondering.

One of the first popularizers of evolution in the modern era was evolutionary biologist Richard Dawkins. His book *The Selfish Gene*[8] captured the spirit of evolution after the revelation of the double helix of DNA. Dawkins emphasized that the higher forms of life are just meat sacks assembled by genes for the purpose of propagating the genes. The genes, of course, did not set out to consciously design an elephant, a sequoia, or a human. The genes responded to the natural selection of their environment to make more of the hosts that propagated most fruitfully. Assembling the cells that hosted the genes in their nuclei into bipedal creatures that used sex for efficient procreation just worked better than other schemes. Yuval Noah Harari promotes a variation on this theme with his suggestion in *Sapiens*[9] that agriculture was invented by wheat as a means of propagating its genes more abundantly. Dawkins discusses in depth the evolutionary basis of instinct and altruism, topics also addressed by Darwin.

An interesting component of *The Selfish Gene* was Dawkins's invention of and advocacy for the concept of a *meme*. In Dawkins's terms, a meme is an idea that propagates amid human minds in a way that is similar to the spread of genes but does not require the machinery of genes. A meme could be an aspect of culture or of religion. In our wired world, the term *meme* has been adopted to represent an image or video, often modified to make a cultural point, that spreads on social media. These memes spread on the internet and flourish or die out depending on their cultural fertility.

CONDITIONS FOR EARLY LIFE

We thus can look back through astronomical, geological, and biological history to trace how life began and evolved on our planet The predominant supposition is that life arose on Earth perhaps using ingredients—amino acids and sugars—that had already formed in space. There remains the possibility that life came from elsewhere, a process of *panspermia*, but that just puts off the problem of where and how life arose. We do know that in the first 0.5 billion years after Earth formed, the environment was inhospitable to life. This was the Hadean Era when parts of the assembling planetary system still whirled and collided in the dissipating proto-solar nebula. Something perhaps the size of Mars collided with Earth and broke off a chunk to become our Moon. This hypothetical but very likely object is called Theia. Calculations based on how long it would take the magma on the Moon to cool after the Theia collision give a rather precise date for when this happened: 4.425 billion years ago.[10] Other fragments of a range of sizes pummeled the surface of Earth, rendering it extremely hot if not molten. The best estimates are that life could have neither formed nor survived in this epoch.

Eventually the bombardment eased. There was a final *late heavy bombardment* that probably wiped out any incipient life about 4 billion years ago. The origin of the water that formed the oceans remains unclear. It may have been delivered by asteroids or comets or another vehicle. We do know that the oceans formed around the end of the Hadean Era. The water may have remained suspended in the atmosphere as vapor until the surface cooled sufficiently. Then the rains came.

Amazingly enough, life arose soon after. While details are debated, there is evidence for early life in the form of fossils dating back to 3.5 billion years ago, only 0.5 billion years, or less, after the end of the late heavy bombardment. These fossils, called *stromatolites*, are lime deposits left by certain *cyanobacteria*. The cyanobacteria were single-celled microbes capable of photosynthesis. Both ancient stromatolite fossils and active contemporary stromatolite colonies are found in Australia.

Even though they appeared quickly, stromatolites were already very sophisticated living, evolving, molecular machines. Prior to their appearance, that is, even closer to the Hadean Era, there must have been a phase when the first biochemical molecules developed the ability to reproduce.

RNA (*ribonucleic acid*) plays a critical role in modern life. DNA (*deoxyribonucleic acid*) cannot do its job of genetic coding and reproducing and manufacturing proteins without RNA and the proteins that provide the supplementary machinery of the *ribosome* that is itself an intricate combination of RNA and proteins. The ribosome splits the double helix, pulls in the proper amino acid ingredients from the environment in the correct order, and spins out two new DNA molecules and RNA coded to produce needed proteins. We will delve into this process in more detail in chapter 10. For now, it is sufficient to note that proteins need DNA to manufacture them but DNA needs proteins to do the manufacturing. How then did DNA arise? Or proteins?

While RNA does not have all the reproductive power of DNA, it can both code and reproduce itself in the absence of other coding molecules.

Studies of this very early era often focus on the possibility of an *RNA world* where RNA formed, evolved, and thrived.[11] In this view, DNA eventually developed and incorporated the RNA molecules to form ribosomes, with this combination proving to be the more robust reproductive unit.

Others argue that there was not a clean separation of an RNA world from a later DNA world. In this perspective, the surface of the early Earth, like contemporary conditions on Titan, was covered with organic material that formed in the atmosphere by photochemistry and then rained down on the surface. The surface could have been a soup of co-evolving hydrocarbons, polymers, amino acids, RNA, and DNA with metabolic processes that transported and stored energy to power the whole development. One variant of this picture is the proposition that cells were not necessary for the first life. The presence of a viscous soup was enough to create gradients in properties—with gradients in energy being a primitive metabolism—so reproducing, evolving life preceded the development of the cell with its isolating membrane and separate interior structure. Picture molasses dripping off a rock. The rotation of Earth could have caused a daily cycle of heating and cooling, of dehydration and rehydration that stirred the mix, enabling complexity to arise. In this view, there was a continuum between abiotic conditions and biotic life. There was no place and time when one could say this is life and this is not. Conditions gradually shifted from a lot of non-life and a little life to a lot of life adapted to every environmental niche.

These conditions might not have required the complete cessation of the Hadean Era. It is popular to consider *impact frustration*—that rocks from outer space banging into Earth at high speed will only terminate life where they land. An alternate view is that the rims of impact sites might be an ideal place to generate energy gradients between hot and cooler regions and to stir the soup to coalesce and come alive.[12] These conditions might well represent Darwin's "warm little pond." Astrobiologist Charles Cockell opined, "The good thing about a hole in the ground is that water seeps into it. So you've got water, organics, and heat all in the same place, which is good for the evolution of life on Earth."[13] Steve Benner, one of the primary proponents of the RNA world, posits that a Hadean impact of appropriate size might have been critical to the chemistry that produced RNA.[14] This hypothesized impact has a name, Moneta, because it might have delivered the surface veneer of metals like silver and gold that avoided being incorporated in Earth's molten core of iron and later became money.

Another environment that is popular to consider for the origin of life is that of the *black smokers*. These are hydrothermal vents that blow hot water and nutrients up through the floor of the deep ocean. While the center of the vents may be too hot to support life, the immediate environment of modern black smokers teems with life from microbes to exotic sea animals. The natural existence of heat and complex nutrients might have given the right environment for biochemistry to arise. A possible impediment would be that too much water will dissolve DNA and black smokers are, after all, underwater. They might be some approximation to a warm little pond, but the verdict is still out.

LUCA: OUR UNIVERSAL ANCESTOR

Genetic studies imply that all modern life arose from a *last universal common ancestor*, LUCA. We do not yet know the zoo of early life forms that might have bloomed and faded before LUCA became established and hogged all the resources, banishing the other struggling variations. LUCA availed itself of all the power of variation, selection, feedback, and recursion. Within a few hundred thousand years, a wink of cosmic time, LUCA had formed, thrived, and given rise to stromatolites and then, by now, the modern tree of life with an estimated (probably under-estimated) 2.3 million species.

Biologists have come to realize that cell division with mutation is not the only way to induce variation in an evolving system. In the contemporary world, *horizontal gene transfer* induces a significant amount of variation. This is a process by which genes escape a cell nucleus, float around in the general environment, and then get absorbed and incorporated into the nucleus of a different type of cell. This mode of gene transfer can give the latter cell new properties of survivability and reproducibility, thus sending it onto a new evolutionary path. Gene transfer might have been an especially active process as life first formed. Horizontal gene transfer may have been the dominant means by which information and complexity were promulgated in the primordial viscous soup. In this context, LUCA is viewed not as a single proficient ancestor bacterium that got the lead on everything else. LUCA is instead interpreted as a collective, system-wide transition from an epoch when horizontal gene transfer was the dominant mechanism of sharing information to an epoch when inheritance by offspring became the dominant mechanism. Technically, LUCA is only traced by ribosomal RNA. Other tracers would and do give a different tree of life. All this trait swapping may have gone on in a sea of many different kinds of metabolisms.

THE TREE OF LIFE

LUCA led to more complex living structures. The first microbes to form and thrive were primitive *Prokaryotes*, cells with free-floating genetic material in their interiors. The early Prokaryotes split into *Bacteria* and *Archaea*, modern versions of which survive in challenging hot, salty, acidic, and radioactive environments. For example, Archaea flourish in the heat vents of Yellowstone. The branch of Prokaryotes evolving toward Archaea stumbled on a structure that formed not just cell membranes but a distinct nucleus where the genetic material resides, protected from the ravages of the external environment. These became the modern *Eukaryotes*. Eukaryotes became us.

Carl Woese was the pioneer of the phylogenetic tree of life with its three main branches of Archaea, Bacteria, and Eukarya and burgeoning substructure.[15] Biologists released a new tree of life[16] in 2016.

AND ALONG CAME HUMANS

The current view of human evolution is not a tree with distinct branches leading to our exalted, if transient, status. A more accurate picture is a bush with many complex branchings, a braided tree. An even better metaphor might be a river delta where the river splits into different flows, but then sometimes the flows recombine. The fact seems to be that there was a lot of interbreeding among hominid species even as evolution was driving the speciation.

We did not descend from monkeys. Both apes and we evolved from a common ancestor. We are all Africans. Traces of the great apes, the hominids, lead to Africa. Genetic studies show that African apes diverged from the ancestors of orangutans about 14 million years ago. Our ancestors and chimpanzees split off from gorillas about 9 million years ago. Our branch parted ways with the chimpanzees about 6 million years ago. We are now separated from the chimps by only 2 percent of our genetic material, leaving 98 percent of our genes in common.

A common hypothesis is that we hominids evolved an upright stance after splitting off from our knuckle-walking chimpanzee cousins. This story has been muddied by the discovery of the fossil of an ape, *Danuvius guggenmosi*, that lived in Germany about 12 million years ago.[17] Evidence suggests that *Danuvius* both walked on two legs and climbed trees long before hominids invented this trick in Africa. Based on her research, University of Tubingen paleoclimatologist and paleoanthropologist Madelaine Böhme argues in the book *Ancient Bones*[18] that while apes may have originated in Africa, some, generation by generation, migrated to Europe. She proposes that *Danuvius* is the last common ancestor of chimpanzees and humans. Böhme suggests that a key genetic mutation then led to human ancestors who migrated back to Africa. This claim remains controversial but is an important reminder that evidence of our roots is still sparse, and our history might be more complex than current pictures suggest even as hominid evolution has proven to be. Perhaps rich sources of fossils are more abundant in Africa than in Europe and Asia. Some of our roots may be in Greece rather than Tanzania, but the preponderance of evidence remains that our genus, *Homo*, arose in Africa over 2 million years ago and that our species, *Homo sapiens*, originated in Africa about 300,000 years ago.

In any case, upright walking came to Africa. *Australopithecus* was an early hominid living around 4 million years ago. *Australopithecus* had the pelvis and leg bones of modern humans and walked erect. One young female born in what is now Ethiopia 3.2 million years ago had no idea of the famous legacy she would leave, the most complete fossil of the era, known as Lucy.[19] Whereas *Australopithecus* was still mostly apelike, a new genus, *Homo*, arose that more closely resembled modern humans. *Homo habilis* lived about 2.8 million years ago and left early evidence of tool making. *Homo erectus* and *Homo ergaster* were the first to use fire and left Africa for Europe and Asia somewhat over 1 million years ago.

Early *Homo sapiens* appeared on the scene in Africa between 250,000 and 400,000 years ago. One powerful technique for tracing the roots of our origins is DNA in the *mitochondria* in our cells. Mitochondria have the vital task of converting food into energy. The mitochondria float around in cells outside the nucleus, having been entrapped and exploited at some earlier stage of evolution. Mitochondrial DNA is only passed from mothers, not fathers, to children. Tracing genetic variations in this matriarchal DNA then gives us a way of tracking our lineage backward. The result has been the suggestion that all of us can trace our roots to a single female who lived about 200,000 years ago in what is modern Botswana.[20] This female has been called *mitochondrial Eve*, the mother of us all, but she was not the Eve of the Bible. There were many other *Homo sapien* females—and males—around as well as others of the *Homo* genus, it is just that their family lineages died out over the ages. There may have been a single male ancestor of us all as well but from a different time and place. Tracing the male lineage through the Y chromosome is a less certain process.

Recent work with more sophisticated technology has allowed the probing of nuclear, not just mitochondrial, DNA. While only a few individuals were studied who lived in sub-Saharan Africa, Tanzania, and Malawi 12,000 to 18,000 years ago, their nuclear DNA recorded earlier generations stretching back to 80,000 years ago. This evidence showed that there were three genetic lineages originating from eastern, southern, and central Africa, and that these hunter-gatherer peoples were not geographically isolated but collectively moved long distances across the continent over the ages.[21] This genetic data put evidence for the changes in architecture, linguistics, and the spread of cultural artifacts in a new context. It was not just ideas, languages, and trinkets that were traded around; people actually moved large distances, interbreeding as they went as the Middle Stone Age gave way to the Later Stone Age. Over the last 5,000 years as agriculture took hold along with slavery and colonization, this vast migration ceased, and the genetic record shows people once again settled down in restricted areas.

When *Homo sapiens* emerged, there was already a variety of other genus *Homo* creatures both in Africa and spread throughout the world due to earlier migrations. *Homo naledi* lived about 300,000 years ago in southern Africa around modern Johannesburg. They shared characteristics of both *Australopithecus* and modern humans.[22] The primary repository of their fossils is the Rising Star Cave. Access to the cave where the fossils were found is through a complex, narrow system of tunnels. It would be impossible for the remains simply to be dropped from the surface. There were no animal fossils. The presumption is that the corpses were carefully transported into the depths of the cave for interment or to avoid the smell.

There were Neanderthals (*Homo sapiens Neanderthalensis*) in Europe from about 400,000 years ago until about 30,000 years ago when they died out, overwhelmed by modern humans. Neanderthals used tools, wore clothes, and probably made fire. Neanderthals were significant predators in their neighborhood, probably using ambush techniques to pursue big game. Their brains were as big as ours, but DNA evidence suggests we developed changes in our cortex and frontal lobe that might have aided our ability to form complex thoughts.[23] The 2022 Nobel Prize in Physiology or Medicine was awarded for studies of Neanderthal DNA.

A cave in Siberia produced evidence for Denisovans (*Denisova hominins*). Denisovans apparently arose about 1 million years ago, perhaps evolved from *Homo erectus*. Human DNA records show evidence for Denisovans in East Asia, New Guinea, and perhaps even Australia. *Homo floresiensis* were small people living in southeast Asia, especially Indonesia. They may have flourished over 100,000 years ago until about 50,000 years ago when, like the Neanderthals, modern humans supplanted them. Their small size, less than four feet tall, led to their nickname "Hobbit." Some argue that they were anatomically modern humans subject to dwarfism, but the preponderance of evidence points to a separate species.

The evidence for when humans spread out of Africa and around the world remains rich and controversial. There might have been several waves of migration from the Horn of Africa into the Middle East, Europe, and Asia driven by climate changes beginning about 100,000 years ago. Alternatively, there may have been a single major migration about 50,000 years ago. Curiously, spoken language evolved about this time, 40,000 or 50,000 years ago.

Genetic evidence suggests that Neanderthals bred with Denisovans. When humans arrived, they likely drove both groups to extinction. In the process, humans bred as well as conquered. Neanderthals, Denisovans, and modern humans were less different than brown bears and polar bears, and they could easily produce fertile offspring.[24] All modern humans have a small percentage of Neanderthal genes so you know there was hanky panky. (Some evidence suggests those Neanderthal genes may be related to susceptibility to COVID-19.)[25] Scientists have also been able to grow human cells containing Neanderthal DNA in a petri dish.[26] The spread of Denisovan genes may be less broad, but evidence suggests that people in southeast Asia and the western Pacific have Denisovan variations in their genes.[27] Those from Africa and Europe show none.

The descendants of *Homo sapiens* who migrated from Africa to Europe and Asia evolved and adapted to their new climate, including an ice age that lasted in the northern hemisphere from 30,000 to 20,000 years ago. In Europe, skin became lighter, and lactose tolerance developed. The Cheddhar Man, remains found in a cave in Cheddhar Gorge, England, dated from about 9,000 years ago.[28] The individual probably had dark skin but light eyes, maybe even blue.[29]

Studies published in 2023 suggest this history is even more complex than suspected.[30] Genetic evidence reveals that several waves of modern human hunter-gatherers migrated from Africa to Europe beginning about 45,000 years ago. There are trails of at least eight distinct populations with light-skinned, dark-eyed people to the east and possibly dark-skinned and blue-eyed people to the west. Contemporary researchers have named some of them the Fournol, the Vestonice, the GoyetQ2, the Villabruna, the Oberkassel, and the Sidelkino. These people co-existed with the Neanderthals and died out with them, suffering the tribulations of the Ice Age before modern Europeans and Asians became distinct.

When in this process did we become human? Was a switch flipped somewhere there in Botswana? When we used tools, developed language, adopted sophisticated social interactions with secondary males and females conspiring to thwart dominant males? In the context of the discovery of *Homo naledi*, Frans de Waal had an insightful opinion piece in the *New York Times* about the implications of the discovery of this branch of hominids.[31]

> The problem is that we keep assuming that there is a point at which we became human. . . . Apart from our language capacity, no uniqueness claim has survived unmodified for more than a decade since it was made. You name it—tool use, tool making, culture, food sharing, theory of mind, planning, empathy, inferential reasoning—it has all been observed in wild primates or, better yet, many of these capacities have been demonstrated in carefully controlled experiments.

LOOKING AHEAD

Homo sapiens emerged on this pale blue dot[32] in the last seconds of Carl Sagan's cosmic year. Modern *Homo sapiens* is only a few hundred thousand years old. Nature did not design *Homo sapiens*. Nature tried every possible variation consistent with the physics and chemistry of DNA and then tested it for survivability. There is no reason to think Nature is not still testing us. *Homo sapiens* is likely still evolving due to basic genetic processes. Our environment evolves. Our genes alter. Mutations happen. A key difference is that in this new age of the Anthropocene, we can drive changes faster than Nature can. That makes for a whole new ball game.

The history of life on this planet and the great vista of time before us yields one firm prediction for the future of humanity. A million years from now, a billion years from now, *Homo sapiens* will not exist. We might die off, we might evolve, but we will not be as we currently are. This prediction is rooted in the immutable laws of genetics. On geological and astronomical time scales, we will drift and change; yet that natural drift—as powerful as it has been in our history—is probably now irrelevant as we will explore in more depth in chapter 10.

10

GENETICS
Designing Ourselves

WRITING THE CODE OF LIFE

Prior to chapter 9, I've focused on the exponentially more rapid developments associated with the digital revolution, but a second topic has been
unfolding that has as much, if not more, capacity fundamentally to alter the
future human condition: biology. Biology has also advanced exponentially,
with knowledge building on knowledge.

Our understanding of biology is sufficient for us to take over from Darwin's Nature and forge our own evolution. Overwhelming changes are
likely to happen over the next few decades even as the digital revolution
races on. Over longer timescales, the mind boggles. It is virtually impossible that we will look and behave the same in the distant future with only
our clothing to differentiate us from our kin today as in so many science
fiction movies.

The subject of biology is vast, complex, and sophisticated. I can only capture some of the high points that have caught my attention as an interested
outsider, an astrophysicist who has dabbled occasionally in astrobiology.
Readers who seek more depth should consult the classic text *Molecular
Biology of the Cell* by Alberts et al.[1]

Antonie Philips van Leeuwenhoek, the father of modern microbiology, marked an important waypoint on the exponential curve of biological
knowledge by using his own microscope invention to study microbes in

the late seventeenth century. In the nineteenth century after Darwin's revelations, the Austrian monk Gregor Mendel established the mathematical foundation of genetics, putting a firm base under Darwin's remarkable insights. In the twentieth century came the revelation of the twisted helix structure of *deoxyribonucleic acid* (DNA) by Rosalind Franklin, James Watson, and Francis Crick. DNA is the very stuff of which our genes are composed.[2] Then we learned to sequence DNA and read the code. Biology could now approach who and what we are from the deepest molecular roots with major ramifications for society. All we need to do is write the code.

THE BITS AND PIECES OF LIFE

Elements of the molecular basis of life were introduced in chapter 9 in the context of the origin of life. Here we will pursue that topic in more detail in the context of modern biology and where it may lead in the future.

There is a close connection between the digital revolution and the biological revolution: both involve code. In the digital world, the basic letters are *0* and *1* (representing Off and On). In biology, the code is written in strings of four molecules: *adenine, thymine, guanine,* and *cytosine*—A, T, G, and C. These ring-like assemblages of carbon, nitrogen, and oxygen are held together by the weak electrical attraction of atoms of hydrogen[3] such that A and T naturally grip each other, forming a *base pair*, as do G and C.

The base pairs can then link up to form the remarkable double helix of DNA. A phosphorous atom surrounded by four atoms of oxygen makes a *phosphate*. Phosphates connect to form two spiraling ribbons that constitute the outer framework, or backbone, of the DNA. Within the framework, the phosphates connect to the molecules of the A, C, G, and T. The base pairs span the distance between the two ribbons of backbone like the rungs of a twisted ladder with an A on one side linked across the gap to a T on the other side and with a G linked to a C. The base pairs are an analog of the *0* and *1* in digital computing or the twenty-six letters of the English alphabet. All can be assembled into instructions made up of what we might call words, sentences, paragraphs, and tomes. The order of the base pairs along the DNA strand constitutes words and instructions.

The molecular structure of DNA has a special capacity that is not available to *0*s and *1*s or to the letters of the alphabet. The twisted helix can be separated into its component single strands, and then each bare A can attract a new T and each bare G a new C. In the other separated strand, each T attracts and binds with a new A and each C with a new G. The result is two strands of DNA that are identical to the single original strand. This is reproduction, the miracle of life. A strand of DNA in a living being has billions of linked base pairs. Segments of the strands constitute specific *genes* that code for the functions of life.

Each strand of DNA is very thin. A human hair is 40,000 times wider. Each is, however, very long—a meter or two. All the DNA in the nucleus of a single cell would stretch several kilometers, and all the DNA in a human body would encircle the outer planets of our solar system. In our bodies, the DNA is not stretched out but tangled in knots of a very specific 3D structure that form the *chromosomes*. Chromosomes are larger, thread-like structures formed by single strands of DNA that are tightly wrapped around a framework of proteins. This construction not only allows the DNA to fit in the nucleus of a cell but gives DNA critical capacities that are not revealed simply by the coded order of the base pairs.

The chromosomes come in two forms labeled by their shapes, X and Y. X-chromosomes meet at a midpoint and then have two branches extending in one direction and two branches in the opposite direction, thus looking like a wiggly letter X. The Y-chromosomes have two upward extensions but only a single downward extension. DNA is replicated during cell division, thereby doubling the number of chromosomes. When the cell divides, half the chromosomes go to each new cell, effectively cloning the original cell. During sexual reproduction, new chromosomes are built half from the mother and half from the father. A new being is created with a mix of properties from each parent, and life goes on.

The *genome* of a living thing comprises every strand of DNA in every chromosome in a cell. We humans have about 25,000 genes in our genome. That large but still quite finite set of instructions dictates all that we are. An approximately 2 percent change in our genes would yield an ape. The genome includes the portions of the DNA strands that represent individual genes and also portions that do not directly code for genetic instructions but have other important functions.

Most of the DNA is in the chromosomes in the nuclei of cells. Some is in the *mitochondria* that float within the cell wall but outside the cell nucleus. The genome in each individual cell is a complete set of genetic instructions for all the complex interacting parts of an organism from single-cell bacteria to a human with hundreds of trillions of differentiated, intricately interacting cells. Every human begins as a single cell with all its genetic information. Cells divide, each new one preserving its information of how to grow individual organs, how to heal injury, when to stop dividing to avoid the runaway growth of cancer, and when to die. In a mature human, about

2 trillion cells divide every day—skin cells more often, nerve and brain cells less so. Each new cell knows the whole recipe of what makes you you.

DNA provides the code for this intricate process of building a living organism, but enabling the construction requires an entire complex cellular machinery built of a dizzying array of proteins. Most of the stuff in a cell is water, but of the material that is not water, most of the mass is in proteins. There are far more proteins than there are genes. Proteins are tiny molecular machines that function by going through cycles of structural change that are finely tuned to their particular biological function. Some proteins instruct the cells when to divide and when to stop dividing. Other proteins provide the mechanical structure of cells, enable cells to move, or allow cells to communicate with one another and with their external environment. Yet others are *enzymes*: proteins that catalyze chemical reactions without themselves being altered in the reaction. Antibodies, hormones, and collagen are varieties of proteins. All the proteins in our bodies constitute our *proteome*.

Proteins are organic polymer molecules built of strings of *amino acids* that are themselves composed of an atom of carbon that is bonded to other molecular components. All the proteins of life on Earth are composed of some arrangement of just twenty-one amino acids. We see evidence of amino acids in space. This suggests that some of these basic building blocks of life formed in that hostile environment and then settled onto Earth to enable life to form, evolve, and flourish. Amino acids can also form when ultraviolet light hits an appropriate broth of molecules or perhaps when lava erupts beneath the ocean.

The structure of a protein is dictated by the order in which its amino acids are linked. Although one can think of that order as a linear sequence, the power of proteins—their specific function—is determined not just by that linear order but by the way the proteins fold into dense, complex networks of atoms like 3D jigsaw puzzles. The order of the amino acids dictates that 3D structure and hence the protein's function. One of the significant problems facing biologists and physicists who study protein structure is exactly how proteins fold from linear sequences into their specific 3D forms. If the arrangement of the amino acids shifts, the protein would have a completely different function or fail to fold correctly and be dysfunctional.

Life as we know it has developed a complex machinery to translate the genetic code of DNA into elaborately functioning proteins. Other key ingredients in this translation process are RNA and ribosomes. The ribosomes are intricately structured combinations of RNA and proteins.

Like DNA, RNA is a *biopolymer*, a long string of molecules composed of similar bases—A, C, and G but with the T replaced by the base U (uracil). Unlike DNA, RNA is usually a single strand with a single backbone that doubles over so that the bases can pair up—A with U and C with G—to form a double helix with the backbone on the outside and the bases linked across the gap. Rather than the long structures of DNA, RNA tends to form compact structures folded into complex, 3D forms analogous to proteins. There is a variety of RNA molecules: *mRNA* is messenger RNA that can read and carry the genetic instructions of DNA; *tRNA* serves to transfer those instructions into the construction of proteins; *rRNA* is ribosomal RNA that does not itself code but forms part of the ribosome that makes the proteins; and *snRNA* is small nuclear RNA that enables *mRNA* to form and function.

The various types of RNA control complicated functions in cells including translating the DNA code into useful proteins; catalyzing certain reactions like cutting DNA or other RNA molecules; or tying strands of DNA and RNA into new configurations. As for proteins, the 3D structure of RNA is critical both to maintain the stability of the molecule and to define its function. This structure allows certain proteins—serving as catalytic enzymes—to modify the arrangement of the bases by strategically adding other molecules along the RNA structure. These modifications allow chemical bonds to form between portions of the RNA that would be far apart on the polymer in the absence of the folded 3D configuration. The bonds are critical to stabilizing the whole folded structure.

DNA along with certain RNA and specific proteins combine in an intricate dance to read the code of the DNA and translate it into new proteins. Nature has evolved a system by which every three rungs on the DNA ladder constitute a *codon* that corresponds to a single amino acid, the building blocks of proteins. In the process of *transcription* of DNA code to protein, a certain type of RNA moves along a specific gene in the DNA, splitting the double helix into its two separate strands. Bases are attracted from the surrounding cell fluid in the nucleus and attach to the open bases on the strands of the DNA; T grabs A, and G attaches to C. The adenine, A, does not attract a T, but instead attaches to a U. The resulting single strand of molecule is a piece of mRNA, that peels away. It is nearly an exact copy of the complementary strand of the DNA but with T replaced by U at strategic intervals. The mRNA carries the instructions for making proteins. Sections of mRNA join together and diffuse out of the nucleus into the surrounding *cytoplasm*.

In the cytoplasm, the ribosome comes into play. The mRNA threads through the ribosome, and the ribosome reads the code of the mRNA three units or one codon at a time to make a specific chain of amino acids. Molecules of tRNA are attached to one of the twenty amino acids that ultimately constitute proteins. The tRNA carries its specific amino acid to the ribosome. As the mRNA is read three units at a time, the tRNA that matches that codon delivers the corresponding amino acid. The amino acids link up sequentially. The string of amino acids then folds to form the specific protein for which the gene in the DNA provided the code.

CHICKEN OR EGG?

This remarkable biological transcription machinery presents a deep conundrum. It requires DNA, RNA, and proteins to function—to make new proteins. DNA cannot function without proteins, but proteins need DNA to form. Ribosomes are critical to transcription, but they are made of RNA and proteins. Which came first? How did this all get started? An active research community has pursued the notion that RNA may have been the first biological molecule. RNA does not carry code as efficiently as DNA, but RNA can replicate itself, and some forms of it have a rudimentary ability to code for proteins or serve as catalytic enzymes. Some viruses have genomes composed entirely of RNA that codes for proteins. Advocates of the RNA world hypothesis propose that RNA was the precursor for all the elaborate biological machinery to come. Others argue that proteins or *peptides*—short chains of amino acids even more basic than proteins—or *amyloids*—long fibers of peptides and proteins—may have been the first fundamental construct in an *amyloid world*. Yet others make the case that an early rudimentary version of the ribosome or a primeval soup of DNA and RNA may have bootstrapped the whole modern process. All the bases that form RNA and DNA have now been found in space, hinting at a *panspermia* that spread life to the Earth and elsewhere.[4] A recent suggestion is that an appropriate mix of cyanide and carbon dioxide might have been the basis to form both the amino acids and proteins that kick started self-reproduction and life.[5]

This chicken or egg problem was solved by billions of years of Nature performing biological experimentation. Scientists are striving to understand the fundamental principles of biological organization that led to the very complex specific machinery used by living things today. Understanding how the 3D arrangement of protein composition and structure accomplishes specific biological functions could lead to the ability to design new proteins and medicines. Rama Ranganathan of the University of Chicago Center for

Physics of Evolving Systems posits that the principles of protein design and function may apply more generally to the network of neurons in the brain and to cells, tissues, organisms, even to entire ecosystems. The aspiration to understand how complex networks form and function is known as *connectomics*. All living systems on all scales, from DNA to the human body, are dense networks of interactions. Researchers like Ranganathan seek the basic rules that underlie this ultimate complexity by studying Nature's solutions. Ultimately the goal is to explain exactly what kind of machines proteins are, how they work, and why they are built the way they are through the process of evolution.

Google's DeepMind team that created the Alpha Go AI code announced a breakthrough in protein folding in 2021, AlphaFold2.[6] This AI computational method predicts the 3D structure of folded proteins at the atomic level given only the order of the amino acids in the protein. DeepMind had presented results from a preliminary version in 2020 but no detail. Postdoctoral associate Minkyung Baek at the University of Washington used that clue to construct her own successful protein-folding AI and released the open-source code.[7] That goaded DeepMind to release the source code for AlphaFold2. The success of these AI systems and their universal access promises a revolution in medicine and bioengineering. The 2023 Breakthrough Prize in Life Sciences was awarded to Demis Hassabis and John Jumper of DeepMind for the development of AlphaFold.

Scientists are probing this biological organization by attempting to emulate Nature's design and constructing artificial proteins. These synthetic proteins can then be injected into cells and their functions monitored. The goal is to select a specific desired function and then design a protein—its constituent molecules and their folding—that will perform that function.

While there is much to learn, the depth of understanding of this protein machinery and our ability to manipulate it was illustrated during the COVID-19 pandemic. Vaccines were developed in a record time of months rather than years. Conventional vaccines—for the flu, for instance—often contain a fragment of viral material. This piece of organic material acts as an *antigen* that the body considers an infectious agent. The artificial insertion of the antigen triggers an immune system response that prepares the body to repel a future infection that contains the same antigen.

By contrast, two of the first COVID-19 vaccines to be authorized for broad public use—one by Pfizer and one by Moderna—relied explicitly on an ability to design proteins. These new vaccines injected mRNA that was designed from scratch. A laboratory in China announced the discovery of COVID-19 on January 9, 2020, a Thursday. The same lab produced a DNA sequence over the next weekend and immediately released the data to the public. Within days of receiving the DNA sequence of the COVID-19 virus, Pfizer and Moderna designed the specific mRNA structures they wanted. Within weeks they had used DNA templates and standard vaccine materials to create special mRNA trial vaccines and had begun to inject them into mice. Within two months, they proceeded to critical human trials.

The artificial mRNA triggered the body's ribosome–tRNA production machinery to produce a specific, made-to-order fragment of the COVID-19 viral protein that in turn served as an antigen. The vaccine specifically stimulated the production of the spike proteins that constitute the welter of protrusions on the surface of the COVID-19 virus that give it its characteristic hedgehog shape. The spike proteins are the mechanism by which the virus sticks to healthy cells, penetrates them, and then replicates. The artificial spike protein antigen produced from the manufactured mRNA prepared the body's immune system to repel the spike proteins on the COVID virus by causing the immune system to make *antibodies* that would cap the spike proteins, rendering them dysfunctional. Unlike traditional vaccines, the mRNA and the artificial spike proteins had no chance of causing the actual disease. The mRNA did not linger in the body but was eliminated by natural processes, thus minimizing immune reactions. The whole process was a miracle of contemporary biological technology.

CRISPR

The CRISPR revolution came to the fore publicly around 2015 after I had taught my Future of Humanity class for a couple of years. My outsider's impression was that a whole new revolutionary science had been invented right under my nose. This became one of my quintessential examples of exponential growth—knowledge feeding on knowledge—driving us into the future.

The roots of this explosive development in biology were planted earlier than I first knew. The name for this biomolecular magic was given to it by Fransisco Mojica, an Italian biologist who studied it in a highly salt-resistant bacteria, but it is an ingredient of ordinary yogurt. CRISPR is an acronym for "clustered regularly interspaced short palindromic repeat." CRISPR is a strand of DNA with two components. One consists of short bits for which the A, C, G, and T arrangement reads the same forward and backward—a palindrome. The palindromic pattern is repeated along the CRISPR strand but interrupted with *spacers* that are different from one another. The CRISPR structure is part of the immune system of bacteria that retains a genetic memory of viruses that previously invaded the bacteria. When a virus first attacks, the bacteria register the viral DNA and preserves a representation of it in the spacers of their CRISPR. The structure of the foreign DNA is registered but is in an inactive state. The bacteria genome often contains a CRISPR-associated (*Cas*) gene that codes for a special Cas protein that aids the process. If that same virus attacks again later on, the bacteria can find the match to the viral DNA in its CRISPR and produce a *guide RNA* corresponding to the invading DNA. The guide RNA attaches to a Cas protein that can cut the viral DNA. The RNA locates the virus DNA by matching base pairs, and the Cas protein cuts the foreign DNA double strand to destroy it.

University of California at Berkeley microbiologist Jennifer Doudna was engaged in a study of how cells use RNA molecules to regulate the genomic instructions to produce specific functional proteins. At a conference in 2011, she met Emanuelle Charpentier, a French microbiologist with a multinational career. Charpentier was studying a single gene that coded for a single protein, Cas9, one of ninety-three known Cas proteins, but did not know the function of the protein. The two women launched a collaboration to understand the function of Cas9. It was the spark that changed the world.

Cas9 has the capacity to generate a double-strand break in DNA by separating the two strands and snipping the DNA at a location that matches the base pair sequence in a guide RNA. The guide RNA interacts with a *tracer RNA* to form a structure that attracts a Cas9 protein floating in the cellular broth and attaches to the protein. The construct of the two RNAs and the Cas9 protein is required for the protein to find the right place on the DNA and cut it. Doudna, Charpentier, and their young associates set out to create a simpler system that linked the two RNAs into a single guide RNA with a single protein. They succeeded in developing a CRISPR/Cas9 system that could find and cleave any double-stranded DNA sequence. A significant body of work had already established that cells have mechanisms to heal double strand breaks by adding or deleting strands of DNA at the break, thus altering the genome at the site of the break. The great success of genetic sequencing told scientists where to find genes that performed certain functions or were involved in certain diseases. The problem was how to manipulate this complicated machinery.

CRISPR/Cas9 gave the solution with a single, programmable system that could find any desired DNA sequence. The RNA was far easier to edit than DNA. The ribosomal transcription process was sufficiently well understood that RNA of a specific base structure to match a strand of DNA could be manufactured in a lab or even ordered through the mail from a commercial laboratory. It was also much easier to program the RNA than to program the snipping protein as researchers had been attempting to do. The guide RNA could be combined with a Cas9 protein to snip the double strand of DNA at any pre-chosen location. The single Cas9 protein could be employed for any section of DNA by simply changing the sequence of the guide RNA associated with the Cas9.

It was as if in a world of artists with brushes, easels, and canvases, someone had suddenly invented paint. The CRISPR/Cas9 combination proved to be a remarkable tool. The technique could be used by anyone with basic training in molecular biology. In 2015, I asked our associate dean of research if he could give me a list of microbiology researchers on campus who were using CRISPR. He said it would be easier to give a list of those who were not. A fundamental, curiosity-driven research project rapidly turned into a technology that enabled an explosion in creative applications of genetic editing in universities and commercial laboratories around the world. Cas9 could be programmed with multiple different guide RNAs to generate multiple breaks in the same cell. Other Cas proteins were pressed into service. Applications were found in basic biology, in targeted genetic changes in a wide range of organisms, in biotechnology, in biomedicine, and in human gene therapy. In 2020, Doudna and Charpentier became the first women to jointly win the Nobel Prize in Chemistry.

Feng Zhang of the MIT/Harvard Broad Institute uses the analogy of CRISPR being like a cursor that can be placed in any line of text. The Cas proteins can then delete letters, words, or sentences and put in new ones. The result is like being able to easily edit, like going from pen and paper to a word processor. Zhang was an early collaborator of Doudna's at Berkeley but effectively became a competitor when moving to the Broad Institute to work with the pioneering geneticist George Church. Zheng was the first to show that the generic CRISPR/Cas technique worked in the cells of animals and plants and not just the single cells of microbes. Zheng, Church, and their team have been prolific inventors of new uses of CRISPR. Berkeley and the Broad Institute became involved in a battle royal over patents to the applications.

Humans have long tinkered with the genetic structures of living things; they just called it *selective breeding*. The difference now is that scientists understand and can operate at the microscopic level of the genes themselves. CRISPR/Cas made this especially easy and accessible. Some changes can be made in the genes of individual organisms, including people, to alter that individual. A particularly powerful example would be treating leukemia, sickle cell anemia, and perhaps even Alzheimer's disease by altering a person's genes. That sort of treatment would affect only that particular individual.

CRISPR has the potential for much more than that. While the results of traditional gene therapy cannot be passed to progeny, CRISPR can be used to permanently edit the *germ line*—eggs, sperm, or embryos—making changes that are inheritable forever. The potential to produce designer babies became a practical reality. Doudna says that this "changes our relationship to Nature."[8]

ETHICAL ISSUES

We are still in the early days of this new technology. Even with CRISPR, biological systems are complex and rife with possibilities for unintended consequences including unwanted or off-target edits. Nevertheless, the direction is clear. Doudna was immediately aware of the ethical consequences of her invention and organized meetings to consider how to guide and control this new technology. In the spring of 2015, she called for a worldwide moratorium on the use of CRISPR to make heritable changes in human embryos. Wise heads cautioned to proceed very slowly. There is a powerful parallel to the discussion of the implications of AI that arose after the advent of ChatGPT.

Both the power and the potential peril involve working with embryos. In the embryonic stage, the cells are *stem cells* with the capacity to develop into any specific type of cell, all of which make up a mature being: skin and nervous system cells; gut and lung cells; bone cells; muscle cells; circulatory system cells; and more. Fredrik Lanner, a Swedish developmental biologist, may have been the first scientist to use CRISPR/Cas9 to edit healthy human embryos in 2016. This work continues with the goal of attaining deeper understanding and tapping the power of stem cells to heal. Responsible research destroys the embryos after the work is done, a practice that itself leads to ethical issues and condemnation in some quarters. Some of this work has shown the frequency with which CRISPR editing in embryos can lead to alterations near the target gene that were not intended or far from the target gene in the complex 3D tangle that is critical for the functioning of the whole DNA complex. Caution is warranted. Lanner declared that designer babies were not on his radar and that he stood against "any sort of thoughts that one should use this to design designer babies or enhance for aesthetic purposes."[9]

Ethical concerns were high on the agenda of the International Conference on Gene Editing in Hong Kong in November 2018. That is why there was shock and consternation when a young Chinese scientist, He Jiankui, announced that he had altered the germ line in the embryos of twin girls, thereby becoming the first person in the world to create genetically edited babies using CRISPR. Dr. He had been trained in the United States at Rice and Stanford. His advisors warned him against prematurely engaging in such activity even if the technical capability were there lest he become an international pariah. He, however, chose to proceed, hoping to render the twins immune to HIV, a disease carried by their father. In that he appeared to have been successful in the short run. The infant twins were apparently healthy. The scientific community was aghast. Among the concerns was that there could be genetic issues that would emerge later in the life of the twins or among their offspring. The Chinese government first subjected He to house arrest and then sentenced him to three years in prison for conducting an illegal medical practice.

Doudna has written a book about her discovery and its attendant promise and peril: *A Crack in Creation.*[10] She describes a dream in which she is shown into a room in which a person is sitting in a chair facing away from the door. She approaches the chair, and the person turns out to be Adolf Hitler asking, "Tell me about this CRISPR."

This tangle of rapid technological progress and ethical constraints continues to foster discussion and debate. Witnessing the demonstrated capacity for unintended consequences, Fyodor Urnov, a geneticist at the University of California at Berkeley, declared, "This is a restraining order for all genome editors to stay the living daylights away from embryo editing." R. Alta Charo is the Warren P. Knowles Professor of Law and Bioethics at the University of Wisconsin and a proponent of embryonic stem cell research. In 2017, she chaired a panel of the National Academy of Science that gave a yellow light to research in the editing of the germ line, meaning, "Proceed with caution." In rare situations under tight regulation, inheritable genetic changes could be employed not just to treat but to prevent disease. This perspective would allow engineering a healthy, genetically modified child for parents who could not otherwise have a healthy child naturally.

Charo acknowledges that new technology might be used for evil but argues that our worst fears usually do not come to pass.[11] She points out that if we have genetic information, we must decide what to do with it; it cannot simply be ignored. The technology is neutral. It is up to humans to make the choice. Ultimately, Charo makes the case for a nuanced view. It is easy to argue that editing the germ line should be forbidden before the technical capability exists. Once the technology actually exists, as it now does with the development of CRISPR/Cas, attitudes and perspectives change. Some restrictions are appropriate but not those so stringent that crucial positive goals cannot be achieved—for example, curing disease and producing healthy children. Charo argues that nothing is absolutely safe. The practical goals are to make a technology like CRISPR as safe as we can and then carefully assess the remaining risks. At the same time we should have mitigation plans in place if things go wrong and a mechanism in place to turn off a genetically altered species released in the wild. She calls for tolerance of individual choices. These are messy decisions in a world that stubbornly refuses to be either-or.

We are at the beginning of a new chapter of human evolution with new tools at the ready. This genie is out of the bottle. There will be more designer babies.

In the meantime, there is an explosion of creative uses of CRISPR/Cas techniques. CRISPR has been used in attempts to write coded information into DNA. CRISPR may also have a role in *xenotransplantation*, the transplantation of organs from one species to another. Pigs and humans are compatible in many ways. Pig heart valves and pig skin are already used in human therapy. Pigs could be used as a supply of organs for humans with diseased hearts, kidneys, or livers. The first pig kidney was successfully transferred to a human in 2021. A practical problem has been that pigs carry viruses that can threaten humans. George Church and his group used CRISPR to create pig embryos from CRISPR–edited cells to grow pigs free of viruses. Future possibilities include growing human organs in pigs by swapping appropriate human genes for pig genes or recovering something resembling extinct creatures like Dodos, Tasmanian Tigers, or mammoths. Other researchers dream of adjusting the genes of future astronauts to make them more resistant to radiation or able to produce their own vitamins. (There are hints that astronauts are subject to shifts in their DNA.)[12]

GENE SWITCHES

Another recent biological revolution concerns the fact that most of the strands of DNA are not genes that code for proteins. For many years, these portions of DNA were known as *junk DNA*, *dark matter DNA*, or *selfish DNA* because they did not appear to contribute to the fitness of the organism. The problem was that there was a lot of it: only about 2 percent of

DNA actually codes for proteins. Why would Nature have developed such a sloppy, inefficient system?

Over the last decade, biologists, chemists, and physicists have revealed the critical role of this non-coding material. The genes in a given living being are all the same in all the chromosomes in its cells; however, not all of the genes are activated at all times. Genes can be turned on and turned off. In one cell, the genes that instruct the cell to become part of the heart muscle are turned on, and those that instruct it to become part of the pancreas are turned off, and vice versa. The non-coding parts of the DNA are vital for that function. One of the aspects that took some time to understand is that a non-coding section in one part of a strand of DNA can affect a gene in a very different part of the DNA; different, that is, if you think of the DNA as a single, stretched-out strand. That is where the 3D structure comes in. With the DNA in practice tightly wound up in the chromosome, a non-coding strand in one place along the DNA can be brought into close, intimate, physical contact with a gene in another part of the DNA.

The non-coding bits are *gene switches*, part of a critical intricate structure that depends on the 3D structure of DNA to activate or inactivate certain genes. This ability to turn genes on and off in a systematic way gives a vastly richer capacity to the DNA complex. Gene switching helps to explain why immensely complicated living things can function with only a relatively small number of genes. A rough analogy is to picture all the light switches in all the buildings of a city. At one extreme, one can picture all the switches on or all the switches off. That behavior would pale compared to the actual situation in which individual switches are turned on or off according to the needs of the occupants of the buildings. The result is a complex pattern of lights that varies moment by moment, transcending the finite number of switches. The genetic machinery is more like that.

The understanding of gene switches triggered another major development in biology and genetics. Jean Baptiste Lamarck, a French naturalist and zoologist of the early nineteenth century, worked roughly a generation before Darwin. Lamarck promoted the important idea that evolution was driven by natural laws. That was well and good, but the particular theory of evolution he advocated was that physical changes in an organism during its lifetime could be passed on to offspring. A classic example given to illustrate Lamarckism is the long neck of a giraffe. The Lamarkian hypothesis is that if an adult giraffe stretches its neck to get to high leaves in a tree, then its offspring will also have longer necks. Darwin's radical thinking about evolution overthrew that notion. For Darwin, there was a bunch of giraffes with necks of various lengths. Those lucky enough to be born with a genetic variation that gave them longer necks could feed more easily and hence survive to reproduce and to pass that gene for longer necks to their offspring. The fundamental driver was natural selection, survival of the fittest. Zoologists have discovered some examples of dwarf giraffes. They are presumably the result of a genetic mutation, not some sin of the parents. For what it is worth, it is not clear that Lamarck ever advocated the giraffe example himself. He talked about why moles are blind and why mammals have teeth but birds do not. In any case, Lamarck was a deep thinker whose central idea led to ridicule in later years. There was some truth to what he advocated.

In the wake of our deepening understanding of genetic functions in general and gene switches in particular, a new branch of biology was born: *epigenetics*. Epigenetics deals with the very Lamarckian concept that the environment can affect genetic structure in a way that can be passed on to offspring. The idea is that in some circumstances conditions in the environment do not change the structure of the genes but can alter which genes are switched on and which genes are switched off. If some of those switches are turned on or off in the genetic material of eggs and sperm, then the effects of the environment can be passed to future generations. In a rough analogy, the genes are the hardware, and epigenetic effects can rewrite the software.

One might think of atmospheric pollutants or harsh chemicals in the environment as triggering epigenetic gene changes, but the effects can be more subtle. One line of study suggests that trauma or fear can reset which genes are active, and those settings can be passed on to future generations. It was important for our ancestors, and it still is for us, to have a reasonable fear of things that threaten us. If an ancestor underwent a terrifying experience and survived, it may be that his or her genes adjusted to maintain a permanent record, and an understandable phobia was passed to future kin. Resetting gene activity in our ancestors may account for why we are intrinsically repulsed by spiders or snakes.

In contemporary epigenetic studies, scientists have found that trauma experienced by parents can be passed down for generations. In one experiment, mice were exposed to the scent of cherry blossoms and then given a mild shock. The mice developed a fear response to the odor even when the shock was not administered. That makes sense. The remarkable aspect is that this fear response to cherry blossoms was exhibited by the next two generations of mice, animals that had never been shocked. This was true even when the next generations were conceived by artificial insemination and raised by mice to which they were not related and that had never been exposed to the cherry blossom odor or to the shocks. Their automatic fear behavior was inherited. Another study showed that similar effects of chemically induced stress lasted for fifty generations in the mitochondria of a certain species of worms.

The ideas of epigenetics have also been applied to humans. Some studies argue that the traumatic effects of the Holocaust have been passed down to subsequent generations, giving those generations a greater propensity for anxiety and depression but also for resilience. Others make the same case for Blacks in the United States whose ancestors were enslaved. That community is deeply affected by ongoing social injustice, but some of the impact might very well be inherited.

Epigenetics thus adds another dimension to ongoing complicated and controversial arguments about nature versus nurture and hence affects the fields of sociology, criminology, and public health. The new pursuit of *sociogenomics* is dedicated to exploring such issues. Perhaps the season of insemination or of birth plays some epigenetic role in the setting of gene switches, giving the tiniest shred of credibility to astrology.

Epigenetics raises the idea that if some of the negative effects of multi-generational stress like heart disease, obesity, and diabetes are the result of gene switching, perhaps we can learn to reverse such effects with clever bio-chemistry. Another strain of research pursues the issue of whether people prone to violence can pass down to their offspring a tendency to be violent, hence yielding a greater propensity to criminality. Epigenetics cannot be the whole story of why someone turns to crime, but it may play a role. Perhaps we can engineer epigenetic effects and reset unwelcome genetic switch settings. This prospect has profound structural and ethical implications for society. Who decides to tinker with gene switches, and who gets tinkered?

DRIFT AND FLOW

Although perhaps not as dramatic as CRISPR and epigenetics, recent ge-netic research has brought a raft of other developments. While too much mutation can be a bad thing, it is important for genes to alter at some level. If there are no genetic changes from one generation to the next, then stasis sets in. Nature has nothing to select to determine that one set of genes is more fit than another. That is one reason Nature developed sex: to ensure some mix of genetic potential between parents and offspring.

Nature has other, more subtle ways of dealing the genetic cards. There is a role for *genetic drift*, where random changes happen in a population by chance rather than by mutation and selection. New genes sometimes ap-pear in individuals randomly not because they are healthier or less healthy, or tougher or less tough, or more or less likely to reproduce. Chance happens, so the gene pool will always drift a little with time even if conditions are otherwise exactly the same.

The gene pool can also change by the general processes of *gene flow*. This can happen when individuals of a certain genetic characteristic move into a population where those characteristics do not exist. Gene flow can happen on a macroscopic level as species migrate but also at the microscopic level. Scientists have discovered that genes are not fixed entities like a certain combination of Lego blocks that have been glued together but that that they are dynamic, as if blocks of different shapes, sizes, and colors were constantly swapping places. The result is that new genetic functions can come into existence, and old functions can cease to exist. Junk DNA can evolve into functional DNA and vice versa. DNA sequences that have no special function in one species can find their way into another species where they transcribe to a messenger RNA and become part of the useful protein manufacture there.

A variation of gene flow is known as *horizontal gene transfer*, by which genetic material—strands of functional DNA—moves laterally from one organism to another rather than vertically by reproduction. Horizontal gene transfer is involved in the spread of antibiotic resistance in bacteria. The little buggers do not just evolve to be more resistant, they swap their defense mechanisms. There is even evidence that genes can be swapped among plants and animals.

Geneticists have learned to emulate these natural genetic processes by experiments in genetic engineering often facilitated by CRISPR/Cas techniques. The discipline of *optogenetics* modifies genes to produce cells containing proteins that are responsive to light. The result is that scientists can use light to precisely control cells in living tissue. Experiments are currently being done in creatures like fruit flies, but scientists envisage the ability to create artificial tissue or control tissue development.

Another application is called *gene driving*, a process that artificially increases the probability that a particular gene or set of genes will be propagated to offspring at greater rates than would be the case according to the standard rules of heredity. Rather than a 50 percent chance that a gene from a parent will be passed on, that probability can be raised to 100 percent. The process of gene driving begins with the release of animals into the environment that have been modified to have a gene in their germ line capable of activating the CRISPR/Cas mechanism. When that animal mates and chromosomes from male and female pair up, the CRISPR in the gene-drive gene recognizes its complement in the natural gene of the opposite chromosome. The Cas protein carried by the CRISPR reader cuts out the natural copy of the gene before the embryo develops. Nature then takes over. The repair processes in the cell use the complement in the chromosome—the gene-drive gene—as the pattern. The cell thus ends up with the gene-drive gene in both strands of the chromosome. The drive is now hard-wired into the genetic structure, and the gene drive, not the natural gene, will be passed down to all future offspring. The gene drive can totally replace the natural gene within a few generations.

One proposed application of gene driving has been to eliminate disease-carrying mosquitoes. The gene drive was designed to disrupt female fertility. Some have proposed using gene drive to eliminate mosquitoes that carry malaria, West Nile, or Zika. There are concerns about unintended consequences if gene-driven species get out of control. There are also moral and ethical issues in deciding to engineer extinction.

GENE SEQUENCING

A significant recent accomplishment of geneticists is *genetic sequencing*. This is the hard-won technical ability to read at first bits, then long strings of linked bases in both genes and junk and then the whole string of DNA in entire chromosomes. It was as if a Martian stumbled across the Library of Congress and had to start from scratch to recognize letters, words, sentences, and then whole books.

The first techniques for reading the sequencing of bases in DNA were developed in the 1970s. Fragments of synthetic DNA could be matched to strands of the DNA being investigated. The larger fragments of synthetic DNA were sorted by size, and the size distribution gave clues to how the original DNA was linked together. In the 1980s, these techniques were made more sophisticated, and advances in computer power enabled the puzzle of how the overlapping fragments must fit together to be solved more rapidly. Automated sequencing machines were developed that allowed first millions then billions of base pairs to be analyzed quickly and cheaply. The first complete genomes of bacteria were decoded. By now, the oldest DNA to be thoroughly decoded is that of a species of mammoth 1.2 million years old.

In 1990, the Human Genome Project was launched, a massive, government-sponsored, international collaboration to decode the human genome. The project cost $3 billion and was expected to take fifteen years; it was accomplished in 2003, a little ahead of time, at least in draft form. A private project led by Craig Venter and his company Celera Genomics started a competing project in 1998 promising a faster, cheaper process in part by capitalizing on data produced and freely released by the public Human Genome Project. Celera's hopes to patent genes were dashed when President Clinton declared that the genetic sequence could not be patented and that the associated data must be made public. Celera published its competing draft about the same time as the public project. The years since have seen great progress filling in gaps in the original drafts. The Human Genome Project collaboration finished the job in 2021: 3.05 billion base pairs and 19,969 protein-coding genes. A more complex *pangenome* based on forty-seven genetically mixed people was completed in 2023.[13]

The question of who owns genetic data is still fraught and unresolved. Several states have laws that prohibit theft of DNA but in many states anyone can sequence DNA that is effectively abandoned by an individual. This would include DNA left on a drinking glass in a restaurant, DNA in the trash, DNA on a discarded cigarette butt, or *environmental DNA* just floating in the air. Any free DNA is fair game for the police, a suspicious spouse, or a political operative. The companies 23andMe and Ancestry.com encourage people to send in a sample of saliva and receive a report on their genetic history. The companies are less open about what they do with that data, including what is shared with third parties. It is ironic that people can be identified not by their genes but by the order of bases in what was originally considered junk DNA. DNA is a powerful tool for identification and for tracing roots. The rights of indigenous people to share or restrict their own genetic information is also a sensitive and evolving topic often ignored in past research.

DNA sequencing and related genetic studies give insight into age-old questions regarding the role and relative importance of nature versus nurture. For example, are people similar or different because of their genetic inheritance or because of the situation or society in which they were raised? I have long hypothesized that people are born with a tendency to be liberal and cherish diversity or to be conservative and tend to distrust or dislike those who are not similar to them. I may have been only partly correct. Relevant studies of this kind involve identical twins raised in separate families. That way, they closely share genetic structure[14] but not the environment in which they were raised. Such twins tend to select the same neckties, for example, when they are older. Twin studies have shown that height is 80 percent genetics—to no one's surprise—but that politics is only 30 percent at best. Genetic sequencing studies also showed that even for something as basic as height, no single gene is involved but a complex mix of tens of thousands of genetic markers.[15]

Tracing DNA has enabled police to solve some crimes, including decades-old cold cases. The solution is not simply to find the criminal in a genetic database but to use DNA as a starting place. Find the cousin, and you may find the culprit. Current studies reach to third cousins, a category that for most people includes about eight hundred individuals. If even one of those people is in a genetic database, sophisticated genealogical tracing of family connections can lead to the person sought. A ramification of this process is that a large fraction of people can be individually identified whether they are themselves in a genetic database or not. This statement currently applies to Americans of European ancestry because those people are the most frequent users of genetic services. Around 60 percent of those people are currently identifiable, and researchers expect the number quickly to grow to 90 percent.[16] Other ethnicities and other countries are likely to follow. Statistics show that any individual of any ancestry can be identified if the genetic database contains even as little as 2 percent of a particular category.

By now even the complete genome of Neanderthals has been sequenced. About 2 percent of the genes of Modern Americans, Europeans, and Asians come from our Neanderthal ancestry. It was first thought that Africans had no Neanderthal genes, but they do. Our history of migration and interbreeding is more complicated than once thought.

The benefit of such genetic methods is that they may lead to personalized medicine and great health benefits. A particularly dystopian implication is that a poison could be designed to affect a specific individual or ethnic group. A more likely looming drawback is a surveillance society of unprecedented intrusion. Private health data might be compromised because even an anonymized genetic profile could be traced to and identified with a single individual.

Genetic sequencing has brought new perspectives to the concept of race. For reasons of which I am still unsure, my class "The Future of Humanity" of about twenty-five students drew a cross section of individuals who were, to me, of delightful diversity: White, Black, Latinx, and Asian. In one class, I got to the point where I talked about the fact that genetics could identify the color of a person's skin. I realized I had just broached the delicate issue of race in this diverse setting and had no choice but to forge ahead and hope that it was a teaching moment. I got little feedback and do not know what individuals were thinking. At least there was no overt outrage.

Geneticists have long established that there is no genetic basis for race as it is commonly used to divide people into categories. People have a range of traits from height to intelligence, but there is no discernable difference in the average or spread of those characteristics that depends systematically on the color of their skin or hair, the shape of their nose, or the fold of their eyelids. In that sense, the concept of race is meaningless. Human genomes are all the same at the 99.5 percent level. The odd remaining 0.5 percent might correlate with the geographical origin of your ancestors, but there is no way directly to map that information into race. The fact remains that genetic studies can, with increasing fidelity, determine the color of a person's skin based on the sequence of their DNA. The DNA can thus discern the properties of a given individual that are currently used to assign a person's race. It is not clear what to do with this understanding. Race among modern humans is essentially meaningless.

Among some people, these developments in DNA sequencing and genetics are reinforcing the notion that race is a biological category. This has complex implications in the context of personalized medicine. Blacks are more susceptible to sickle cell anemia. Using that information is a reasonable choice to target particular therapies. How is selecting African ancestry to aid therapy any different than using race, the color of one's skin? Dorothy Roberts explored these difficult issues in her book *Fatal Invention*.[17] Roberts argues that race is not a biological construct but a social, even political, construct disguised as a biological category to justify injustice. She points out the paradox of pursuing race-based genetic variation in medicine while paying insufficient attention to the health disparities and racial inequalities in American society and elsewhere. She advocates framing the issue of genetic structure in medical treatment and other contexts not as how it affects racial groups but how it affects individuals.

Another aspect of genetic science that garners much attention from researchers and the public is genetic modification of our food. Humans have been doing genetic modification of plants and animals for a very long time by the process of selective breeding. This process calls for the selection, often visual, of a property deemed desirable and then breeding to enhance that aspect. This process is slow and steady, but it results in gene modification. An interesting recent example is the experiment done in Siberia in which Russian scientists carefully selected foxes with some hint of positive response to humans and managed to breed tame foxes in fifty generations. The difference with modern gene modification is that it works at the molecular level; it can be done quickly in embryos or seedlings; and its effects might spread rapidly. There is fear in some quarters of unintended consequences.

FRONTIERS

With knowledge of the genetic code and other technical advances, other potential applications arise. Three-dimensional printing began as a way to make small plastic parts. One could feed a digital design into the "brain" of the printer, and the printer would use a laser to sinter tiny layer after layer until the 3D object was constructed. This field, like so many others, has advanced exponentially over the last couple of decades. Three-dimensional printers are widely used by both manufacturers and hobbyists. People print plastic guns that escape detection by metal detectors. Others print rocket engines or houses. Applications have been developed in the printing of body parts, bones, and prosthetics. Material science engineers at the University of Wisconsin have 3D-printed arteries that can monitor the pressure of blood flowing through them. Dutch physicists have microprinted a tugboat smaller than a human hair that could float in that artery. The shape is somewhat fanciful but complex enough to challenge the state of the art of microscopic printing. Variants could carry medicine or diagnostic tools. Scientists from Tsinghua University and Drexel University have gotten closer to the root of things by 3D printing a structure for growing embryonic stem cells.[18] Round clusters of embryo cells are more relevant to natural clusters of embryonic cells than flat arrays of cells grown in petri dishes and hence give researchers more precise control of the stem cell development. This 3D structure could be used to grow tissue and organs and test medicines. Printing the stem cells themselves remains a distant potential.

Why build a conscious robot when you can type in the genome and print a designed human on a 3D printer?

There is yet another realm of technological development that complements biology and encroaches upon it: *nanotechnology*. Nanotechnology involves the manipulation of matter on the scale of atoms and molecules with a characteristic size of a nanometer, 1 billionth of a meter. Nanotechnology brings the skill to manipulate individual atoms and build them into larger structures. Nanotechnology has an immense range of applications, but some relate to biology. Messages can be written using the DNA code—an example of *nanocomputing*. *Nanobots* the size of cells could crawl through our bodies, monitoring functions and delivering medicine in precisely targeted ways. Nanoparticles have been used to kill brain-eating amoebae while leaving healthy human cells unaffected. MIT engineers have embedded carbon nanotubes in spinach, yielding a plant that can detect traces of explosives in groundwater and send a wireless message to a cell phone—an early development in *plant nanobionics*. By breaking down the plant-human communication barrier, scientists hope to tap into the sophisticated environmental detection system inherent in plants. Israel has become a center of nanotechnology research and manufacturing. One company, Olfaguard, is developing a nanobiological system to scan food products and detect pathogens like salmonella and *E. coli* in real time.

Chemical biologists at the University of California at San Diego have employed nanotechnology to create a close representation of a human cell with an outer coat of plastic rather than a natural cell membrane and an inner compartment that contains DNA like a natural cell nucleus.[19] The synthetic nucleus releases RNA to the rest of the cell, thus stimulating the creation of proteins. These synthetic cells can then pass these proteins to other synthetic cells and thereby enable communal behavior. Synthetic cells have the potential to deliver drugs, detect cancer cells or toxic chemicals, improve diagnostic testing, or even form artificial tissue.

Microelectromechanical systems (MEMS; also known as *smart dust* or *smart motes*) are a variation of nanobots. These tiny robots or other devices are light enough that they can float, wafting in the air like dust. They can detect light, temperature, pressure, vibrations, magnetism, humidity, and chemicals such as pollutants. MEMS can be powered by heat or vibrations and wirelessly relay the information they collect. If they are as big as a grain of salt, they can even incorporate cameras. They can be 3D printed in precise detail and mass produced. There are visions of scattering these particles to collect data and empower smart cities. Others consider this the path to dystopian surveillance cultures.

A contemporary example of the marriage of AI and biology is the development of *xenobots*. Xenobots are a new type of life, an artificial, living, self-replicating, programmable organism that serves to redefine the meaning of "robot," "machine," "organism," and "program." Xenobots were assembled from bundles of cultivated stem cells of the African clawed frog.[20] The program is the shape of the bundle. The bundle can move around gathering up stray stem cells in the medium and constructing new copies of itself. A team from Tufts, the University of Vermont, and Harvard used AI technology to study billions of potential bundle shapes and find the one that most effectively reproduced. Bundles of first fifty stem cells were studied then bundles of three thousand or more. Spherical bundles could make one new generation of themselves but a *C*-shaped bundle reminiscent of a Pac-Man could efficiently gobble up stray cells and produce new *C*-shaped bundles that would repeat the process for up to four new generations. The bundles of frog cells replicated not by growing and dividing but by moving, finding building blocks, and compressing and combining them into self-copies.

The original xenobot research required the existence of already rather sophisticated stem cells, but it points the way for the development in the lab or in the wild of other organisms that can form and propagate without either natural selection or genetic engineering. Xenobots show how self-amplifying processes can emerge spontaneously. They may give new insight into the origin of life, perhaps an amyloid world as mentioned in chapter 9 that preceded the RNA world.

Xenobots or the inventions that follow from them may be used to address issues like repairing injury and birth defects or curing disease and aging. Team member Michael Levin of the Allen Discovery Center at Tufts University and the Wyss Institute for Biologically Inspired Engineering at Harvard foresees broad applications of this biotechnology, saying it may help to "better understand and control the goals and behavior of swarms of active agents—in this case cells, but those same lessons will help us make sure that the IoT, swarm robotics, and many other technologies actually have beneficial outcomes."[21] Kirstin Peterson, a roboticist at Cornell gives an optimistic assessment: "Robots that can copy themselves are an important step toward systems that don't need human operators."[22] The technology is sweet and seductive, but this is also a path to gray goo.

ETHICS AND RELINQUISHMENT

There will first be a trickle and then a flood of new biological developments. How will we as a society decide how to use this powerful technology to shape our own ends? What will be allowed? What prevented? The technology exists now to tell parents how tall and smart their precious infant will be and also the likelihood he or she will be prone to alcoholism, tobacco addiction, heart disease, or cancer. What parents will not be tempted to tinker with that future?

Stem cells can be manipulated to create artificial embryos. This has so far only been done to grow mice,[23] but someone, somewhere, will try it with human embryos. Recent research has shown that using CRISPR/Cas to modify the genetic codes of embryos can lead to unexpected negative effects in the far-flung complex of the 3D DNA structure laden with tricky gene switches of unknown implication. Some feel it is ethical to pursue modification of genes to control severe diseases but not to design babies. It may not be possible to walk that fine line.

One government might develop one set of ethical guidelines, another government quite something else. China has set a goal of collecting blood samples from 700 million males to construct an elaborate genetic map of its citizens. This complement to China's elaborate system of facial recognition is another deliberate step in the development of a powerful, controlling surveillance state. Chinese scientists are attempting to construct a process of DNA phenotyping by which genes can be analyzed for external traits like skin color, eye color, and facial structure that can, in turn, be correlated with ancestry. The goal is to create artificial facial images that are sufficiently realistic to identify criminals or victims—or Uighurs.

Even Russian President Vladimir Putin understands the lessons of this chapter. Addressing the 2017 Sochi World Festival of Youth Conference he said,

> Genetic engineering will open up incredible opportunities in pharmacology, new medicine, altering the human genome if a person suffers from genetic diseases. All right that is good. But there is another part to this process. It means—we can already imagine it—to create a person with the desired features. This may be a mathematical genius. This may be an outstanding musician. But this can also be a soldier, an individual who can fight without fear or pain. You are aware that humankind will probably enter a very complicated period of its existence and development. And what I have just said may be more terrifying than a nuclear bomb.

ANTHROPOCENE
The Age of Humans

CLIMATE CHANGE

Alongside the digital revolution, the probing of our own brains, and the capacity of biology to alter our evolution, a fourth imminent exponential challenge to humanity is us. The number of *Homo sapiens* on the planet has been growing exponentially for millennia. There are signs of that growth saturating and leveling off, but we have well entered the age of the *Anthropocene* in which the very presence of hordes of humans affects the ecology in which we survive and thrive.[1] In his encyclical *Laudato Si* of March 2015, Pope Francis addressed the issue by saying, "Clearly, the Bible has no place for a tyrannical anthropocentrism unconcerned for other creatures." A principal implication of the Anthropocene is the prospect of global warming, or more generally, climate change.

For decades, climate scientists have mapped the dire possibilities of human production of carbon in the air around us and the associated effects of global warming. It is very difficult for many individual people and for societies in general to address issues of future threats that are not directly and manifestly present now. Meanwhile, the planet has gotten more carbon choked and warmer. The choices now are bad or worse. The climate conditions we see now are not just worse than the past; they are the best we can expect in the future even with heroic measures of mitigation and adaptation.

The signs of damage from climate change are all around us. We ran out of cute hurricane names in 2020 and had to employ basic Greek letters: we got to Hurricane Delta. Fires in California not only turned the sky orange but the smoke threatened the famous Sonoma Valley wine production. The California forest fires of the summer of 2020 burned the crowns of giant sequoias for the first time in recorded history. Other fires devastated Siberia. Among people paying attention, there are new symptoms of *eco-anxiety* and *climate grief*. In 2018, Jem Bendell[2] self-published a paper titled "Deep Adaptation: A Map for Navigating Climate Tragedy" after it was rejected in a peer-reviewed journal. The premise is that as global warming affects social, economic, and political systems, society may collapse. Bendell predicts starvation, destruction, migration, disease, and war. While criticized for lack of rigor, the paper triggered a broader conversation about the devastation that might ensue and how values might need to shift to adapt to extreme changes in what is regarded as civilization.

When I first heard about climate change several decades ago, one of my first thoughts was whether the changes being debated were part of the natural cycle. As an astronomer, I'm aware that our Sun goes through sunspot cycles every eleven years. The star is hotter now than when it was born about 5 billion years ago. Earth has gone through periods of warming and freezing. There have been ice ages that sent glaciers almost to the equator. There is discussion of a *snowball Earth* when ice covered the whole surface of the planet more than 650 million years ago, the so-called *Cryogenian period*.

Two things shook me from this predisposition. One, purely anecdotal, was when preparing a lecture on the issue, I ran across a photo of some Inuit children playing golf near the Arctic Circle. My immediate thought was, "This ain't right!" More seriously, I started paying attention to the periodic reports from the UN Intergovernmental Panel on Climate Change[3] (IPCC) that began in 1988 and the complementary US Government Climate Report that is mandated every four years by Congress.[4] As I started this book, there had been five IPCC general assessment reports—the sixth was released in August 2021 and February 2022 and finalized in a synthesis report in March 2023—and a number of special-purpose update reports.

One of the raps on climate change predictions is that the underlying methodology is uncertain. That is true. Climate involves issues of mind-bending complexity that incorporate the heating and cooling and flow of air and water. Computational methods have gotten ever better, but there are intrinsic uncertainties. Those uncertainties cut both ways. Over the time that I have been paying attention to the topic, with each new report, conditions have been worse than the average predictions of the previous report. Temperature data have landed at the upper end of possibilities sketched by the older data. That is not a good trend. The trend continued with the sixth IPCC report that recorded higher temperatures than anticipated by the fifth report and concluded that some effects of climate change were locked in, irreversible.

Some people argue that given the intrinsic uncertainties of the methodology, things might not be as bad as the climate alarmists think. The record is not on their side. There are others who fall in the category of having their livelihoods affected by proposed responses to climate change, a situation that promotes an understandable negative mindset. I sympathize with disrupted lives, but Nature will do what Nature will do.

The Greenhouse Effect

A fundamental notion underlying the concept of climate change is the *greenhouse effect*. Greenhouses designed to grow plants have abundant clear glass windows. Window glass has the important physical property that it is transparent to optical radiation, hence ordinary sunlight. That is actually a quite remarkable property given that glass is hard and solid and that it is easy to design glass that is totally opaque. Most of the energy in sunlight is concentrated in the optical bands. That is why humans and animals have evolved to see such radiation. Optical radiation floods in through the greenhouse windows and deposits energy as heat in the plants, soil, and benches within. Those components re-radiate the absorbed energy but as lower energy infrared radiation, not optical, since the interior of the greenhouse is not as hot as the surface of our Sun. A less immediately perceived property of glass is that it's opaque to that longer-wavelength infrared light. The infrared radiation cannot escape through the window glass so the energy of the original optical light is trapped within the greenhouse, providing a comfortable, even hot, interior compared to the external environment. Hothouse plants can thus be cultivated even in the winter. Anyone with south-facing windows knows this effect.

Earth is not surrounded by window glass, but its blanketing atmosphere provides the same effect. Due to the intrinsic properties of the atmosphere's dominant constituent components—nitrogen and oxygen—the atmosphere is transparent to optical radiation. The peak power of optical sunlight thus floods in through the atmosphere. The surface of Earth is heated and re-radiates that energy as infrared radiation. As for greenhouse windows, the atmosphere of Earth is opaque to infrared radiation. The incoming heat of the sunlight is trapped, warming the planet and keeping it habitable. We can survive and thrive on Earth because its thin blue sky acts to provide a greenhouse effect. A simple observation shows the fundamental truth of this. Our Moon has no atmosphere to provide a greenhouse effect. Its surface is barren and cold even though it is just as close to our Sun as Earth is.

We can look elsewhere in space for relevant wisdom. Mars may have once had a denser atmosphere that trapped heat and allowed some form of life to develop and survive. That is why Mars is an enticing goal for those seeking life elsewhere in our Universe. Mars apparently lost much of its atmosphere by evaporation into space. It no longer has an effective greenhouse effect, and its surface is not habitable to us. The lesson is that you do not want your planet to lose its atmosphere.

Venus provides an opposite and more visceral lesson. As a planetary atmosphere heats up, gases can be released from the rocks into the atmosphere. Those gases increase the capacity of the atmosphere to trap incident solar energy. The result is more heat, more release of greenhouse gases, more trapping, and more release: a *runaway greenhouse effect*. All evidence suggests that Venus may have once been more temperate but that being closer to our central star than Earth or Mars, it suffered a runaway greenhouse effect. The result is a suffocating, sulfuric acid–infused atmosphere hot enough to melt lead. The lesson is that you definitely do not want your planet to undergo a runaway greenhouse effect. Earth is headed in that direction, and humans are contributing.

A tricky aspect of this issue is that the gases that are effective in trapping the infrared radiation and warming the surface of Earth are minor constituents of the atmosphere. There is relatively little carbon dioxide and methane in the atmosphere, but these molecules are significant absorbers of infrared radiation and hence are labeled *greenhouse gases*. For eons there has been just enough of these greenhouse gases to keep the surface temperate. Carbon dioxide is released from oceans and emitted in volcanic eruptions, the decay of plants, and the exhalation of animals. Carbon dioxide is removed from the atmosphere by absorption in oceans and by living plants. Methane is released by oceans, decaying wetlands, and, interestingly, termites as they chew up and degrade cellulose.

A molecule of carbon dioxide comprises a carbon atom attached to two atoms of oxygen. Methane consists of an atom of carbon surrounded by four atoms of hydrogen. The common element is carbon. When discussing global warming, it is thus popular to refer to the problem as excess carbon in the atmosphere, *carbon* being shorthand for carbon dioxide, methane, and other greenhouse gases for which carbon is a key constituent. About

95 percent of the carbon dioxide in the atmosphere and 36 percent of the methane are produced naturally. The remaining 5 percent of carbon dioxide and an impressive 64 percent of the methane are produced by human activity. That is where the plot thickens.

About 90 percent of the human contribution to the carbon dioxide in the atmosphere comes from burning coal, oil, and natural gas to light and heat our homes and power our industries and automobiles. About 10 percent comes from deforestation: the loss of trees that would otherwise absorb and reclaim carbon dioxide. Human methane production also comes predominantly from the use of fossil fuels with a significant minority coming from cattle and pigs.

Hockey Sticks and Butterflies

Humans have been burning fuel for heat and raising animals for a long time. Studies that track the abundance of carbon dioxide in the atmosphere show there was a change in the nineteenth century, the start of the Industrial Revolution. That is when we began the massive use of fossil fuels that powers our modern lifestyle. Both the amount of carbon dioxide and methane and the average temperature of Earth's surface and oceans began to track upward. Graphs of these quantities as a function of time show a roughly level value that began to trend upward in what is often described as a hockey stick plot (see chapter 2) with the bend from the horizontal handle to the rising, puck-slapping part occurring about the time of the Industrial Revolution. There are bumps and wiggles and controversies about the details (even the existence of the hockey stick pattern in some quarters) but the overall trend is upward. The climate is changing, the oceans and atmosphere are warming, and human activity is a major contribution to the change. Global carbon emission was up by 1.6 percent in 2017 but 2.7 percent in 2018.[5] That is only a small interval in time, but the result suggests a growth that is even more rapid than exponential.

The surface of Earth is a complex place with currents of air and ocean water and exchanges of energy and constituents between them. In contemplating the nature of climate change, it is important to differentiate climate and weather. *Weather* refers to local conditions both in position and time. *Climate* refers to the overall average properties on Earth's surface. They are related, but they are not the same. The weather is subject to chaos,[6] formally known as sensitivity to initial conditions. The fundamental notion is that there are systems for which a tiny change in the conditions ultimately can lead to substantial changes later. This is related to the *butterfly effect* mentioned in chapter 9 wherein the metaphorical flapping of the wings of a butterfly in Brazil can trigger a tornado in Texas.[7] Formal studies of the effects of chaos on the weather show that it is impossible to predict the local weather more than about ten days in advance. Intense study and more powerful computers have extended that range, with the results being less and less accurate with time. In some instances, the overall trend in weather in a locality, hotter or colder, can be predicted with some accuracy for months in advance but the specific temperature or the likelihood of rain cannot be. Earlier in the twenty-first century there was a trend for TV weather reporters to deny or minimize the effects of climate change. That tendency has moderated as more people come to understand the science behind both weather and climate.

The average global warming that accompanies a buildup of carbon dioxide and methane in the atmosphere is complicated by increases in volatility of the weather and climate. Average heating can disrupt the flow of jet streams that normally trap frigid air in the Arctic regions. This disruption can allow that cold air to surge southward, leading to unexpected spells of cold weather as a result of overall heating. Heating can, of course, also lead to a rise in temperatures, drought, and forest fires.

The heating process can lead to something like exponential change where the rate of change is related to the accumulated amount of previous change. One example is the melting of tundra in Siberia. The frozen tundra is a substantial trap, sequestering carbon. When the tundra melts, as it is doing now, the frozen ground relaxes. Methane leaks into the atmosphere, increasing the rate of heating. Large craters in Siberia indicate that sometimes the methane ignites and explodes. The drying of soil and vegetation has led to massive fires in the Siberian boreal forest, or *taiga*, that in turn release massive amounts of carbon dioxide, thus enhancing warming trends. Analogous effects may happen in the tropics—without the explosions. A vast amount of carbon is trapped in soil where it feeds microbes and fungi. Some of that carbon is used for growth, and some is released into the atmosphere. Global warming may lead to more active microbes that increase the rate of injection of atmospheric carbon.[8] Deforestation and forest fires both release even more carbon dioxide.

As Earth warms, the oceans also respond. Warming oceans kill coral reefs and the teeming life that depends on them. The current that flows north and east across the Atlantic delivering relatively temperate weather has been interrupted, threatening colder winters in Europe. Ice melts, replacing surfaces that reflect sunlight with dark water that absorbs sunlight, thus increasing surface heating. The melting ice causes the ocean level to rise, threatening populations on coastlines around the world. The Arctic is heating twice as rapidly as other regions on the planet. Dramatic changes in Arctic ice can be seen in satellite movies taken by NASA of the polar regions.[9] The Arctic ice goes through natural cycles of forming in the northern winter and melting somewhat in the summer. In recent years, the forming has been less robust, the melting more extreme. The difference is palpable as the movie progresses. Scientists speak of a new Arctic as snow and ice are replaced by rain and open water. Greenland is losing ice that shears from vertical cliffs at the water's edge one hundred times more rapidly than outmoded estimates based only on melt water from deep in the glaciers.[10] In the Antarctic, coastal glaciers act as a dam, holding back a continent's worth of ice. Those glaciers are breaking off, threatening to allow the rest to slide into the ocean.

The climate of Earth is thus rife with feedback effects: cooling can lead to more cooling, and heating can lead to more heating. Given the human impact on the climate, climate scientists are focused on the latter. Their concern is that there is a tipping point, an average temperature resulting from myriad collective effects that will mean the changes are irreversible. According to current climate models, Earth is unlikely to suffer a runaway greenhouse effect until our Sun burns out its hydrogen fuel in the center and begins to expand in a few billion years. Before that, however, water could have evaporated from the surface into the atmosphere and led to a *moist greenhouse* effect. The result would still be the extinction of life of all kinds, including humans or whatever they have evolved to become.

Pay Me Now or Pay Me Later

Some people argue that climate change and global warming is a hoax. It is difficult to argue with such a position since evidence holds no sway. Others acknowledge that there is some warming but question whether humans contribute substantially. The evidence for that is slowly turning the tide. The deepest arguments concern cost: the cost of doing too little if climate change is real and the cost of doing too much if it is not. In 2020, the US Futures Trading Commission, which normally considers the regulation of sophisticated financial instruments that govern the price of corn, wheat, oil, and other commodities, issued a report on the threat of climate change to US financial markets.[11] The report concluded that the costs of wildfires, droughts, catastrophic storms, floods, and the loss of biodiversity will propagate through a variety of financial institutions, ultimately affecting everyone. Such studies undercut the arguments of people like William

Nordhaus, winner of the 2018 Nobel Prize in Economics, that industries that work indoors will be exempt from the effects of climate change.

To those who accept the reality of climate change, the cost is either mitigation and adaptation or suffering. Some aspects of suffering are rather obvious: fires and drought. As glaciers melt—Mt. Kilimanjaro is no longer snowcapped—and oceans rise, there will be devastation along coastlines around the world and on low-lying islands. Saltwater is already bubbling up in suburban neighborhoods of Miami. One can envisage, at the very least, a collapse of coastal real estate markets. For a long time, it was thought that trying to move coastal communities was too expensive and disruptive. People were encouraged to insure and rebuild. Consciousness has grown that rebuilding in flood-prone areas makes little sense. The Army Corps of Engineers has begun conversations with city officials that they will have to force people from their flood-prone homes or be denied federal funds for flood protection. In 2020, the Federal Emergency Management Agency (FEMA) began a new $500 million program to encourage and pay for large-scale relocation. The Department of Housing and Urban Development (HUD) began a related program with a projected cost of $16 billion.

Other ramifications of climate change are more indirect but no less destructive. An example is migration. An argument can be made that climate change contributed to the drought that drove farmers from the countryside into the cities of Syria. Disruption and primarily peaceful protests brought a violent crackdown by the government. That led to civil war involving not just Syrians but many countries in the Levant. The turmoil drove migration from the Middle East into the countries of Europe. The backlash triggered repression in some European countries and resistance in others. Out of such reactions arose illiberal, right-wing movements that threaten governments and even democracy itself. Future social reactions to climate change could be like that only more so. The flooding of Bangladesh is likely to bring migratory trauma throughout the Indian sub-continent. There is a variety of causes for migration from Africa to Europe and from Central and South America to the United States, but the threat of drought driven by climate change has the potential to severely exacerbate the situation in both areas of the world.

The US Department of Defense (DoD) perceives these changes as a threat to national security. In January 2016, the DoD issued Directive 4715.21 titled "Climate Change Adaptation and Resilience"[12] that assigned responsibilities throughout the DoD to assess and manage risks associated with a changing climate. In 2019, the DoD released a congressionally mandated report[13] assessing significant vulnerabilities to climate-related events in an attempt to "identify high risks to mission effectiveness on installations and to operations." Both of these reports were produced during an administration that was openly hostile to the notion of climate change.

The cost of mitigating and adapting to the effects of climate change is imposing. A principal one is the need to wean civilization from the widespread use of fossil fuels, the consumption of which contributes so much to the addition of carbon to the atmosphere. Such change means disrupting the global extraction industry, and the wealth that goes with it, and many of the industries that rely on fossil fuels, from heating and power to the automobile industry. Such a move would mean a substantial loss of associated jobs. Individuals and industries that gain by polluting will resist. As Upton Sinclair noted, "It is difficult to get a man to understand something when his salary depends upon his not understanding it." Optimists see new opportunities for industry and employment in green energy solutions including wind, solar, and nuclear power. Eventually we might replicate the thermonuclear power of the Sun itself by converting the hydrogen in seawater into helium.

In weighing the costs of mitigation versus suffering, it may be useful to consider the cost of mitigation as a form of insurance. No one knows if their house will burn down, but it is prudent—even mandated—to purchase fire insurance. The issue is what is the reasonable cost of insurance against the likelihood of climate change. An advantage of mitigation is that it holds the promise of a cleaner, healthier environment.

Some relief may come from an unexpected quarter: the insurance industry itself. Climate-related natural disasters cost money, and much of that comes from the pockets of insurance companies. The giant French insurance company AXA leads the Net-Zero Insurance Alliance, a consortium of large insurers and reinsurers (those who insure the insurance companies) that have committed to insuring only companies that promise net-zero greenhouse gas emissions by 2050. This is not an idle exercise. If insurance companies stopped insuring coal companies, the coal companies could not get financing. The whole coal industry might wither within a decade.[14]

The need to address climate change has introduced a new term: *managed retreat*.[15] The notion is that rather than an admission of defeat or a one-time step, adjustment to climate change should employ a managed, strategic approach that incorporates retreat from rising oceans and tinder-box forests with long-term societal development goals.

John P. Holdren, science advisor to President Obama, invoked the metaphor of driving a car with bad brakes toward a cliff in the fog.[16] It is prudent, Holdren maintained, to tap the brakes of human carbon production to avoid

a cliff-like, point-of-no-return tipping point even if the fog of scientific uncertainty prevents us from knowing exactly where the cliff is. The science
is getting better, the tipping point more assured and closer if not passed.[17]
Some signs are the Greenland ice sheet collapse; the West Antarctic ice
sheet collapse; the collapse of ocean circulation in the polar region of the
North Atlantic; coral reef die off; the sudden thawing of permafrost; and
the abrupt loss of ice in the Barents Sea.

It is a defining aspect of an issue like climate change that the ramifications will be manifest most severely in the future, but there is a growing
consciousness—especially among young people who perceive that they
will bear the brunt—that that future is now. Greta Thunberg, the young
Swedish woman who emerged as a strong proponent for addressing climate
change, captured this spirit at the 2019 UN Climate Action Summit when
she addressed the audience of her seniors who were placing hope for the
future on young people, saying, "How dare you!"

Another significant aspect of an issue like climate change is that it is
global. It affects large countries and small, developed countries and third
world, polluters and the polluted. Trying politically to engineer the international agreements to address the problems is a complex, unprecedented
task. People of color are significantly affected but have little current role in
attempts at solution.

The United States, China, and India are major contributors to the
production of carbon and global warming, and they have all taken steps
to address the problem. While still dependent on coal, China is a major
manufacturer and user of wind turbines. India is ahead of its goals to have
a significant fraction of its energy generation from renewable sources. The
American Clean Power Association has brought together conglomerates of
utility companies to engage in clean energy solutions.[18] Even oil companies
like Royal Dutch Shell are considering plans to expand their remit to include cleaner energy sources.

New Techniques

Mitigating climate change is likely to involve new technology. In some
cases, wind and solar power have become economically competitive with
fossil fuels even setting aside the overriding issue of the contributions of
fossil fuels to climate change. The technology for converting solar energy
to electrical power has gotten both more sophisticated and cheaper. The
cost of solar energy production is now less than that of natural gas. Gigantic
wind turbines are sprouting up around the world. Having the advantage

of controlling its own internal electrical grid, the state of Texas initiated a substantial effort in wind production of electricity.[19] Issues of global warming have driven automobile industries worldwide to shift production to all-electric vehicles with the underlying presumption that the electricity will come from sources that contribute little to the warming. General Motors and Volkswagen have pledged to produce only electric vehicles within a decade or two.

There is a side effect of this move to electric vehicles. What are we going to do with all the batteries? I have been asking myself this question for well over a decade and despairing at the lack of an answer in the literature I come across. Piling batteries up in landfills is not a sustainable solution. Fengqi You of Cornell University remarked, "What to do with all these retired electric vehicle batteries is going to be a huge issue. There's very little discussion right now about the environmental dimensions of improving battery design for recycling or reuse." You and colleagues proposed a procedure to reuse vehicle lithium batteries to store wind and solar energy, reducing the carbon footprint by 17 percent.[20] Then the battery goes into the landfill. There need to be ways to extract and reuse the raw materials in the batteries. Little is in prospect now.

There are other ways to produce energy without making greenhouse gases. In 2022, the Department of Energy announced a new Enhanced Geothermal Shot program to reduce the cost of geothermal energy by 90 percent by 2035.[21] Fuel cells that chemically combine hydrogen and oxygen and exhaust water have been in use for some time and may play a more substantial role in the future to power both land vehicles and even aircraft.[22] Miniature versions of fuel cells have been designed that make energy from sugar (glucose). Such devices might be used to power implants.[23] Techniques for producing energy as our Sun does by the thermonuclear fusion of hydrogen into helium have seemed a decade away for decades. In fact, by a technical measure of net energy production, progress has been steadily made at the same rate as Moore's law for the growth of transistor density and capacity. There is progress both to compress hydrogen with bolts of laser energy and to confine hot hydrogen plasma in a magnetic torus.[24]

The National Ignition Facility (NIF) in Livermore, California, uses the world's largest laser to compress pellets of deuterium—known as *heavy hydrogen* because it has a neutron in addition to the single proton that defines a hydrogen atom—in a process called *inertial confinement fusion*. In 2021, the NIF achieved *burning plasma*, a condition in which thermonuclear reactions converting deuterium into helium produced more energy than did the laser pulse. In 2022, the NIF announced they had achieved a state in which the thermonuclear burning produced more energy than both the laser and multitudinous loss mechanisms.[25] The goal of producing more energy than that of the total laser pulse is close. The goal of commercial viability remains in the future but is ever closer.

Some private companies think they can produce toroidal magnetic confinement of a thermonuclear burning plasma faster and more efficiently than large international research consortia by employing the new technology of high temperature superconducting magnets that can be cooled by tractable liquid nitrogen rather than exotic and demanding liquid helium.[26] One prominent initiative is Commonwealth Fusion Systems, an MIT spinoff led by cofounder and CEO Bob Mumgaard.[27] Efforts are underway to ensure community buy in and environmental justice as this new technology is developed and employed on a large scale.[28] In September 2022, the Department of Energy announced a program to enhance the US fusion energy industry by initiating a program based on milestones similar to the effective NASA program that developed commercial resupply and crew transportation services for the International Space Station (see chapter 14).[29]

Another approach to the threat of climate change is to remove the greenhouse gases that have been produced and introduced into the atmosphere. There are techniques for removing carbon from the exhaust of fossil fuel power plants. Experiments have shown some success in using carbon scrubbers to remove carbon from the atmosphere although it is not yet clear that this technology can be made economical and operated at a sufficiently large scale to affect the atmosphere globally. At the UN Climate Summit in Glasgow in 2021, the Department of Energy announced a major new effort to develop the technology of carbon renewal and to make it more economical.[30] One idea being explored is to dissolve carbon dioxide in water in a manner similar to making fizzy soft drinks. Instead of being sold, the carbonated water is frozen, thus trapping the carbon dioxide. A critical third step is then to drive that frozen water underground where the carbon dioxide will permanently, chemically bond with basalt rock.

There are also natural solutions to carbon removal. Many of these focus on trees, Nature's carbon scrubbers. Appropriate steps are to halt deforestation and promote *afforestation* (the creation of forests where there were none before) or *proforestation* (protecting current forests and promoting them as intact complex ecosystems that maximize carbon retention). An important starting point would be to protect big trees such as sequoias. The 30x30 Initiative aims to have preserved 30 percent of the world's land and oceans in a pristine state by 2030.

Other studies have outlined the steps needed to address severe changes associated with the warming planet. While the target remains somewhat uncertain, climate models show that to avoid the type of catastrophic collapse envisaged by Bendell's "Deep Adaptation" paper, the average global temperature must be kept to less than 2 degrees Celsius compared to conditions at the start of the Industrial Revolution. The planet is already halfway there, having warmed about 1 degree Celsius since 1850. In 2020, climate scientists at Princeton released a study titled "Net-Zero America"[31] that outlined the steps needed for the United States, individual states, and even particular counties to de-carbonize the economy by 2050. The report proposed extensive but potentially affordable changes. In 2021, the National Academies released a complementary study titled "Accelerating Decarbonization in the US Energy System"[32] that also presented steps for the United States to achieve net-zero carbon emissions by 2050. The recommended steps were projected also to make the economy more competitive, to add jobs, and to reduce social injustice in the energy system. The development of new technologies may bring new jobs and expanded employment, but the transition from our current system that is so dependent on fossil fuels to one that is not will be long, complicated, contentious, and difficult.

MALTHUS AND LEBENSRAUM

It will perhaps not escape the reader's attention that when we are contemplating climate change, a key component is humans. This raises another contentious issue: population. People beget people, and the rate of production of people is roughly proportional to the extant number of people. As for other quantities discussed in this book, for a long time, the number of people rose roughly exponentially (albeit with interruptions like the Black Death).

Estimates for the deep past are highly uncertain, but there are guesses that there were already 1 to 10 million people 10,000 years ago. Reasonably accurate estimates of population exist for only the last few hundred years. There were about 1 billion people by 1800. The time to double the population shrank from about 300 years in 1800 to about 50 years in 1975, showing a growth somewhat more rapid than exponential. Since then, the growth has slowed to a doubling time of roughly 100 years. The current population of Earth is about 7.9 billion and growing at about 2 people per second.[33] Various estimates suggest a slow rise, even a peaking of the population going forward. Picking a doubling time of 100 years as a possible example, 1,000 years from now, Earth will host 32 trillion people. That likely cannot hold. Departures from steady exponential growth, however, bring other issues.

There are advantages to the growth in population. Every human has an intrinsic value. One never knows where in a burgeoning population a special person might arise who will change the life of all for the better so the more people, the better the chance. A fundamental premise of much of economic theory is that growth is healthy, even demanded. Economic growth arises from growth in productivity, but traditionally growth has been positively associated with demographics—growth in population.

There are also problems associated with too many people. Human-induced carbon production with associated climate change is one. As population density grows, the urban-natural boundary expands, leading to deforestation. More people live in forests and closer to wildlife, bringing threats of fire, viruses, pathogens, and pandemics. Too many people in a given region can lead to starvation and population decline.

One of the first people to think deeply about the nature and implication of population growth was the English cleric and economist Thomas Malthus. In his book *An Essay on the Principle of Population*, published in 1798, Malthus started a conversation that continues today. Malthus proposed that human society was doomed to boom-and-bust cycles. When conditions were good, the population would grow until people overwhelmed resources. That would lead to a shrinking of the population through war, famine, and disease until resources were adequate. Rather than stabilizing, the population would then tend to overgrow again, repeating the cycle. The inevitable crash when population exceeded resources foreseen by Malthus became known as a *Malthusian catastrophe*. Malthusian cycles are well known to scientists who study animal population dynamics. Species will tend to overuse the resources available to them, leading to overpopulation and a crash. The crash is followed by a typically slow recovery that leads once again to the overuse of resources. Animal populations rarely return to their original peak. Are humans immune to such dynamics?

The leadership in Germany in the 1930s could not foresee that technological solutions would allow the feeding of a growing population. They concluded that the solution must involve more room in which to live. They called this *lebensraum*. Seeking lebensraum was part of the motivation for Germany to invade the rest of Europe. In this sense, one could argue that the catastrophe of World War II was sparked by a drive to avoid a Malthusian catastrophe. Buckminster Fuller argued that we have far more energy and resources than we need but that we use them inefficiently.[34] He pictured the battle between capitalism and communism as a struggle over resources as if we did not have enough—effectively a global lebensraum argument.

Malthusian notions were later brought to public attention by Stanford biologist Paul Ehrlich who in 1968 published a popular book, *The Population Bomb*, co-authored with his spouse, Ann Howard.[35] Ehrlich foresaw problems with famine and disease and warned of issues with climate change. He predicted that hundreds of millions of people would starve to death in the 1970s because of overpopulation. Ehrlich was wrong in that prediction, but the issue remains as to whether he—and Malthus—was wrong in general or just early. Financial expert and online commentator John Mauldin argues that Malthus and Ehrlich were not just wrong but that Malthus was "the most dangerous economist to ever live, more so even than Karl Marx."[36] Others think the Malthusian experiment is still underway, with climate change a warning sign.

Some feel that each human life is sacred and that any suggestion of artificially limiting populations is anathema. Another view is that technological solutions will arise to forestall Malthusian catastrophe. It was the latter that immediately tripped up Ehrlich. There was a threat of food shortage in the 1960s when Ehrlich wrote. Even as he published, however, the so-called green revolution in agriculture was underway. The green revolution involved new methods of farming, use of fertilizers, and especially the development of new, high-yield species like dwarf wheat. Today the world has plenty of food. Any issues of malnutrition are related to politics and the distribution of goods. A trillion people might change that story.

These considerations lead to questions of how many people Earth can sustain. The United States has a density of 90 people per square mile; India, about 1,000 per square mile. Individual cites are more densely packed. Karachi in Pakistan and Mumbai in India have about 60,000 people per square mile. Manila has over 110,000. If all the land surface of Earth were packed at the density of Macau and Monaco with about 50,000 people per square mile, that would correspond to about 10 trillion people. Macau and Monaco are currently rich and thriving places, but it is difficult to believe they would be so if the entire land surface of Earth were packed at the same density with no room for forests and farms. Each of us can ask whether we would want to live in a place with the population density of Manila or even greater. If that seems a distasteful prospect, how do we avoid it without some limits on population?

Perhaps we could expand to occupy the surface of the ocean, thus increasing the habitable area of Earth by a factor of about 2 and allowing 20 trillion people on the planet. As noted above, if the population doubled every 100 years, we would hit that limit in less than 1,000 years. Perhaps technology would arise that allows us to build up on the land surface or down under the oceans. If we could sustain more people that way, would we want to? Could we do so without irreparably damaging the oceans that are already under threat from pollution and dangerous warming?

We will eventually populate Mars, but any migration will be slow compared to historical rates of growth of the Earth's population. If similar rates of population growth applied to Mars, we would soon fill that planet as well. If we have a population problem on Earth, migrating to Mars is not the solution. Earth is a finite home. Surely it can successfully host only a finite number of people.

If we are growing toward a Malthusian catastrophe, we have been doing so with some success. Global poverty and illiteracy have declined in the last few decades. In 1953, 35 percent of the world's population lived in extreme poverty, making less than the equivalent of $1 per day. By 2015, that fraction was only 14 percent.[37] More women are now educated, upward of 60 percent even in South Asia and sub-Saharan Africa, where rates are lowest. The death rate of children is down. This good news is not universal in either time or place. The COVID-19 pandemic of 2019 set back much of the positive news. In 2020, extreme poverty increased globally for the first time in two decades.

If there is to be a goal of limiting the population of Earth to some finite value, what should that number be? What is the population that Earth can successfully sustain? The answer is perhaps not much more than the current number and maybe even less. An informal poll of the students in my class came up with a number of only 3 billion, a number last seen in about 1950 and far below the current population. While estimates range widely, other more formal studies have given similar numbers. In contrast to Buckminster Fuller's 1981 claim of ample resources, the Global Footprint Network[38] estimates that humans already use the equivalent of 60 percent more of Earth's ecological resources than the planet can supply, more in developed countries, less in underdeveloped countries. No country on the planet currently uses resources at less than the rate at which supplies can be replaced.

People in developed countries currently use far more resources per capita than does the average human on the planet. If we were to hold Earth's population roughly constant but bring everyone up to the level of quality of life enjoyed in the United States and Europe, the demand on already taxed natural resources would be strained even more. While energy consumption is not the only measure of resource demand, it is representative. If the whole world were to consume energy at the rate per capita of the United States, the world's energy production would have to increase fivefold. Doing this with fossil fuels would have a large, negative impact on climate change. Whether such a demand could be met with non-polluting renewable energy remains to be seen.

How Many Is Enough?

To defend ourselves from ourselves, we thus may need to limit, even shrink, our populations. To bring things into balance, we will need to consume less, produce more without harming the environment, or limit the population. We could, and probably should, consume less meat and use less plastic, but it is not clear that will solve the ultimate problem. There are speculations that we could engineer our bodies to be more efficient and need less food. In the absence of such a technological solution to change us, there are ways to produce more food. The green revolution showed that food production can be made more efficient. There are current efforts to grow produce in vertical farms that require less acreage and water. There are many initiatives to manufacture synthetic meat and even to grow meat from molecular scratch without needing animals as intermediaries.

Another approach is to face the population issue head on. Given some evidence that we have already overpopulated Earth and are making unsupportable demands on its resources, one can ask how the population could be shrunk and what would be the implications of that decrease? How would we get to about 3 billion people if that were the goal? How do we achieve a state where wealth and opportunity are spread uniformly in a stable manner that avoids the Malthusian catastrophe?

The problem is that if one considers the means to limit, never mind reduce, the global population, the solutions tend to be traumatic. Wars, disease, and starvation contribute.[39] So do policies to restrict the birthrate. China foresaw problems supporting its gigantic population and enforced a one-child policy. China also educated women and gave them opportunities. There became less impetus to have more children because it was too expensive. There were more draconian outcomes. Illegal second children were regarded as non-persons with no government papers and thus were excluded from ordinary commerce. There was a tendency to seek male children and even to abort females or resort to female infanticide. The result is now a generation of single males who might not be able to single-handedly support their aging parents as has been typical of Chinese culture. There are now about 120 males for every 100 females. Men are having trouble finding mates, and there are reports of two men marrying one woman.

One aspect of the modern world that has the potential to contribute unhappily to population decline is the spread of *endocrine disruptors*, chemicals that can alter cell development. These are often found in plastics, shampoo, cosmetics, and pesticides. Endocrine disruptors may be associated with a global decline in sperm count and even weird outcomes like two-headed sperm.

Whether a declining birthrate should be viewed as a positive development or a negative one, it seems to be happening spontaneously. There is evidence that the rate of growth of the global population has slowed compared to that of the last fifty years. In 1950, no countries had a birthrate so low that its population could not be supported. Birthrates have since decreased by about a factor of 2, and now half of all countries are experiencing a baby bust with shrinking and aging populations. Studies in 2017 and 2020 by the Global Burden of Disease collaboration[40] predicted that the global population would peak at about 9.7 billion around 2064 and fall to the range 7 to 12 billion by 2100. Spain, Portugal, Italy, Japan, Thailand, South Korea, China, and other countries were projected to have their populations shrink by a factor of 2 by 2100. Some people may welcome this as a relief to strained resources and lessened threat of catastrophic climate change; others may regard a decline in population growth as a disaster unto itself. In this view, a growing population of young people who produce new ideas is critical for a healthy economy.

The fall in birthrates seems not to be due only to war, pestilence, or a decline in sperm count but also to demographic and sociological shifts. The fact that the childhood death rate has declined substantially means that pressure for large families has declined, and women have fewer children. Women also have greater access to contraception. A dominant factor is that more women have gotten educations and sought work. As illustrated in China and throughout the developed world, one of the gentlest and most effective ways to reduce the birthrate is to educate women. That has worked in most developed and many developing countries. It is not yet universal.

During the period of transition in which we seem to be living, populations will probably continue to grow in some parts of the world while shrinking in others. The United States is predicted to double to about 570 million by 2100. It is conceivable that some regions of the world will be faced with traumatic sociological adjustments to declining populations while others will be subject to Malthusian disasters, especially those beset by droughts and floods induced by climate change. It is a sad irony that the developed world with stabilizing populations is driving the climate change that will most severely impact the parts of the world still growing their populations.

The population growth rate in Africa has doubled since 2000 and is projected to double again by 2050, representing about two-thirds of the world's population growth. Half the world's population may be African by 2100. To some people, merely contemplating that future with the implication that something should be done about it is tantamount to racism, even genocide. Limiting the birthrate anywhere to promote the quality of life is not intrinsically racist. Instead, the projections can be accepted as possible, even plausible, and efforts expended to bring the quality of life upward for those Africans. The goal is balance, with all Africans sharing our quality of life. Educate African women, give them economic security, and let Nature take its course as it has in developed countries around the world. There is nothing racist in that.

A transition to declining populations, whether by Nature or design, will bring challenges. The National Intelligence Council is charged with producing a Global Trends Report every four years. The 2021 report addresses trends in demographics, human development, environment, economics, and technology over the next twenty years. The report foresees shrinking and aging populations. A decline in confidence that governments and institutions will be willing and able to address peoples' needs may lead to growing social schisms.

Impacts

As populations decline, the balance of old and young people will shift. For the first time in history, there are now more people over sixty-five than under five. Shrinking workforces will cause economic disruption. An aging population will put more burdens on mechanisms of health and social support. Dramatic declines in population in India and China would lead not only to economic disruption but also to a shifting of the centers of global power. Immigration could limit the effects of declining working-age populations, but it is difficult to see Japan or China embracing a sufficiently large immigration from Africa to address that decline. Christopher Murray, professor of health metrics science at the University of Washington who led the Global Burden of Disease collaboration pointed out that the problem of an inverted age pyramid—more on top, fewer on the bottom—is a serious issue for how societies are organized, how economies work, and how taxes get paid. He said, "What we really need to figure out is how to transition from the state we're in now [to steady or declining population]." Jane Goodall advocates a process of *voluntary population optimization.*

Growth is the positive mantra of classical economics. Economic growth is traditionally strongly linked to population growth. Economists may need to revise their basic conceptual framework to characterize the positive aspects of a steady or shrinking population rather than regarding that as an unmitigated disaster.[41]

Perhaps greater formal cost needs to be assigned to pollution of the commons. The British economist William Forster Lloyd introduced the notion of *the tragedy of the commons* in 1833. Lloyd hypothesized a situation where grazing on shared or common land would benefit individual owners of cattle. He projected that in the absence of regulation, a significant number of owners acting in their own self-interest would lead to overgrazing and by doing so harm everyone. In modern terms, the commons would be our atmosphere, oceans, rivers, and forests that are subject to pollution with little direct cost to the polluters. Elinor Ostrom won the 2009 Nobel Prize in Economics by showing how communities with regulated access to shared resources can cooperate to exploit resources without tragic overuse. There are more Nobel Prizes to be won for economists who show how to thrive in a world of constant or shrinking populations.

It is relevant to ask what would happen if Earth's population settled into a steady value at about the number we have today. What economic and social constructs would change with a constant population? The goal need not be the quantity of people but the quality of life for individuals and societies as measured by options in life choices, satisfaction, and contentment. Rather than total economic growth, the economic growth per capita may be more relevant. Countries with shrinking populations and numbers of workers like Germany and Japan are still relatively content because their economic growth per capita is still positive. In India, people tend to use meager resources to purchase cell phones rather than toilets and sanitation that could to lead to healthier lives. Can we design and engineer a situation where they have both without more people?

There is a school of economics that addresses these issues. Romanian economist Nicholas Georgescu-Roegen trained at the Sorbonne and later worked at Harvard and Vanderbilt. He is regarded as one of the founders of a study known as *ecological economics*. In his masterwork *The Entropy Law and the Economic Process*,[42] he argued that natural resources are finite and steadily degraded when employed in economic activity. His successors include Henry Daly, emeritus professor at the University of Maryland School of Public Policy, who has explored steady-state economics; Tim Jackson, director of the Centre for the Understanding of Sustainable Prosperity at the University of Surrey;[43] Kate Raworth, senior associate at Oxford, who works on the balance between human needs and planetary limitations; Giorgos Kalas, research professor at Universidad Autónoma de Barcelona, who advocates a theory of *degrowth* to reduce *social metabolism*; and Dietrich Vollrath, chair of the Department of Economics at the University of Houston, who has argued that a stagnant economy is a sign of success. The married couple Esther Duflo and Abijit Banerjee, professors of economics at MIT, shared the 2019 Nobel Prize for Economics for their related experimental work on alleviating global poverty.

A related branch of ecological thinking is the *rights of Nature movement*. Advocates argue that rather than thinking Nature (here and on the asteroids and on Mars) as property to be owned, we should allot Nature inalienable rights to be included in humanity's moral structure. In keeping with this philosophy, Bangladesh has given legal rights to all the country's many rivers. Ecuador, New Zealand, India, and Colombia have granted rights to rivers and forests. Ohio gave Lake Erie the legal right to "exist, flourish, and naturally evolve" and then gave Ohio citizens the right to sue on the lake's behalf were it to be polluted. The Amazon rain forest has rights, if often abused and ignored. This nascent movement faces difficult questions of the rights of people versus the rights of Nature and perhaps even the fundamental nature of capitalism and the basic Western reductionist mode of thinking of Nature as a separate thing to be studied.

A possible solution to the resources of a finite planet is to expand into space,[44] the topic of chapter 14. There are abundant materials there though they will be difficult to access and extract. Robert Ayres, currently at the International Institute for Applied Systems Analysis in Laxenburg, Austria, has proposed the development of a "spaceship economy" on Earth analogous to a well-provisioned, isolated ocean craft or spaceship. This notion might be expanded to include the resources of the entire solar system, but of course these are also finite because in 5 billion years our Sun will die. If we cannot beat the speed of light, we will always be restricted to a finite reservoir of resources although if this reservoir is sufficiently large, a growth-oriented economist can always continue to kick the consumption can down the finite-resource road.

Solving the Aging Problem

Yet another factor germane to the issue of population is garnering attention. This aspect again comes from the realm of biology, our deepening ability to know and engineer ourselves. Do we consider aging a disease and attempt to treat, cure, even eradicate it? There is an active community of researchers and advocates who are seeking to do so. One of the principal advocates is Ray Kurzweil himself, who promotes a strategy of living long enough to live forever. The idea is that with an exponential growth in medical and biological knowledge, people alive today may live long enough for the problem of aging to be solved.

The prospect of curing aging is real enough in principle. We are designed by natural selection to live long enough to reproduce, to pass along our genes. Already modern medicine has allowed us to live decades longer than our reproductive shelf life. Our selfish genes no longer have a need for us after we have adequately reproduced so evolution has not designed a system to keep our cells fresh and our *telomeres* capped.

Telomeres are non-coding bits of DNA comprising repeating sequences of nucleotides at the end of our chromosomes. In humans, the telomeres are thousands of base pairs long. They serve somewhat like the plastic tips of shoelaces to keep chromosomes from fraying or from tangling with one another. The telomeres prevent the DNA repair mechanisms in the body from interpreting the tips of chromosomes as double-strand breaks and attempting to fix them. The telomeres thus protect the genetic information in our chromosomes and hence ensure against cell malfunction, disease, and death. In a sense, however, the telomeres are like a fuse burning down. Each time a cell divides, its telomeres become shorter. Eventually, telomeres become too short to function properly by allowing healthy cell division. When cells are unable to divide properly, they become inactive, die, or continue to divide in an abnormal way that can lead to dangerous diseases like cancer and Alzheimer's. It is not the fall that kills you; it's the landing. Likewise, it may not be aging per se that gets you, but the shortening of your telomeres.

While the length of telomeres may only be a symptom of aging—like gray hair or wrinkles—rather than a direct cause of death, telomeres are a focus of study in the anti-aging community. Another focus is on *telomerase*, an enzyme that acts to lengthen telomeres and hence keeps telomeres functioning properly. With continuing cell division, telomerase levels become depleted. There is thus some thinking that if means were found to increase the production of telomerase in our bodies as we age, our telomeres would remain long and healthy, and aging and death could be postponed, perhaps for a very long time. It is also possible that longer telomeres and increased telomerase could induce cancer.[45] More research is needed.

In the meantime, advocates of hanging in there until a robust cure for aging is found suggest we should maintain our telomerase levels by exercising regularly, eating a diet high in antioxidants, and perhaps mindfully meditating. It is also recommended to decrease stress. Best practices would suggest avoiding being born into poverty.

Another avenue to reset the biological clock is to return aging cells to vigor by converting them to stem cells. David Sinclair of the Paul F. Glenn Center for the Biology of Aging Research at Harvard Medical School and colleagues have reprogrammed gene expression in mice. The mature cells became stem cells that could subsequently grow into fresh, healthy cells.[46] Stem cell biologist Jacob Hanna and his team at the Weizmann Institute of Science in Rehovot, Israel, created fully synthetic mouse embryos from stem cells without eggs, sperm, or a uterus.[47] A related experiment created mouse embryos from stem cells that lived for 8.5 days (one-half a mouse pregnancy) and developed a yolk sac, digestive tracts, rudiments of a central nervous system, beating hearts, and brains with well-defined subsections.[48]

The lure of special waters from a fountain of youth or some other magic elixir to preserve or restore youth runs throughout the history of humanity. How about a magic protein? Researchers at the Harvard Stem Cell Institute surgically joined a young and an old mouse so they could share blood. The heart of the old mouse was rejuvenated. Regrettably, the young mouse aged more rapidly. The team, led by Lee Rubin, codirector of the neuroscience program at the Stem Cell Institute, determined that the special ingredient in the blood that helped the older mouse was a protein called GDF11.[49] Rubin and his team wondered if the same ingredient (without traumatic surgery and sacrifice) would work in humans. In 2022, they founded the company Elevian to develop therapies to slow or even reverse aging with GDF11.

In an especially startling revelation, a team led by Nenad Sestan, professor of neuroscience, comparative medicine, genetics, and psychiatry at the Yale School of Medicine revivified a pig that had been dead for an hour by infusing it with a special solution called OrganEx.[50] The pig did not get up and dance around, but it showed evidence of brain and cell activity. This development leaves us a long way from applications to humans, but one cannot help but reflect on the immortal words from the 1931 film *Frankenstein*[51]—"It's alive! It's alive!"

The lure of a long life is strong. I'm very curious myself to see where exponential change takes us in the future. Advocates of curing aging, however, tend to avoid a central issue. What will we do with all the people?

A presumption is that long-lived people will be healthy and robust and hence can continue to contribute to the economy, society, and culture rather than being a burden. Granting that aspiration, if we recognize that too many people challenge our ecosystem, what do we do about babies?

Do we regulate births or even prevent people from having them? What will young people do for jobs if the old people won't retire and go away? Will old people become paranoid, living in careful isolation because only an accident can kill them?

What about sociological changes? The reality of death is woven into our cultures, our religions. Death promotes the concept of renewal. Futurist Peter Diamandis advocates abandoning acceptance of death as an inevitable part of the human condition but instead adopting a "longevity mindset." He presents an optimistic case for extending the "healthspan" of humans and acknowledges the sociological disruption that would have to occur to accommodate such a change.[52] He does not address how we would manage this change. Enabling longer lives with advanced biological technology would almost surely begin among wealthy people in developed countries. That alone is likely to lead to social turmoil.

Of course, just because it is complicated does not mean it will not happen. Modern medicine has extended our expected lifetimes far beyond the use-by date wired in by evolution. If the capacity to extend lifetimes further—even preventing death—proves possible, then it will be done. There will, however, be ramifications.

There is a Venn diagram that encompasses three attitudes, or mindsets: (1) Malthus and Ehrlich are somewhere between misguided and dangerous, even evil; (2) the disease of aging should be solved so everyone can live forever; and (3) the prospect of reducing or even of limiting populations is a threat. The only way to hold these points of view simultaneously, as some do, is to embrace unlimited growth of population. That is not possible on the finite surface of this planet.

The estimated timeframe within which climate change will cause irreparable damage or solutions will be put into place is several decades. By coincidence, that is also about the time it will take for the effects of artificial intelligence to go from interesting to ubiquitous and overwhelming—the onset of the singularity. Several decades is also the timeframe within which exponential progress in neuroscience, medicine, and biology may alter the course of human evolution. May your life, and that of your grandchildren, be interesting.

12

ECONOMICS

How Does Capitalism Work in a Hyper-Automated World?

THE INVISIBLE HAND

As we move into our new world of exponentially rapid progress in robots, AI, neuroscience, medicine, and biology, there are likely to be massive sociological changes. Adam Smith said, "It is not from the benevolence of the butcher, the brewer, or the baker that we expect our dinner, but from their regard to their own interest." In our technological future, who or what will be the butcher, brewer, or baker that serves their own interest but through that self-interest might enrich us all? One way to gain perspective on this question is through the lens of economics. How is wealth generated through extracting raw materials, manufacturing things, or intellectual property, and how will that wealth be distributed through society after the singularity?

Economics is the study of how people determine the value of everything from dirt to data. More specifically, economics concerns the production, distribution, and consumption of goods and services. The current economic system in much of the world is a form of liberal capitalism. This system has complex roots in the Enlightenment but it flowered in the eighteenth century with *liberal* implying an alternative to hereditary aristocracy, feudalism, and dictatorship. Fundamental principles are private ownership of assets and free markets that minimize the use of tariffs and subsidies by which mercantile nations seek power over rivals. The roots involved an unfettered laissez-faire that was gradually moderated by combining the power of free enterprise with appropriate regulation and levels of social support. Other experiments were done. "From each according to his ability to each according to his need" had a simplistic appeal but top-down, command-and-control communism by and large failed. China is challenging Western liberal capitalism with an autocratic capitalism, the ultimate success of which remains to be seen.

Liberal capitalism has been remarkably successful with Adam Smith's *invisible hand* globally raising the quality of life, but it has its rough edges. There is a constant struggle to find the proper balance between the power of capital and the rights of labor, between winner-take-all capitalism, appropriate regulation, and the needed level of support for those who slip through the cracks. Our burgeoning digital world is putting liberal capitalism to yet another test.[1]

Talent and drive—qualities needed to thrive in a capitalist system—are distributed in roughly a bell-shaped curve: most of us are in the middle, a few are short-changed, and another few are especially gifted. In a capitalist system, the results are distributed in a very different way. Those who won the talent-and-drive lottery end up with a disproportionate share of the rewards. Jaron Lanier calls this the "star" system, or winner-take-all.[2] There are many garage bands but only a few end up selling millions of records, dominating streaming sites, and filling stadiums on concert tours. There have been many aspiring high school basketball players but only one LeBron James in his era. There have been many startup dreams of digital riches but Amazon, Apple, Google, Microsoft, and Facebook sit atop immense wealth and wield correspondingly immense power reminiscent of old feudal systems.

There is nothing intrinsically unfair about rewarding success, but if too many people are left behind, social disruption results that can harm even those at the top. The Roaring Twenties of the early twentieth century led to the crash and the Great Depression, with even formerly wealthy bankers jumping from windows. A century later, similar dislocations are occurring. In his book *Capital and Ideology*,[3] French economist Thomas Piketty summarized 250 years of the distribution of income and wealth, including that in the recent era. Over the last several decades, the tenets of free trade led to the export of jobs to low-wage countries and record levels of wealth inequality with CEOs making on average about three hundred times more than the average employee salary—a recent record—and some much more.[4] At the risk of vastly oversimplifying, this led to recessions, great and small, disgruntlement, and a turn to autocrats in parts of the Western world.

A healthy society is one in which the maximum number of people have the maximum number of options. A basic assumption of economics is that a healthy society demands a healthy economy. So far, so good. A fundamental tenet of economics is that a healthy economy requires growth, not so much as to fire up inflation but sufficient to avoid deflation, recession, and depression. There are arguments over the best ways to measure economic growth, but traditionally one speaks in terms of *gross domestic product* (GDP). A typical target to avoid the Scylla of inflation and the Charybdis of depression is a GDP growth rate of about 2 percent per year.

If we are thinking far into the future, as we are attempting to do, there is a deep issue right there. Recall from chapter 2 that a fixed percentage increase per year is equivalent to exponential growth, with the only formal limit being infinity. Given that we cannot really expect an infinite GDP, there must be some limit to this economic scheme. Again traditionally, growth in a quantity such as GDP requires growth in the combined product of the productivity per person and the number of people. We cannot and should not depend on the population of our planet to grow without limit or even to be much greater than Earth supports now. Climate change suggests we should have fewer people. That would suggest that inasmuch as economic growth continues, it must depend on the increased productivity of each individual. For a long time the increased sophistication of manufacturing provided that increased productivity. These days much of the economy depends on consumption and services that are not as amenable to growth in productivity per person. In addition, technology in the form of AI and robots will first tend to displace people in service, gig, and other low-income jobs. That threatens the notion that a healthy economy should benefit all members of society.

Within the loose bounds of liberal capitalism, there are three big issues: the future of capital, the future of labor, and the future of money. An immediate central question that arises is how to maintain a strong middle class in an exponentially developing, digitally dominated world.

THE FUTURE OF CAPITAL

Let us first look at the status of capital. After the development of the internet, a dominant winner-take-all system developed. This outcome was not inevitable, but in the absence of an alternative structure, it was a natural development. A principal underlying driver was again the power of exponential growth. A company that got off to a good start, even by a small margin of time, could exploit the new opportunities of the internet and grow exponentially. Success would breed success, attracting more customers and more investment that would allow more success. A competitor that started a little later would find it very difficult to catch up. There are many other factors including better ideas, technical prowess, ambition, and aggression, but with the opportunity of exponential growth, being first out of the slot helped a lot to produce a few overwhelming stars in the new digital world. That world may, however, be unstable and not the best environment for a content and productive middle class.

Early in the development of the digital world, there was a strong dominance in some quarters of the ethos that "information wants to be free."[5] That has a certain seductive, democratic-sounding appeal but not if you are a member of the creative class—an artist, writer, or musician—producing that information. People recording concerts, plays, or movies on their cell phones even for their later private use, never mind selling those recordings, violate the creative rights of the artists involved. Sharing that information for free directly undermines the aspirations of those creators to join the middle class if not to become its stars. This trend also pervades the digital world in the form of piracy of books that—while not necessarily actively promoted by the stars of the digital world like Google and Amazon—is not robustly prevented. Even in the absence of outright piracy, streaming creative product has tended to reward its creators less than previously under a system of physical musical records and tapes and traditional publishers and bookstores. More broadly, information may not want to be free if that information is personal data that are blatantly or secretly collected, massaged, and exploited, and your only reward is convenience. Those data are wealth, and a star system is an inadequate system to appropriately distribute that wealth and maintain a healthy society.

When Tim Berners-Lee developed the standards for locating and linking information online that led to the World Wide Web in 1989 and the fabulous wealth that devolved from that, there were basically no digitally driven businesses in the mode we see today. There were computers and software—IBM, Apple, and Microsoft—but the ability to link them all via the internet was a qualitatively new aspect. Businesses had to be developed from scratch as people recognized the potential and rushed to exploit it. Some, like Apple, built new hardware. Others, like Amazon, sold stuff. Yet others—Google and Facebook—dealt in the data, the information, and access to it. Things might have been different and might yet change, but the structure we have now is what it is. The original intent of the internet was that it would allow everyone to communicate with everyone, yielding a flourishing of creativity, democratic exchange of ideas, and commerce. Besides vast new wealth, we learned a lot about people who gain an unfettered ability to communicate.

In hindsight, there was one factor glaringly absent in the early *Star Trek* universe: private enterprise. Gene Roddenberry envisaged ambient computing. Captain Kirk could look vaguely off into the command deck of the Enterprise and say, "Computer" then request some information or issue a command to the ever-aware and ever-responsive Enterprise computer. There was no hint that every word spoken by Captain Kirk and the rest of the crew, all their facial expressions, all their body language, and the activity of every bit of machinery on the Enterprise was recorded and beamed back to Earth to companies that declared ownership of all that data to be sold to other companies in order to manipulate captain and crew into purchasing certain widgets. That, however, is the direction in which we are currently heading.

The spectacular growth of the World Wide Web and the companies that dominate it today was not an accident. The companies worked hard to establish their business and to grow. Businesses that got a head start on the World Wide Web took advantage of the exponential growth in opportunity by invoking the *network effect*. The idea is to establish and control a network of users of a specific enterprise. Once you're established, other people flock to the network because others are already there. It is thus critical to be first, to establish that beachhead of users so that other users will sign on because the first are there. Facebook is a classic example. Once Facebook had a decent collection of users, others turned to Facebook because that was where their friends and family were already exchanging gossip and recipes. Next thing you know, nearly 3 billion people are using Facebook, almost half the population of the globe.

There are techniques to grow and exploit the network effect. One is the use of powerful user agreements that users virtually never read. Those user agreements give permission for companies to use and sell customer data; it's hardwired into the agreement. You cannot use the service without promising to share everything you do, to whom you talk, where you shop, what you browse. The result is a "sharecropping" model for social networks. The users get free use of the network but they also do the work of providing the data while the platform owns and monetizes that data. In order for each company to build and support its platform, the algorithms that control it are deliberately written to be addictive. If you like this, try that. The systems function with the use of AI, and AI—as we discussed in chapter 3—requires vast amounts of data to predict what a given person will want to do.

It is imperative that data be scooped up and hoarded. Google and Facebook make their money by selling advertisements. The advertisers who invest their budgets in this way want to be able to target customers with high specificity so the more information they have on the user, the better. Facebook takes this another step. Rather than just providing information on users, Facebook actively steers users to activities that enhance its profit. The result is that Facebook is not just selling advertisements but deliberately engaging in behavior modification.[6] Harvard emeritus professor Shoshana Zuboff gives a chilling description of how Google designed Pokémon GO to be a specific exercise in high-efficiency behavior modification that would drive customers to stores that paid to be PokéStops.[7] TikTok is but the latest example of AI-enabled behavior modification.

Another technique for growth and dominance is to eliminate competition. If another business comes up with an idea that threatens your dominance in your network, you undercut or otherwise eliminate them, or you buy them. That is why Google came to own YouTube; Microsoft took over Skype; Facebook devoured WhatsApp; and Amazon enveloped Goodreads and ate Whole Foods for good measure.

The result has been the development of what Lanier calls "siren servers," in analogy to the Sirens of Homer who lured sailors. The siren servers are winners of all-or-nothing contests. They may have extensive physical plants, but they live on digital networks. Lanier argues they are characterized by "narcissism, hyperamplified risk aversion, and extreme information asymmetry."[8] A complex array of smaller enterprises is subject to creative destruction—an important ingredient of a healthy capitalism—and is thus relatively robust. The result of the network effect and siren servers is enterprises that are powerful but brittle where all the risk of something going awry is put on the user. This creates a situation of intrinsic moral hazard. We can move fast and break things because someone else will pay the price. Things have broken.

THE ATTENTION ECONOMY

The *Kevin Bacon effect* seems innocuous enough. The notion is that because Kevin Bacon was in so many films with ensemble casts—for example, *Animal House* and *Footloose*—early in his career, you can start with Bacon and within six steps link to every person of influence in the film industry. The Bacon effect is an example of how a network works. In our new digital universe, there are also networks, but the connections are not idle. The algorithms of social networks actively guide links. The goal is the feeding, growing, and empowering of the network.

What developed was an *attention economy*[9] where the goal was to go viral. The algorithms actively promoted likes and re-posts that spread the original post. People learned to game this system, and the algorithms helped them to do it. People got a kick when their meme went viral and their base of followers grew. Social media benefited because it could get links to targeted ads in front of more people. The problem was that some people and the algorithms quickly learned that negativity is likely to go viral more quickly than positivity. Outrageousness, even if it is false, spreads quickly, thus generating attention, clicks, and revenue. Simple truths and rebuttals languish. The result was that under the active guidance of the recommendation algorithms, the amusing Kevin Bacon network effect led in only a few links from cat videos to hate speech.

A parallel effect was the spread of disinformation. The old news system, for all its flaws, attempted to disseminate "all the news that's fit to print" according to the masthead slogan of the *New York Times*. The crucial word is not *all* but *fit*. The news should be edited and curated. All that went out the window when people began to turn to social media for their news, and Facebook could publish links to articles in the *New York Times* without adequate recompense. Chatterings about friends, acquaintances, and hobbies became primary sources, corrupting the traditional structure of curated news. Comment sections became cesspools of adamant claims not backed by numbers, data, or sources and hence that were mostly free of fact-checking or independent moderation. People act and talk online as if no one is watching when millions of Kevin Bacon connections allow a vast network of people to do so. Other people keenly understand this system and actively exploit it to spread disinformation.

Fake news spreads and is especially helped by people protesting fake news. As disinformation propagates, people become suspicious of all news not just that on social media but in the traditional curated media. Social media flourishes as traditional media withers. Facebook ducks responsibility, hiding behind claims of free speech even as users are steered to ridiculous and dangerous conspiracy theories. Governments in Russia, India, and the Philippines assault the free press and promote violence and authoritarian policies or even genocide as in Myanmar and Ethiopia.[10] Avaaz,[11] a US-based nonprofit organization that promotes global activism on issues such as climate change, human rights, animal rights, corruption, poverty, and conflict reported that hoaxes spread on Facebook are "a major threat to public health" by making it much harder for doctors to engage in best practices with their patients.[12] The attention economy has not just generated new ways to make money—which it certainly has—but also threats to our political system by undermining confidence in a free press and systems of governance. The motto of the nascent Google was "Do no evil." The new motto of Google's current umbrella company, Alphabet, is "Don't be evil," a subtle difference worth contemplating.

WHO OWNS THE DATA?

The economic system that grew on the internet was not inevitable. Lanier[13] argues that the winner-take-all system did not have to take root. He imagines a system in which the value of personal data is determined—the companies profiting from it know the value full well—and that micropayments are made to the individuals who produce those data. That might reduce the value of the companies from fabulous to spectacular but provide a built-in means to redistribute the wealth represented by the data and support people further down the wealth distribution curve. That system did not originally develop, and we are—temporarily at least—stuck with the system we have as we race exponentially into the future. For better or worse, the current system reflects aspects of our humanity. What was dreamed to be a marketplace of ideas became a rich source of confirmation bias. Conspiracy theories are driven by our craving to understand. Fear leads to inappropriate reactions. Just as people learned to game the attention economy, surely people would learn to game Lanier's system of micropayments should something like that come to pass.

The advent of AI based on LLMs gave us another chance to rethink how our current system works. The vast data on which ChatGPT, DALL-E, and related processes work was scraped from the contributions of a large number of writers and artists who gave no permission and were not compensated. Protests over that process are giving rise to discussions and even demands that some recognition and recompense be paid to the originators. The result might be a version of Lanier's micropayments.

I confess I was suspicious of Facebook from the beginning when they asked me to post where I went to high school. Why would they want to know that? I did not understand who would see my comments or whose I would see so I kept a very minimal profile. In hindsight, I am glad I was conservative in that regard. For all their liberal bias, I still prefer to get my curated news from the *New York Times* and NPR and marvel that anyone, anywhere, turns to Facebook or TikTok for news.

In 2020, Charlie Warzel of the *New York Times* spent a few weeks monitoring the Facebook news feeds of two middle-of-the-road people in their sixties. What he found was "an infinite scroll of content without context. Touching family moments are interspersed with Bible quotes that look like Hallmark cards, hyperpartisan fearmongering and conspiratorial misinformation. . . . In a word, a nightmare."

The fact that Facebook evolved from a friendly social network into a global, corruptible, and corrupted news source resulted from deliberate policy and is the reason Facebook has drawn increasingly negative attention. There are strong psychological forces that make people amenable to sharing and believing misinformation, but Facebook has built its structure to foster those tendencies. It is hard to change the fundamental psychology of people, but reshaping corporations is within the bounds of possibility. Politicians, news organizations, and users of the internet are all beginning to clamor for change.

There are other chilling aspects of our wired world that are woven into the fabric of the targeted-advertising, data-hoarding structure. Zuboff refers to this as "surveillance capitalism."[14] One component of this is the mobile phone that so many of us routinely carry. The phones are tracked through the cellular signal towers that enable the system. These data are nominally anonymized but in a pair of striking articles[15] in the *New York Times*, Charlie Warzel and Stuart Thompson were able to show that it was relatively straightforward to link the tracks of a given phone to an address—typically, where the phone routinely spends the night—and then to other personal information that identifies an individual. These data are routinely, massively, and obscurely collected on all of us, every day.

As Kevin Roose of the *New York Times* remarked, "It's long past time for the world on our screens to be managed as thoughtfully, and with as much accountability, as our roads and schools and hospitals."[16]

The internet has done many positive things. It vastly expanded access to information, making the writing of books like this much easier. It broke down barriers where curators of information evolved into self-serving gatekeepers. It fostered an explosion of wealth, some of which trickled down. The trick going forward is to build something better without destroying the golden goose.

In order to limit the spread of disinformation, fake news, hate speech, and conspiracy theories and to limit the traditional monopolistic tendencies of big companies, there are movements to invoke antitrust measures, even to break up companies like Alphabet-Google, Facebook, Apple, and Amazon.

The latter two are not directly accused of spreading disinformation, but when companies have more wealth and power than some governments, attention must be paid. There is a traditional argument that large companies naturally stifle competition and creative destruction.

Social media companies like Facebook, Twitter, and TikTok operate in a different economic mode than those like Apple and Amazon that sell products or Google and Microsoft that sell software and services. In classic economics, goods and services ideally are sold for the marginal cost of producing them. If you charge less than that, profit is being lost; if you charge more, people will not buy. Amazon and Apple roughly operate by that rubric. The social media companies are very different. Because they make their profit entirely from selling finely targeted advertisements, the marginal cost of adding another customer is essentially zero. That drives the value of content toward zero independent of the cost to produce it. That is why traditional news venues are in economic peril. The same logic drove online trading to zero transaction fees and at least temporarily allowed individuals who coordinated online to catch some hedge funds in a short squeeze costing billions of dollars. The long-term effect on stock markets remains to be seen.

Social media companies have introduced a digital version of the tragedy of the commons. An individual's right to free speech, to say any damn thing they would like, must be constrained if that speech causes broad, deep, societal damage. Shouting on a soapbox to people within earshot is one thing; shouting into the amplified, reverberating, echo chamber of the internet is quite another. This is a hugely complex and controversial issue, but if an individual's interests conflict with those of society, then society gets to have a say.

REGULATION

Both excess power wielded by large companies and over-regulation by governments can limit innovation in a healthy economy, but intervention by government properly done can promote creative destruction. In the 1980s, the US Department of Justice brought an antitrust suit against IBM. Although the suit was finally dropped as the world shifted from mainframe to personal computers, it helped to clear the way for the explosive growth of Microsoft and Apple. In 2000, the Department of Justice accused Microsoft of being an abusive monopoly. Once again, ultimately there was no action to break up Microsoft, but the reaction of the company to avoid severe punishment gave room for Google to grow to dominance. Now Google, Facebook, Apple, and Amazon are accused of stifling competition. Even the threat of antitrust action may promote an evolution that will lead to innovation and perhaps a repeat of the cycle.

Europe got ahead of the United States by passing the General Data Protection Regulation (GDPR) in 2018.[17] This law gave users of the internet more control over what data is collected on them and how it is shared, aspects at the heart of the ad-driven business model of the social media companies. The GDPR requires companies to get permission from the user before employing their data, a significant change from the previous default system in which giving up personal data was automatically written into user agreements and employed in a deliberately opaque manner. Under the GDPR, anyone can inquire about the personal information a company acquires and ask that it be deleted. Interestingly, this right applies not just to technology companies but to banks, grocery stores, even one's employer. The GDPR also contains concrete means to penalize companies that are deemed to have violated the law. Companies can be fined up to 4 percent of their global revenue and not just their European revenue. The very existence of the law can nudge companies in different directions. After passage of the GDPR, Facebook announced that it would offer the privacy controls required by the law to all users. The GDPR was first used in a significant punitive way in 2019 by fining Google 50 million Euros for not properly disclosing how data was collected in its various services. In 2022, Europe added to its armament by passing the Digital Services Act (DSA) to combat disinformation and to limit targeting online ads based on ethnicity, religion, or sex; and the Digital Markets Act (DMA) to limit anticompetitive behavior in app stores, advertising, and shopping.[18] The European Parliament passed a draft AI Act to put restrictions on that technology, including facial recognition software and requirements for the creators of LLMs to disclose more about their data bases. The challenge now is effective enforcement. The ultimate effect of the GDPR, DSA, DMA, and AI Act on defensive tech companies and the global population of users remains to be seen.

The United States, home of the target companies, has continued to lag in terms of regulation with no coherent action by Congress. In 2020, the Department of Justice filed an anti-monopoly lawsuit against Google. Also in that year the Antitrust Subcommittee of the Congressional Judiciary Committee released a report[19] that recommended changes in federal laws that would facilitate imposing limits on Google, Facebook, Apple, and Amazon including restructuring or spinning off component divisions. The FCC has considered new privacy rules by which customers, not companies, would own their personal data but has vacillated on imposing such rules. There are proposals for a special regulator for the big tech companies that would stop them from favoring their own services just as railroads were once prevented from owning businesses that required railroads to ship goods. Amazon is currently a model for just that sort of aggressive, self-dealing behavior.

The big technology companies do not just allow us to browse the internet and buy goodies on a whim. They provide the basic infrastructure of the internet from maps to operating systems to cloud services. Even their competitors must rely on their services. Several decades ago, the United States had strong antitrust laws that might have judged illegal recent practices of predatory pricing or buying and dissolving smaller rivals that have been employed in the growth of the big tech companies. Although not much has yet happened in the United States, all this rumbling may perhaps signal the "end of the antitrust winter" in the words of Tim Wu of Columbia University.[20] A bipartisan AI Caucus formed in the US Congress and the policy research AI Now Institute[21] was created to address the concentration of power in tech industries. The Federal Trade Commission, the Consumer Financial Protection Bureau, and the Food and Drug Administration have all pondered associated AI issues. Our past experience with monopolies and government regulation of railroads, automobile safety, tobacco products, and the opioid crisis gives lessons for the future.

While attention in the West tends to focus on the well-known giants of Silicon Valley, there are parallel, but significantly different, developments in China. The giant tech conglomerate Tencent operates the WeChat app that is effectively Google, Facebook, and Amazon rolled into one.[22] WeChat serves 1.3 billion Chinese both behind the Great Firewall in China and in the Chinese diaspora throughout the world. By facilitating browsing, chatting with friends, digital payments, and ordering goods, WeChat has become nearly indispensable in modern Chinese life. While in the West there are stirrings of resistance to the overweening influence of the tech giants, an important difference in China is that the autocratic government controls WeChat and uses it in turn as a means of social control. The government keeps a grip on the population by monitoring and censoring the myriad activity on WeChat. Even Chinese nationals living abroad who nominally have access to Western technology tend to remain in the grip of WeChat through the network effect. All their friends and family at home are on WeChat so that is the most effective medium by which to keep in touch. In order to be competitive with Western technology, WeChat does not censor overseas Chinese to the extent it does those at home; however, the app runs AI on the communications from the overseas community and learns how better to censor at home. The insistence on operating the Great Firewall that prevents Western companies from bringing uncensored news to China and resistance to the reach and effect of WeChat in the West may lead to a balkanization of the internet that is far from the original dream of a free-and-open global system.

THE WINDFALL CLAUSE

How do we construct a system in which news and information are spread efficiently and nonsense is filtered out? How do we balance the rights of companies to run their business as they see fit with the desire of politicians to spread their message when those two imperatives are in conflict? What if the companies are in the United States but the totalitarian governments that seek to control the messaging to their populace are a world away? Who decides who is allowed access to social media and who is barred? How are big social media companies brought into harmony with healthy societies without destroying their business models of selling attention? How do we construct a system in which both individuals and professional curators contribute successfully to the circulation of news and information by editing, filtering, judging, and checking information before or as it flows? Is there a role for AI in striking this balance between freedom of speech, healthy inventive economies, and healthy societies? Kevin Roose of the *New York Times* had a somewhat tongue-in-cheek suggestion that the first step should be to ban Share buttons that are all too often invoked before the brain is quite engaged or—also common— despite the fact that the juicy item is likely to be false. There are already attempts to limit the number of people with whom one can share in order to control the runaway virality of a completely unrestricted system. A more practical suggestion might be to evolve to a subscription model. A subscription model would greatly weaken the incentive for algorithms to steer users to addictive content. Subscription models have the negative effect of being biased against people of low income, namely most of the population. Perhaps this could be balanced by government subsidies, but this is a global problem requiring a global solution.

The cycle of growth, dominance, and limitations will undoubtedly continue. A century from now, Facebook, Apple, Amazon, and Google will likely be gone or at least be only tottering remnants of their current structure. New technologies, new opportunities, and new exponential growth will come to the fore.

There may also be new ways to distribute wealth within the economy. Some AI companies have signed the Windfall Clause[23] by which they pledge in advance to donate a significant portion of any extremely large profits they might accrue from developments that yield extreme economic and societal disruption. This may not be as bold as it sounds since three-quarters of businesses based on AI have failed to date. It is unlikely, but not too late, for the current tech giants to adopt this philosophy. There may be more plebian means to limit the reach of tech giants and redistribute their wealth to those whose data provides that wealth: taxes. The State of Maryland took a small step in that direction by passing a controversial new tax on digital advertising in 2021 that was nominally aimed at the business model of targeted advertising.[24] A state circuit court judge later struck down the law.

More radical changes have been contemplated. Since birthing the World Wide Web, Tim Berners-Lee has been deeply involved in efforts to maintain the internet as a vehicle for egalitarian forming of connections and sharing of information. Like Jaron Lanier, he thinks that a wrong turn was taken with the growth of the siren servers, which he refers to as silos that function as surveillance engines with tight control of innovation.

Berners-Lee advocates the development of personal online data stores or "pods" in which each person can control their own data.[25] In stark contrast to the current scheme by which companies like Acxiom, Datalogix, and Epsilon collect data and sequester it for their own enterprises, companies would be given permission to access data or to deliver a personalized ad but could not themselves store the data. Already in about 2014 a spate of companies had formed around the globe—in the United States, Israel, Spain— proposing some version of this scheme. A survey in 2016 found that half the respondents and 63 percent of millennials would exchange their data for cash; another survey of teenagers found that 43 percent would rather receive cash for their data than to take a job.[26] The data control initiative has not yet flourished, but it percolates along. One current version is ODE.[27] Another is UK company Gener8.[28] Founder Sam Jones decries the apathy of users toward their data and the lack of transparency of the companies that exploit it, noting that data legally belongs to the user under European GDPR rules. Gener8 provides a browser extension that enables people to monetize their data while browsing.

There are many technical and conceptual details to sort out. These schemes might provide more transparency and give individuals more control of their data, but the flip side is that you may have to pay or give up recompense to maintain your privacy. This may mean only the wealthy can have privacy and the poor are subject to a default of data exposure. Private smart homes might cost more than ones that report on their users. Landlords might pocket the revenue from data on their tenants. Nevertheless, with these goals in mind, Berners-Lee is developing associated software called Solid in a startup, Inrupt. From such seeds, giant killers might emerge.

THE FUTURE OF LABOR

The digital revolution did not just alter the nature of capital and the means to accumulate wealth, it also threatens profoundly to affect the complementary nature of jobs, labor, and work. Social media companies, despite their wealth and influence, employ far fewer people than traditional manufacturing enterprises. Digital technologies—whether apps like Uber that depend on flexible labor or platforms like Airbnb that are built on a *trust economy* and shift accountability to users—have helped a small number of players accumulate wealth and influence. The lack of skilled workers drives companies to employ more robots and AI even as the adoption of those productivity-boosting technologies promises to displace significant numbers of working people. During the COVID-19 epidemic, some restaurants short on human employees turned to robots. Will they ever turn back? The gig economy—Uber is a prime example—potentially provides more flexible employment but also allows companies to shed the responsibilities of the steady employment, health insurance, and retirement benefits that define comfortable lifetime employment. The net effect is to shift wealth and influence to a small number of people at the top. For the rest, there is a growing sense of *precarity*.[29]

At the same time, the digital economy provides brand new ways of making income that were inconceivable a few decades ago. A difficult practical problem is that those displaced in the new economy may not be those who can benefit from it. An important issue is how to ensure that everyone benefits from the exponential advances of technology and not just the lucky, driven few. A counterforce is the perception that a pool of low-wage labor must be maintained to avoid runaway cost and spiraling inflation.

As I mentioned earlier in this chapter, an economy can grow if productivity grows. This can—but probably should not—ultimately depend on growing the number of people. The alternative is to grow the productivity per person. The growth of productivity in the United States has been rather slow for the last decade but robots and AI promise to boost productivity in an exponentially more dramatic way. A decade or century from now, the perspective may be quite different. There is a joke about the limit to that.

In the future, essentially everything is automated and only two biological entities have jobs: a man with an On-Off switch, and a dog to keep the man from touching it. Funny, but what about all the people rendered unemployed or unemployable in the process? Not even highly trained professionals like doctors, lawyers, and college professors are likely to be immune to the encroaching tidal wave of AI.[30]

One near-term, practical solution is for humans and robots to learn to work together in a collaborative way. That must and will work for some time, but it is not clear that it is a long-term solution in the face of exponential advances in AI. Another suggestion is that humans must learn to engage in constant learning. That will work for some but may not be a universal panacea. The capacity for an individual human to learn is finite, and the exponential advance in technology may race ahead. My Austin colleague Byron Reese is confident that machines will not become conscious and hence that humans will always have an edge over machines.[31] He also points out the practical reality that the adjustment to new technological opportunities will not occur by a coal miner becoming a brain surgeon but by everyone along the economic ladder moving up a rung as the ladder extends. The brain surgeon may learn to master new skills employing AI, and the coal miner first may learn to collaborate with robots to mine the coal and then transfer to a new area where those collaborative skills are in demand, maybe a wind farm just down the road.

UNIVERSAL BASIC INCOME

The prospect of jobs eliminated by technology has also raised issues of how equitably to distribute the benefits of technology and to maintain a healthy society. Two principal notions are a *guaranteed income* and its more extreme cousin *universal basic income* (UBI). The idea of a guaranteed income is to provide a modest amount of support perhaps to a specific group of people, and perhaps for a limited time. Martin Luther King Jr. called for a guaranteed income in 1967. The more radical notion of a UBI is to provide a basic subsistence income to every adult. The idea of a UBI has been around for some time but it was brought to wider public attention by Andrew Yang in his 2020 campaign for president of the United States.

A forced experiment in guaranteed income was instituted by the COVID-19 pandemic of 2020. Millions were suddenly thrown out of work by the need to quarantine. They could neither work nor serve as consumers, resulting in a severe disruption of the economy. The burden fell especially hard on people at the bottom of the economic ladder and among those especially on people of color. The US government responded by giving cash payments to virtually all the adults in the country. For those who were reasonably well off, the payments were welcome but not critical. For those of more moderate means who were already living paycheck to paycheck, the payments and new, extended unemployment benefits mitigated poverty, starvation, and despair.

The idea of the UBI is to support people at the bottom end of the economy and build a middle class to provide a strong democracy based on that foundation. There are several important problems with UBI. It would be very expensive, and there is great disagreement on how to pay for it. Taxes would have to go up in some fashion. There are also fears that guaranteed payments to people, even at minimal subsistence levels, would render them indolent layabouts, sapping their initiative to seek gainful employment. I asked my class what they would do if they received a UBI. A few in the class, all male, promptly chimed up that they would play video games. Fortnight was popular at the time.

Experiments with UBI have been done, mostly funded by governments but sometimes by wealthy private citizens. Sigal Samuel, who writes on this topic for Vox Future Perfect, provided details on guaranteed income programs worldwide.[32] Some are genuine UBIs with support given to the whole community. Alaska has given some of its invested oil income to every citizen since 1982. The Cherokee natives in North Carolina give some casino income to every member of the tribe. Iran has run the only nationwide UBI program since 2011 as a replacement for subsidies it had given for household necessities. Kenya has given twenty thousand people spread over nearly 250 villages a little less than $1 a day guaranteed for twelve years. Pierre Omidyar, who made his money with eBay, provided about $500,000 dollars to help Kenya start the program.

Other projects have been more focused in time or population and so are not universal in the broadest sense but relevant nevertheless. In 2010, India did an experiment providing 6,000 poor individuals a few dollars a month for 18 months. In 2017, Finland instituted a pilot program to give 2,000 unemployed people chosen at random a monthly income of about $600 per month for two years even if they became otherwise employed. Namibia gave everyone over sixty in a certain region a few dollars a month of private donor money. Brazil gives money to poor families if they keep their children in school. In Stockton, California, Mayor Michael Tubbs started a program in 2017 funded by private donors to give $500 per month to 125 people. He got support from Twitter founder Jack Dorsey and Chris

Hughes, one of the founders of Facebook. Tubbs has since started the organization Mayors for a Guaranteed Income. An offshoot is that in 2021 the governor of California included funds in his budget that would enable local governments to run their own guaranteed-income pilot programs aimed at low-income families. There are guaranteed-income pilot programs in Compton, California; Richmond, Virginia; and in Ulster County in the Hudson River Valley. Many cities, including Chicago and Austin, are now using direct-cost payments as a means to curb poverty, replacing or complementing traditional support programs.[33]

Studies of these programs show that fears are unfounded that they will foster communities of laggards. Most recipients did not play Fortnight. There is little or no evidence that people work less and some evidence that they are more motivated to seek employment. Indian villages where even only some received support reported significant increases in food sufficiency and decreases in sickness. In general, the guaranteed support tends to increase nutrition levels, physical and mental health, education, and sanitation and to decrease crime and addiction. Stress goes down and general happiness goes up not just among the recipients but in the broader community. Trust in other people and in institutions goes up. Recipients feel liberated of financial burdens and free to pursue their talents and interests. The birthrate in Alaska has trended upward since Alaska's UBI program began.

The charity GiveDirectly gives cash via cell phones directly to poor people in sub-Saharan Africa and, experimentally, in the United States. GiveDirectly was started in 2011 by a group of MIT and Harvard economics graduate students. They were convinced that giving cash directly and unconditionally was preferred to giving goods because it gives people more flexibility. The recipients know what they need and use the cash appropriately. By 2021, GiveDirectly had delivered over $165 million to more than 534,000 households in eight countries under twenty-five different programs.[34]

While negative sociological expectations have not been realized, true subsistence-level UBI is still expensive. Andrew Yang's attention-getting proposal that every adult in the United States should get $1,000 per month would cost about $3 trillion per year. At the same time, the average worker in the United States makes about $1,000 per week so it is not clear that Yang's proposal would really be adequate subsistence for those most in need. In much of the rest of the world, the cost of subsistence is paltry compared to that in the United States, but then there are also a lot more people.

Estimates of cost must consider potential positive feedback. In 2018, a randomized experiment was performed in the Vancouver area. Fifty homeless people were given $7,500 in a lump sum to use however they wanted, a total of about $375,00. Another sixty-five homeless served as a control group. The latter group were not given funds but were also monitored over the next year. At the end of the study period, most of those supported had used their funds for food, clothing, and rent and had saved a little. The receipt of a lump sum seemed to trigger constructive, long-term planning. Because these people moved into housing, the Vancouver shelter system saved $400,000. The net cost was essentially zero.

The net cost of UBIs or some version of a conditional guaranteed income must reflect these positive financial returns and must be balanced against broad positive benefits that lead to a happier, more trusting society. Suresh Naidu, an economist at Columbia University, imagines a future that aids both individuals and our fragile democracy.[35] He proposes that every citizen receive some remuneration for participating in the democratic political system. He envisages an Aristotelian *eudaimonic* democracy reinforced from the ground up.

The internet did not just promote new, powerful, data-driven companies. It also provided new ways for individuals to make money. Little businesses could acquire raw material from throughout the world and sell in that vast market. Small craftspeople could sell their handmade wares on Etsy. There were new opportunities to exploit the wisdom of crowds. Crowd-work enterprises like Task Rabbit, Upwork, and Amazon's Mechanical Turk provided some income but relied primarily on small-scale work with low, piece-rate pay. This structure assumes that the workers are unskilled and all basically the same. Real people are much more complex. The demands of professional work are also more complex. Crowd work could potentially become a profession worthy of aspiration. Crowd work should evolve so that workers can develop their skills and become more engaged, satisfied, and well paid, and so that employers can find the complex, creative, skilled work force they need.[36]

A whole new economy of *influencers* arose. People who could attract attention by being stars, or by being especially well-spoken, or by being especially good at attracting attention could connect with vast new audiences and monetize that attention by promoting products and posting clickable ads for which the companies behind the ads would pay. Some people who were famous mostly for being famous and some teenagers who, for whatever reason, appealed to other teenagers made impressive fortunes overnight. Whether this sort of income is stable for a lifetime or volatile and transient remains to be seen, but its reality cannot be denied.

Any new medium, from moveable type to film to digital data, finds early use for sex. The internet was no different. Individuals could set up basic video production facilities in their bedrooms and perform for subscribers. Others could film their efforts and sell them to website hosts. Sex workers could be trafficked on the web.

THE FUTURE OF MONEY

A necessity in any modern economy is a means to efficiently facilitate the exchange of capital and labor. Most economic exchange has long been backed by money rather than direct barter of goods. Money itself is a complex issue. It can be represented by a chunk of gold or silver or it can be represented by a piece of paper the exchange value of which is backed by the good faith and credit of some government. The internet and the digital machinery behind it enabled a whole new type of money in the form of digital payments and *cryptocurrencies*. These currencies exist only in digital form (as does your bank account if you think about it) and are rendered secure by various cryptographic techniques—hence the name.

Whereas cash is still king in much of the world, especially huge amounts of US$100 bills, we are in the midst of a move to cashless economies. An early version was PayPal that allowed digital transactions. More recent versions are Apple's Wallet, Venmo, and Amazon's Just Walk Out shopping. These are cashless procedures, but they all must be backed up by government-backed money in a bank or through a credit card company.

Narendra Modi, the somewhat autocratic prime minister of India, shocked the country in 2016, when he removed all large-denomination currency from the market in an effort to eliminate illicit cash transactions in crime and politics. The move devastated small businesses that depended on cash. The result was the rapid growth of an instantaneous, digital, scan-and-pay network that revolutionized the Indian economy. There were two key components to this change. One was providing every citizen with a unique biometric identification number, the *Aadhaar*, despite manifest privacy concerns. An associated component was the widespread use of QR codes connected with the Aadhaar that allowed merchants and beggars to facilitate payments. The Aadhaars are the foundation of the instant payment system, the Unified Payments Interface. This system was an initiative of India's central bank. It offers services from hundreds of banks and dozens of mobile payment apps with no transaction fees. The result was that by 2023, 99 percent of adult Indians, 1.3 billion people, had a biometric identification number and their own personal QR code. The Unified Payments Interface is used by 300 million individuals and 50 million merchants with about 50 percent of the transactions classified as small or micro payments. This system has expanded banking services like credit and savings even to the poor, and it has aided government tax collection. These changes have shown how developing nations with inadequate infrastructure can produce substantial economic growth. India now aspires to export this public-private model to the world's poorer nations.

China has also widely adopted digital payment systems. Between India and China, the use of digital pay networks dwarfs that in the West and especially in the United States.

Cryptocurrencies are units of financial exchange recorded by an open public ledger that efficiently, verifiably, and permanently registers ownership and transactions between two parties. Such currencies are differentiated from fiat currencies produced by governments or central banks because no central authority is in control. Like ordinary currencies, cryptocurrencies are identical to one another and thus fungible. Since they are identical, cryptocurrencies can be used as a medium for commercial applications.

The philosophical goal of cryptocurrencies is to get governments out of the loop by providing a medium of exchange that cannot be controlled by centralized authorities. One advantage of paper or metal money is that once minted, it is anonymous. It can be exchanged peer to peer without government control. Cryptocurrencies promise the same direct peer-to-peer transactions. This eliminates intermediaries who might take a piece of the action, avoids inflation fostered by an inept government, and limits surveillance by an autocratic government—as most in the world are.[37] Cryptocurrencies may have the greatest promise in developing countries where means of exchange are inefficient and denied to many.

The ledgers that provide the secure public record of transactions are typically based on a *blockchain*, which is a list of records called *blocks* that are linked to one another and rendered secure by cryptography. Each block typically contains a *cryptographic hash function* or *hash pointer* that links to a previous block; a timestamp; and transaction data. The hash pointer is a basic cryptographic tool: a mathematical algorithm that serves as a one-way function pointing to a previous block. It is virtually impossible to invert the pointer to go the other way and alter the previous ledger. A network of anonymous, mutually distrustful peers manages the blockchain by collectively defining the protocol that validates newly added blocks.

The most common validation schemes are based on a *proof-of-work concept*. Individuals or groups must use computers to solve increasingly complex numerical problems before a block can be added to the ledger. Solving the problem is awarded with a certain number of cryptocurrency *coins.* The people engaged in this process of generating blocks that validate and timestamp transactions are called *miners.*

The first and most famous cryptocurrency, *Bitcoin,* is based on open-source software that was released in 2009 by an individual or group of people known by the pseudonym *Satoshi Nakamoto.*[38] Bitcoin established the anti-establishment libertarian philosophy of cryptocurrencies and the scheme by which a ledger comprises blocks linked by hash functions and the computational problems that must be solved to mine new Bitcoins. The reward to miners for producing a new block is cut in half every 210,000 blocks. The total number of Bitcoins that can ever be produced is 21 million with the final Bitcoin rewarding nothing to whoever produces it. Half the total number of Bitcoins were produced in about four years. It is estimated that it will take 120 years to make the rest. Although miners are rewarded by fewer Bitcoins per new block as time goes on, Bitcoin's history since its birth has been a radically volatile but spectacular rise in value so miners continue to scramble to produce them.

Nakamoto registered the domain name *bitcoin.org* and started the first block of the Bitcoin chain, known as the *genesis block.* He mined about 1 million Bitcoins before vanishing from the network in 2010. For the first few years, Bitcoins were worth between pennies and tens of dollars with the price fluctuating by factors of two or three over the course of weeks or months. By 2013, Bitcoins were worth $1,000 each but that price plummeted to $150 by 2015 only to rise again to $1,000 in early 2017 and to $13,000 by the end of the year. Bitcoins fell from $10,000 to $4,000 between February and March 2020 as the COVID-19 pandemic set in. From October 2020 to April, the value rocketed from $10,000 to $60,000. The price moves by gut-wrenching amounts depending on the actions of governments or the whims of famous individuals; for example, is Elon Musk in favor of Bitcoin or down on it? Those who caught the upside made fabulous fortunes, millions or billions of dollars, but the volatility trapped others. Nakamoto's 1 million Bitcoins are formally worth about $50 billion today, and the total worth of Bitcoins in the world is about $1 trillion—comparable to the wealth of some nations. As the value of a single Bitcoin has gone up, a new digital coin was invented, a *satoshi,* worth one-hundred-millionths of a Bitcoin, or about 0.03 cents in today's value.

With that sort of wealth on offer, there has been a race to mine Bitcoins and other cryptocurrencies with computers that are bigger, cheaper, and more efficient. Miners have made substantial investments in software and server farms and increased the demand for graphics cards dedicated to mining. In addition to the cost of the hardware, there is a considerable cost in electricity to power it. The cost of mining Bitcoins is estimated to be about $400 million per year, which is still affordable given the price of Bitcoins but ever less so. Running all the servers also produces a great deal of heat. Many miners are moving their enterprises to cold climates like Iceland to reduce the cost of cooling. Mining Bitcoins remains profitable because some of the cost is externalized. The price of the electricity to run and cool the computers yields a significant carbon footprint comparable to that of a small country. The carbon cost of aggressive Bitcoin mining may eventually create a backlash as moves to limit global warming become more stringent.

An offshoot of the development of cryptocurrencies is the invention of *initial coin offerings* (ICOs). Startup companies can each create their own ad hoc currency and associated coins or tokens. These coins are offered to early backers in exchange for investment funds or sweat equity while costing the startup very little.

While exchanges employing traditional money are anonymous, cryptocurrencies are pseudonymous. In principle, the owners of cryptocurrencies are not identifiable even though all the transactions are publicly posted in the blockchain. The original Nakamoto document explained it this way:

> Privacy can still be maintained by breaking the flow of information in another place: by keeping public keys anonymous. The public can see that someone is sending an amount to someone else but without information linking the transaction to anyone. This is similar to the level of information released by stock exchanges, where the time and size of individual trades, the "tape," is made public, but without telling who the parties were.

Cryptocurrency exchanges allow customers to trade cryptocurrencies for other assets such as US dollars, real estate, or other forms of digital currency. Cryptocurrency exchanges are often required by law to register the personal information of users, thus putting a chink in the armor of secrecy surrounding them. Jordan Kelly founded the company Robocoin with the goal of providing a medium to exchange cryptocurrencies. He installed the first ATM-like kiosk in the United States in a bank in Austin, Texas, in 2014.[39] The kiosk has scanners that read driver's licenses or passports to identify users.

Cryptocurrency exchanges flourished for a while but became notorious in 2022, when one of the largest, FTX, went bankrupt and its founder Sam Bankman-Fried was arrested for fraud and double dealing with investors' funds. The value of various cryptocurrencies again plummeted.

Cryptocurrencies are designed to avoid centralized control but that does not mean that entities of central control are not interested in maintaining that control. At the very least, the widespread use of cryptocurrencies might make it difficult for governments to monitor and hence modulate economic activity. Cryptocurrencies are also more difficult than ordinary cash or financial holdings for governments to impound in their role of legal enforcement. The legal status of cryptocurrencies is still undefined or evolving in many countries. Some countries have allowed the use of cryptocurrencies while others have banned them. China has been especially aggressive, outlawing the purchase of goods with cryptocurrencies since the invention of Bitcoin and preventing any financial institution from dealing in Bitcoins. China extended these restrictions to more institutions and to all cryptocurrencies in 2021, causing a prompt drop in their value by about 30 percent in the space of a week. The United States has declared cryptocurrencies to be an investment like a stock and hence subject to capital gains taxes. This contrasts with a traditional currency for which you pay no taxes if, for example, the dollar rises against the British pound sterling. This ruling came as unwelcome news to some Bitcoin billionaires who possessed a lot of Bitcoin value but not a lot of dollars with which to pay the taxes.

There are concerns that cryptocurrencies may foster criminality. Although all transactions are recorded in blockchains, it is difficult in practice to identify participants and account for their transactions. Much illegal business is done on the Dark Web, an encrypted portion of the interconnected World Wide Web that cannot be accessed by standard search engines like Google. The Dark Web primarily runs on *virtual private networks* (VPNs) linked by the browser Tor, thus yielding a world-wide, encrypted network of relay sites that connect users. The encrypted VPNs ensure that internet service providers—and governments—cannot tell that a user is using Tor. Tor anonymizes the traffic. There is nothing intrinsically illegal about the Dark Web, but it clearly enables users who prefer to keep their business as private as possible.

The Dark Web got a dose of infamy in 2013 with the arrest of Ross Ulbricht. Under the pseudonym *Dread Pirate Roberts*, Ulbricht had set up and operated a black market on the Dark Web dealing in illegal drugs that was enabled by VPNs and Tor and fueled by Bitcoins. He is now serving a life sentence, but others undoubtedly seek to emulate him. There are lingering concerns that the Dark Web and cryptocurrencies offer an unregulated way to evade taxes and launder money. Some studies have argued, however, that analysis of blockchain activity is an effective means of fighting crime and gathering intelligence and that fears of illegal finance schemes are overstated. This concept was given some credence in 2021, when the

US government was able to track and recover a substantial number of Bitcoins that had been paid in a ransomware attack. A new industry sprang up represented by Chainalysis and other startups that analyze blockchains to reveal who is transferring cryptocurrency to whom and where.

The debate over the significance and role of cryptocurrencies in the future is as volatile as the currencies themselves. While some see a relief from the tyranny of fiat currencies, others see a Ponzi scheme or a manic bubble analogous to the Dutch tulip mania of 1637 or the more recent cases of the Beanie Baby and dot-com bubbles. Lawrence Fink, head of the gigantic investment fund Blackrock, opines that, "Bitcoin just shows you how much demand for money laundering there is in the world."[40] Another problem is that the decentralized nature of cryptocurrencies means that there is no means to limit losses if digital coins are lost or stolen or an access password is forgotten, all of which have occurred.

In the United States, the Federal Reserve System is examining the possible impact of cryptocurrencies amid concern that private money risks runs on the currency and the introduction of risks to consumer protection and financial stability.[41] One Federal Reserve governor notes that in the nineteenth century, private money yielded inefficiency, fraud, and instability in the payment system.

While some are suspicious of cryptocurrencies in general and Bitcoin in particular, others see intrinsic merit in the basic concept of blockchain ledgers to validate transactions. Many banks are beginning to explore the utility of blockchain ledgers. *Smart contracts* are computer programs that automatically execute contracts when defined conditions are met, thus reducing the need for trusted administrators. The cryptocurrency Ethereum, a popular alternative to Bitcoin, has a smart contract capability based on blockchains that allows decentralized transactions.

Non-fungible tokens (NFTs) are another digital product but one with a unique characteristic. NFTs can have value but they are not interchangeable with other NFTs and hence in that sense are non-fungible. NFTs contain ownership details so identification of owners is transparent and enables ready transfer between owners. NFTs allow tangible assets to be traded with a minimal probability of fraud. They can also be used to represent the identities of people and property rights. NFTs can remove intermediaries, provide simpler transactions, and lead to new markets. NFTs can be split into parts so they can have more than one owner and combined to create an entirely new NFT with its own value. NFTs are vaguely connected to Lanier's micropayments but one is paying for someone else's data and not getting paid for one's own.

The value of an NFT is whatever someone is willing to pay in some fungible currency: dollars or Bitcoins. Cryptokitties is a game that was created in November 2017. Each cartoon kittie is represented by an NFT based on the cryptocurrency and blockchain Ethereum. The NFTs can be combined, or bred, to form unique new kitties with their own value. The average cryptokittie was worth about $70 in 2019, but some have sold for $300,000. NBA Topshot is an NFT exchange based on video shots of basketball players making dramatic plays. An NFT honoring deceased star Kobe Bryant sold for nearly $400,000. Jack Dorsey, the founder of Twitter, created an NFT on December 20, 2020, representing the first Tweet he ever sent that said, "just setting up my twttr."[42] It was sold on March 22, 2021, for $2.9 million.

Also in March 2021, the artist Beeple caused a stir, and brought broad public attention to NFTs by creating a collage of five thousand days of his electronic artwork. An NFT of that work was sold at auction by Christie's for $69 million. The buyer said that he thought it was the beginning of a new age of digital art and would be worth much more in the future. Even the University of California at Berkeley got into the game by auctioning NFTs of documents related to work that led to Nobel Prizes for cancer research and the development of CRISPR. The first sold for $50,000 to a group of Berkeley alums.[43] As a means of exploring the NFT phenomenon, Kevin Roose of the *New York Times* created an NFT of his article about NFT. It sold for about $560,000.

The potential utility of NFTs is clear and is being thoroughly explored, but how they are valued in practice is tangled in the mystery of human thought and emotion—like any other currency. Just who are the people who are willing to pay millions of dollars for digital ephemera? Where did they get their money and why do they think this is a better investment than other vehicles? It will be interesting to see whether this phenomenon is stable and sustainable in the future.

The Beeple case engendered some deep thinking about what art means and whether digital art will change the meaning and value of traditional art. Traditional art is a piece of work that can be owned and displayed. It is not just a digital file with bragging rights. In an article in the *New York Times*, critic James Farago argued that NFTs threaten to destroy art as a reflection of human values and that Beeple committed a "violent erasure of human values inherent in pictures."[44] To add insult to possible injury, NFTs, like all cryptocurrencies, require humongous computing power to generate, thus demanding large stores of energy and exacerbating climate change.

AI AND ROBOTS

The digital revolution has had a widespread impact on capital, labor, and money itself. Where will all this take us as exponential growth in technology races ahead? What will be the nature of economics in 100 or 1,000 years? In 10,000 or 100 thousand years? Over the near term, humans will adapt to the changing technological environment, but as technology evolves more rapidly than individuals, governments, and societies can adapt, the tensions will grow.

AI and robotics will change the nature of work by augmenting the abilities of humans but also by introducing autonomous decision making that is the traditional role of humans. As this occurs, we need to ensure that the working environment is inclusive and equitable and that workers have the power to influence their conditions. Those goals are likely to lag an ever-changing reality.

In 2020, MIT released a report addressing the future of work.[45] The report noted that productivity and wages in the United States had been diverging for forty years, and it addressed structural changes that could bring things back into balance. The report forecast that for another several decades there should be plenty of jobs but beyond that, things get murky. For the time being, robots can do incredible things like carefully picking up a glazed donut and placing it in a box without disturbing the glaze, but that same robot cannot pick up a book in an Amazon distribution center. That flexibility and adaptability remain the prowess of humans. Amazon will continue to employ many humans even as it automates as much as possible. The MIT report said that maintaining a healthy economy and society "will require innovating in our labor market institutions by modernizing the laws, policies, norms, organizations, and enterprises that set the 'rules of the game.'" The report advocates raising the minimum wage, augmenting unemployment insurance, and expanding labor laws to enable collective bargaining for gig, freelance, and domestic workers. The report also called for changes in corporate tax laws that currently favor spending on machines. It specifically recommended an employer-training tax credit focused on the need to directly link skills training to the practical demands of business.

Some jobs will be lost: call center staff, proofreaders, and translators now and also drivers as autonomous vehicles become more prevalent. People adept at working in partnership with AI will have new opportunities. This includes data scientists, those involved in telemedicine, and technomechanics who ensure that the AI and robots function properly. Training will need to be specific to tasks but also interdisciplinary and multidisciplinary.

Other issues are open as AI becomes more prevalent and powerful. How can bias be eliminated from the system whether controlled by humans or AI? The adoption of AI needs careful thought, regulation, and auditing. Many AI and machine learning algorithms are open-source and public but the data they draw on is often not despite containing intimate private details harvested from the public. This may have to change, disrupting current business models.

Who or what will decide to hire and how? Who or what will evaluate performance and how? Who or what will decide to fire and how? AlphaGo Zero cannot currently make hiring and firing decisions, but close cousins are beginning to do so. The nonprofit organization Worker Info Exchange is attempting to tilt the balance of power between gig workers and the companies that hire and control them. Their report "Managed by BOTS: Data-Driven Exploitation in the Gig Economy" addressed the issues of gig workers who are managed by AI algorithms.[46] The AI can find a problem with the worker causing them to get fired or penalized, and the worker often cannot find a human to whom to complain or appeal. If the worker can contact a human, the human often does not know why the algorithm flagged the worker and does not know how to find out. These issues currently apply mostly to gig workers, but in the near future many people may find themselves reporting to *digital control supervisors*.

What happens if there is something like a singularity with the development of general AI and machines that are faster, better, and more creative than us? If the current winner-take-all system fostered early in the digital revolution is sociologically fragile and unstable, what happens if a fantastically capable AI emerges and no human or human corporation is at the financial apex? Does the concept of economic exchange even make sense in such a world? There will be power and striving to achieve maximum efficiency and utility, but traditional notions of capital and labor are likely to vanish.

There will be abundant changes in traditional notions of economics as technology races ahead. Bitcoin is an early and volatile experiment, but it might be the seed that totally changes the global economy as some form of decentralized digital currency based on blockchain comes to dominate. Currency exchange fees and transaction fees would vanish. Governments could not control their fiat currencies. There would be less need for stock markets, banks, and lawyers. This could happen sooner than you think if the use of Bitcoins continues to grow exponentially.

One measure of an economy is the *velocity* of money: how fast funds are exchanged. If everyone in the world had a bar of gold but that gold could not be exchanged, we would not be rich but starving. A new age of *decentralized finance* (DeFi) promises to increase the velocity of money and hence to boost the economy. Smart contracts will likely be an intrinsic component. Smart contracts allow holders of cryptocurrency to accrue interest on their holdings or to borrow against them. Smart contracts also enable the creation of new cryptocurrencies and NFTs.

The construction of our digital infrastructure has created great wealth, but that wealth is arguably not dispersed equitably. The growth of DeFi has introduced the issue of how to institute governance of new digital exchanges or, more specifically, the structure of *decentralized autonomous organizations* (DAOs). Issuance of governance tokens and profit-sharing tokens by DAOs promises a means to more broadly share wealth and decision-making power in our digital economy. Peter Diamandis foresees deep structural changes in society as wealth and power are distributed in a more democratized way in a "new global commons."[47] Fold that into projections of machine learning and AI.

There are likely to be significant changes in the functioning of stock markets. High-frequency, flash trading brought a battle for milliseconds as the speed of light dictated that traders' computers that were physically closer to those of the exchanges had a minute advantage over those that were more distant. AI is already used to monitor market conditions and make rapid trading decisions. That will become more common as the AI becomes more capable and sniffs out opportunities that no human could even comprehend. Imagine an enhanced version of AlphaGo Zero playing not the game of Go but the global market by implementing and even manipulating smart contracts. Nasdaq and the London Bourse could be rendered irrelevant. Imagine a day trader in New Jersey trying to compete with an AI. On whose behalf would the AI be bidding? Some person? Itself?

All these changes will be colored, if not dominated, by a likely flat or even declining population: fewer people and much, much more AI. Automation has its own momentum, but businesses will be driven to automate to increase productivity in the absence of population growth. Economic policy and theory will have to become more focused on how machines and humans interact as machines come to more strongly affect human behavior, even human nature. Decisions will need to be made about what technological developments are encouraged, which are relinquished, and who is to make those decisions.

Even as the population stabilizes, megacities are likely to grow, requiring the expansion of smart technologies, the IoT, AI, and robots. These technologies will diffuse into rural communities. If that diffusion is effective, there may then be less demand for cities as work can be done from anywhere. Having a few big information technology companies in control of this process may prove inefficient and may impede the most efficient distribution of capability. We may require a *robonomics*[48] built on blockchain technology in which processing and storage of data, provision of robotic services, and allocation of bandwidth are delivered through a distributed spectrum of individuals and small- and medium-sized enterprises. Blockchain technology may also help in enforcing privacy regulations that will arise with ubiquitous IoT and edge computing. Our current economic policies and trusted institutions will evolve into distributed institutions that allow effective decentralization. Blockchain technology may then help to scale and redistribute the coming massive interaction and collaboration between humans and AI and facilitate the capacity to audit, insure, and litigate smart contracts between humans and AI.

THE FUTURE OF GROWTH

Classical economics postulates that humans make strictly rational economic decisions. Behavioral economists have long since concluded that this is nonsense. Humans bring complex emotions to economic bartering. They have a sense of fairness ingrained by millennia of evolution. They boil with resentment if they think their portion of the hunt is not equitable. While it could be programmed in, blockchain software and AI have no such intrinsic constraints. As distributed computing comes to dominate our economic enterprises, the nature of economics may change, becoming fundamentally rational with none of the human aspects that pertain today. Or perhaps an AI will demand equal reward for equal ability and effort.

Money in the form of US greenbacks, gold, or NFTs has no intrinsic value, only the value invested in them by people who have the confidence that they can be exchanged for other things of value: food, housing, trinkets, vacations to the South Pacific. The value of money arises from a collective belief in its worth. What if AI does not have that belief?

The opposite also operates in modern economies. A basic principle is that the customer—with their individual subjective feelings—is always right. No matter how efficient to produce and how delicious and nutritious, if a given cereal on the grocery store shelf does not sell, then it is an economic failure. Yuval Noah Harari argues that this economic principle arose with the growth of humanism, the notion that it is individual humans who bring meaning to a meaningless world.[49] Will that power of the individual customer's right to decide survive in an AI-dominated world? Will we evolve from humanism to AI-ism as we collaborate or even merge with ever more capable machines?

Modern economies rely on a belief in the future. That belief is necessary to foster a system of credit. Credit is necessary to finance new enterprises that drive growth secure in the confidence that growth will allow a repayment of that credit—with interest. That means we do not have to share the pie in some zero-sum game that characterizes most balanced ecologies but can anticipate that, over time, growth will yield a bigger pie.

The pie that drives growth has three ingredients: raw materials, energy to forge those materials, and knowledge of how to do so. On our finite planet, raw materials are ultimately finite. Perhaps our expansion into space will alter that reality. We constantly strive for new sources of energy. We need to slack off on fossil fuels, but solar power and nuclear fusion promise immense new sources of energy. Knowledge is the magic ingredient that separates humans from animals. We are now adding a new ingredient with the exponential growth of computers and AI. AI may add to human knowledge and hence economic growth, but what of AI knowledge? Will AI know things that are beyond the human ken? If so, what does that do to the human economic enterprise?

Economic growth is in tension with our threat to the environment that ultimately sustains us. Battling climate change will undoubtedly sap some of our resources and limit economic growth. On the other hand, total anthropic-driven ecological collapse—like nuclear war—can really ruin your day. Will our AI be smart enough to help us find the right balance of economic growth and ecological protection?

The US Supreme Court has ruled that companies are effectively people. They have a right to own property and indulge in free speech. What about AI? Can an AI be a person in the same legal sense? Can an AI own a corporation through a blockchain smart contract? Why not?

The economy can be viewed as a system that gathers data about the wants and needs and capabilities of people. Those data are infused with the power to make decisions. In this sense, capitalism can be viewed as a distributed data-processing system and communism as a centralized processing system. In a time of rapid change such as will be driven by exponential technological progress, capitalism is more flexible and resilient. Stock exchanges rapidly evaluate the impact of new declarations of profit and loss but also of the outbreak of war or the onset of peace. This capitalistic system demands growth. If growth must ultimately be limited to protect the planet we inhabit and the economy needs to adjust to, if not stasis, then some steady state, perhaps a more centralized control structure will be needed. Who or what would control that? A human? A super-powerful intelligent AI?

It is clear that robots, AI, and automation in general are going to displace traditional sources of labor and jobs for people. In eras of cyber warfare, even militaries will need fewer people. How our society in general and the economy in particular adjust to those changing conditions will be a critical question over the next several decades. What will happen as people lose their value as workers and soldiers? Will the people at the top of massive corporations and governments and AI value every human being as having intrinsic value if that value has no economic dividend? Will the loss of that economic measure undermine the very system of liberalism and capitalism that has dominated the last several hundred years?

We are rapidly moving into an information-driven economy, *datalism* rather than capitalism. Instead of the invisible hand of the free market, the information economy depends on the invisible hand of data flow. That information wants to be free, or at least free to flow into the giant corporations that process it for profit. Humans who efficiently process data are good. Death halts the flow of information and hence should be avoided. In a data-driven world, internal subjective human experience has little value. The point of existence is to share it. Only shared experience has value. Private journals become a thing of the past, replaced with the need to record, upload, and share. I share information, and therefore I am. Facebook, Instagram, Reddit, TikTok, Twitter all exist to facilitate that all-important information flow.

In this new world, AI is likely to prove a much more efficient broker of information and data than any human. In a world where information flow is paramount, individual human experience may lose the traditional value ascribed to it in a humanist society. Subjective human experience may no longer be relevant. A conscious AI will not be needed to wreak disruptive changes in our economics and society.

Conscious AI may nevertheless arise.

13

DEMOCRACY
Who Votes in an Age of Artificial Intelligence?

THE INVISIBLE HAND OF THE ELECTORATE

The humanist philosophy that overthrew monarchies, that with some success held autocracies at bay, and that gave rise to capitalism also gave birth to a complementary development in politics and governing: democracy. There are parallels between capitalism and democracy. Rather than the invisible hand of the market, a central idea is that when all vote their self-interest, the process should sort out the result to give the will of the people: the invisible hand of the electorate. Ideally, the individual is paramount. Every person should have a vote either directly or through an elected representative.

In the early days of World War II, Carl Becker wrote an essay on democracy[1] in which he said, "Democratic government, being government by discussion and majority vote, works best when there is nothing of profound importance to discuss. . . . When these happy conditions no longer obtain, the democratic way of life is always in danger." This is one of those latter times.

As for capitalism, reality can bring distortions to the democratic process. Many people do not vote just as many do not participate in stock markets. The rich are the siren servers of the political process, gaining undue influence by lobbying and pouring money into elections. Money is declared to be speech. Companies are ruled to be de facto people. There are efforts to both promote and suppress the vote. As Mark Twain cynically remarked, "If voting made any difference, they wouldn't let us do it."

In addition to the distortions of a complex, turbulent process, there is overt corruption. We find the occasional outright bribable politician, and there are sham elections in autocracies. The vote has been denied to people declared to be less than human. As Winston Churchill said, "Many forms of Government have been tried, and will be tried in this world of sin and woe. No one pretends that democracy is perfect or all-wise. Indeed, it has been said that democracy is the worst form of Government except for all those other forms that have been tried from time to time."[2]

Just as the exponentially developing digital world is changing our economic system, it has also brought challenges to governance. In a short time, digital technologies evolved from being a means to spread democracy to providing platforms to undermine democracy. It is important to realize that widespread notions of democracy have only been with us for a relatively short time, a few hundred years. Changes are happening rapidly, likely too fast for our political processes to respond in careful and thoughtful ways. Already finely tuned computer systems absorb vast amounts of voter data and produce highly contorted boundaries of electoral maps by which parties in power pick their voters and minimize their opposition.

Ubiquitous digital media have simultaneously shattered old political structures by data mining the preferences of individual voters and creating new tendencies to form voter blocs. The 2016 and 2020 presidential elections in the United States showed that even foreign governments could manipulate US elections by meddling with social media. While social media can facilitate protest movements, autocratic governments can use them to organize and promote counterprotests and to surveil and harass opposition activists and journalists.[3] Social media were employed in the Brexit vote for Great Britain to leave the European Union; in the rise of the far right in France, Germany, Sweden, Poland, Hungary; and in the rise of autocrat Rodrigo Duterte in the Philippines. They were involved in the ethnic-cleansing campaign against the Rohingya in Myanmar; Buddhist attacks on Muslims in Sri Lanka; the spreading of false rumors that gangs were kidnapping children to harvest their organs that led to lynchings in Indonesia, India, and Mexico. Jair Bolsonaro used social media to spread COVID-19 misinformation in Brazil.[4] There must be an effort to ensure that social media do not subvert democracy, but it is not clear that democracy can respond sufficiently rapidly to defend itself.

The big technology companies, especially social media companies like Facebook, Twitter, YouTube, and TikTok, have the power to determine who gets to speak on their platform and hence to deliver political messages. The algorithms that gather personal data and target shoppers can also be used to promote conspiracy theories that are antithetical to informed voters and healthy democracies. Facebook may already know you better than do your friends and family. Facebook can predict how you are going to vote with reasonable accuracy if you have pushed the Like button more than about three hundred times on what may seem to you to be random topics.[5] The 2016 presidential campaign of Donald Trump employed campaign techniques pioneered by the Obama campaign of 2008 by employing Facebook exactly as it was designed to facilitate exploitation by advertisers. The campaign used voter data that had been scooped up by Facebook, tested political messages on large numbers of potential voters, and then microtargeted individual voters with messages that were designed just for them. Suppose algorithms can identify a few key voters in closely contested elections and deduce how to nudge them by targeted electioneering. That election could be won not by the politician with the better message but by the algorithm with the best targeted messaging to the correct handful of people.

You may think you share points of view with your neighbors, but with online microtargeting, the commercial and political messages that you are getting are aimed at your proclivities and could be quite different from those received by your neighbor. Politicians have figured out how to push just the right buttons on you and your neighbor in the background with few knowing what is going on. This process promotes a personal rather than public politics. As noted by Harvard professor emerita Shoshana Zuboff, surveillance capitalism and democracy are mutually exclusive.[6]

At the same time, social media have tapped into the innate tendency of people to root for the home team in new ways that distort the political process. Sociologist Zeynep Tufekci of the University of North Carolina at Chapel Hill puts it this way:

> The problem is that when we encounter opposing views in the age and context of social media, it's not like reading them in a newspaper while sitting alone. It's like hearing them from the opposing team while sitting with our fellow fans in a football stadium. Online, we're connected with our communities, and we seek approval from our like-minded peers. We bond with our team by yelling at the fans of the other one.[7]

Unscrupulous actors have exploited this capacity. In the 2016 presidential election campaign, Russian trolls working for the Internet Research Agency in Saint Petersburg created social media sites and then posed as activists across the political spectrum. By seeding extreme posts, they excited the opposition and provided evidence for how bad that opposite team was. The goal was to polarize, fracture, and weaken trust rather than to convince anyone of anything.

THE CORRUPTION OF DISINFORMATION

The founders of social media originally intended to link people and provide information. The creation of an attention ecology was not the goal, but it has been the result. In practice, it is attention—not information—that drives the digital world. Each individual has a limited attention span and capacity. Garnering that attention has become the goal. Attention is effectively the currency that underlies power in our online age. Individuals, politicians, and giant tech companies all strive to extract, wield, and profit from attention. Both the scrupulous and the unscrupulous seek to garner and manipulate it. People who feel they are not getting their share of attention are ripe for political exploitation. The political rise of Donald Trump was, to a great extent, due to his uncanny ability to command great gouts of the currency of attention. Michael Goldhaber, the Berkeley physicist who helped to bring attention (that word!) to the attention economy,[8] is concerned that the attention economy and a healthy democracy may not be able to coexist. As the writer Howard Rheingold said, "Attention is a limited resource, so pay attention to where you pay attention." Also pay attention to those who are deeply resentful of their perceived lack of attention.

Andrew Marantz writing in the *New Yorker*[9] argued that the attention structures associated with social media have changed notions of the public square. Incentives to promote content that titillates, radicalizes, or shocks keep us clicking on links so that is the content that is promoted and that flourishes. Traditional news media governed by traditional libel laws present curated coverage that limits misinformation. Social media drove a Darwinian evolution to draw attention. Their profit incentive was aligned with those who seek to promote outrage and spread misinformation. This evolution has been cloaked in principles of free speech, but its structure is the result of the choices of the users and the attention-driving algorithms of the social media. It was not inevitable but carefully constructed. Marantz concludes that "just as our industrial economy has led us to the brink of climate collapse, our social-media economy is accelerating our political collapse." Even TikTok, which once seemed an innocent repository of short, self-made music videos, threatens to become a major source of mis-information in elections.[10]

While rapid advances in AI have driven surveillance capitalism, the at-tention economy, and the rapid spread of disinformation, we have not yet seen the maximum challenge to democracy that these factors can bring. We have yet to experience an election where deepfake videos (see chapter 3) and powerful AI language programs capable of communicating in idiomatic conversation are broadly employed to propagate disinformation.[11] The com-bination of deepfakes and natural language programs that mimic ordinary human speech could have candidates for office saying things that the origi-nal human would never think or utter. Trust is critical for democracy, and AI has the power to undermine that trust.

Law professors Robert Chesney and Danielle Keats Citron (The Uni-versity of Texas at Austin and the University of Virginia, respectively) sum-marize the range of threats of deepfake technology: distorting democratic discourse; manipulating elections; eroding trust in institutions; weakening journalism; exacerbating social divisions; undermining public safety; and inflicting hard-to-repair damage on the reputation of prominent individuals including elected officials and candidates for office.[12]

Bad guys are out there. On the dark side, generic *ransomware* tools are widespread, allowing digital blackmailers anywhere in the world to lock computers and demand payment lest they destroy your data. We have al-ready seen the widespread public release of computer programs that allow individuals and research groups worldwide to explore the power of machine learning. Technologist Aviv Ovadya warns of the potential for a coming "Infocalypse" in which tools to manipulate perceptions and falsify reality become widely and easily available.[13]

Some research groups have begun to war game the release of programs that make it appear that anything could have happened and anyone could have said anything. The results tend to be frightening and depressing. These techniques could lead to widespread deepfake *astroturfing*, or *polity simulation*: an artificial attempt to create the impression of a real grassroots movement in support of some political goal. AI-driven algorithms could flood the offices of legislators and regulators with demands for attention and the recipients could not readily tell the difference between a real and a fake campaign. If social media can target individuals to sell them stuff, then AI can create fake political messages to individuals that seem to come from people they know. False comments generated by AI using natural language processing could appear real and corrupt the comments ecology that is widespread on the web today. Anyone could do this if the tools become readily available. Imagine a clip purportedly showing North Korean leader Kim Jong-un declaring nuclear war on Japan. There are few current defenses against an aggressive attempt to engage in a willful distortion of the truth for political ends.

The result could be *reality apathy*, a condition when people stop paying attention to or believing anything they see, hear, or read because they do not know what to trust. Even the knowledge of the potential existence of deepfake tools can be damaging.

The mere existence of these techniques gives us a plausible reason to deny a video or post even if it is true. If the threat of fake information causes people to question the integrity of real information, the damage is already done. A single well-publicized political hoax could convince broad swaths of people that nothing is real. This would undermine the solid base of an informed public that is necessary for a healthy democracy.

Some see in this potential not just a threat to democracy but an attack on liberal philosophy. Actions that deliberately attempt to create a post-truth society undermine the foundations of civilization.[14]

All is not yet lost. At the very least, there must be a strong effort to develop cryptographic techniques to verify images, audio, and text. We need to be able to distinguish what is real from what is a manipulated fake. AI threatens to lead us into this morass; perhaps the right AI can provide a defense.

Efforts to counter deepfakes and their kin are underway. Facebook has a policy of removing deepfakes created by AI but not *cheapfakes* produced by cruder manipulation. By declaring cheapfakes insufficiently dangerous and easy to detect, the social media companies allow large-scale misinformation to propagate. Although formally against AI-generated deepfakes, such policies pave the way for potential broader use of more malignant technology. Many big tech companies are wrestling with the issue while trying to maintain a business model that is constructed on techniques that encourage the use of deepfakes and other modes of misinformation. Other companies are building their business by directly contesting deepfakes. One of the goals of the startup Deeptrace Technologies[15] is to develop deepfake detection and associated anti-virus tools. Truepic[16] has developed what they call controlled capture technology that records and verifies origin, pixel content, and metadata in photos and videos and hence ensures they are trustworthy.

A practical problem is that organized deepfake efforts are often produced in countries that are beyond the reach of the legal systems of the countries in which these techniques are deployed. Many deepfakes originate in Russia, Iran, and North Korea.

A TV station in South Korea employed a deepfake representation of a real newscaster. China, on the other hand, has made deepfakes illegal whether used for news or parody.

FREE SPEECH

Closely related to the effect of our digital world on democracy is its effect on fundamental issues of free speech. Freedom of speech in public venues is a basic right spelled out in the US constitution and valued in much of the world. One cannot, however, falsely shout, "Fire!" in a crowded theater. What, then, can one shout on the internet?

Free speech is both a fundamental and a fraught topic in the United States. Consider a law that permits the deportation, fine, or imprisonment of anyone deemed a threat or who publishes "false, scandalous, or malicious writing" against the government. This does not describe new rules imposed by the government of China on Hong Kong but the US Sedition Act of 1798. A century of legal rulings on slander and libel were required before a modern sense of comportment by news media and the meaning of freedom of the press were established. Now we wrestle with new issues of free speech in the context of ubiquitous and rapidly evolving social media.

The American Civil Liberties Union (ACLU) has long been a bastion of free speech, including speech that some find reprehensible. In one of its most famous cases, dating to the 1970s, the ACLU defended the rights of Nazis to march in Skokie, Illinois, despite the fact that the town was home to many Holocaust survivors. The ACLU argued that it was necessary to defend even hateful and offensive speech in order to defend the free speech right of everyone. Even the ACLU has found it difficult to remain a defender of free speech in an era when hate speech is recognized as a form of psychological violence that can lead to physical violence.

As in so many other regimes, the exponential development of social media, including deployment of deepfake technology, threatens to outpace legal and justice systems. Section 230 of the US Communications Decency Act declares that internet companies cannot be held liable for content posted on their sites as a newspaper would be for its published content. That has allowed room for a lot of misbehavior.

Traditional news media are subject to laws regarding slander and libel. Those laws vary in different countries, but they encourage curation and self-censorship regarding published material. Being currently governed only by the liability exemptions of Section 230, the social media companies have broader power and fewer responsibilities than traditional media. The result is that the social media companies basically own and control the public square. That might be manageable if there were vibrant competition among many companies, but in the current environment, the policies of the social media are controlled by a few siren servers and hence by a handful of individuals at the top of these companies. The result has been growing calls for anti-trust action and other restrictions on the big tech companies. Should they be declared public utilities and thus regulated in a similar fashion? Must they open free discourse to lies and slander? Should there be a means to verify that postings come from real people, even if anonymous, or at least a labeling system to differentiate human posting from AI natural language and deepfake posts?

US citizens can say almost anything they want in a public venue, including things that others consider abhorrent. The social media companies are, however, businesses, and there is no right to be heard on their platforms. These companies can thus define restrictions on posted content as defined in their terms-of-service agreements. In the words of attorneys Chesney and Citron, that makes these agreements "the single most important documents governing digital speech in today's world."[17] This situation gives the companies immense power to control and limit speech on what is otherwise an explicitly public platform with global reach. This also gives these companies a responsibility that they did not necessarily seek but nevertheless have. They struggle to write reasonable terms-of-service agreements, yet nothing but their conscience and quest for profit forces them to do so.

If a terms-of-service agreement specifies that only content espousing Nazism or only content supporting racial justice and immigrant rights is allowed on the platform, then if you accept the rarely read agreement, those are the rules to which you accede. If the rules more reasonably say that lies are not allowed, then you can be booted off the platform for lying regardless of your political clout. Some argue that the social media platforms are de facto the new public square demanding unfettered free-speech access. Digital media are so woven into our work and personal lives that access may feel like an inalienable right. The First Amendment, however, says, "Congress shall make no law abridging the freedom of speech, or of the press." The Constitution is silent on the degree to which Facebook and Twitter can regulate speech on their platforms. They are free to ban all political speech if they should so choose. Even presidents can be banned.[18] On the contrary, a government attempt to force a social media company to propagate or suppress certain political discourse would be a violation of the First Amendment rights of that company.

Just as it took a century to assimilate and ameliorate the US Sedition Act, terms-of-service agreements will continue to evolve. These agreements might be modified to give users more power over their personal data including the ability to call for the removal of any deepfake that affects them. Another possibility is that Congress will step in and modify Section 230. That will be an interesting discussion, but at least one held by elected representatives and not a handful of unelected CEOs.

Business writer John Mauldin points out that while Facebook, Twitter, and YouTube garner much of the attention, any attempt to regulate free speech in the context of social media involves a complex ecosystem of companies that need to be considered in any policy changes.[19] To complicate matters, that ecosystem is evolving exponentially rapidly but policy does not. Restrictions on free speech can come from either the political right or the political left. Freedom of religion and freedom from religion are also important for freedom of opinion and expression. Finding the right balance will be tricky and contentious. There must be free and open discussion in the public marketplace of ideas, but how to accomplish that without trampling on other rights is far from obvious.

ALTERNATIVES

In China, the goal is not to promote democracy but to maintain social and political control.[20] China has instituted a social credit system that gives people positive points for good behavior and demerits for unwelcome behavior. This system is bolstered by China's broad use of facial recognition software so that miscreants can be caught in the act and identified by authorities. The determination of what is positive and what is negative is up to government leaders. Positive behavior includes volunteer work or giving blood. Among behaviors to be penalized are jaywalking, illegal parking, not visiting aging parents often enough, failing to keep restaurant bookings, and cheating in video games. Penalties can be denial of the ability to fly or even use of public transportation.

There are other efforts underway to eliminate any central authority by employing blockchain technology to host websites and create social media networks. This would make it more difficult for any company or government to control anti-democratic influences by deleting content or banning certain accounts. The same developments would undermine efforts by autocratic countries of the left or right to control their citizenry by manipulating access to the internet.

An important issue is thus how to work within the framework of our democracy and the current structure of social media to design a new, better system. A foundational principle of our democracy is a free press with the power to investigate and hold accountable both private and government enterprises. That structure—especially local news organizations funded by traditional, non-microtargeted advertising—has been undermined by the ubiquity of misinformation-filled news feeds on social media. How do we restore the severely decimated, traditional, local news structure? Breaking up social media giants into smaller companies using the same click-bait incentives that drive misinformation is not a healthy solution. We need new checks and balances, new guardrails for both social media and our political structure.

Amidst these uncertain conditions, there are glimpses of how technology might be adapted to strengthen rather than undermine our democratic processes. An interesting experiment is proceeding in Taiwan.[21] Taiwan has appointed a digital minister, Audrey Tang, who helped Taiwan navigate the COVID-19 epidemic with sufficient community cohesion that no restaurants closed and there were no mass protests against masks and vaccines.

Tang got her start during protests in 2014, when students occupied the parliamentary building demanding greater transparency in the governing process. In order to communicate what was happening within the building Tang brought in about one thousand feet of ethernet cable and set up a Wi-Fi network of cameras and microphones connected to a large projector on the street outside. The result was that the public could witness in real time exactly what was going on within, what topics were being debated, and what individuals were saying while providing feedback from the crowd outside the parliament building and from vastly more people online. The result was a transparent communication system that was responsive to public opinion, what Tang calls a "listening society." After three weeks, a consensus formed, the protest ended peacefully, and a new vision of how to operate a democracy was born.

The roots of this were already in place in about 2012 when a group of hackers, including Tang, developed open-source software for which everyone involved could offer their own opinion and modifications to the code. In the elections of 2014, candidates who supported open transparent government won and those who did not lost. Voting works! Tang began working with the minister for law and ended up training a thousand government workers in the digital techniques of building consensus and was ultimately appointed the first digital minister.

One technique employed by Tang and her cohort was to establish an interface where citizens could register their opinion on some issue. The interface allows others to upvote, downvote, or pass on each opinion. There is no reply button so no avenue for trolls to hijack the discussion. An algorithm does a principal component analysis to identify the most contentious sentiments. Another algorithm does a machine learning cluster analysis that identifies people with shared sentiments and the ideas that unite them. In the beginning of the discussion of a given issue, sentiments might be all over the map, but by this feedback process, the wisdom of the crowd tends to bring the best ideas to the top and clear supermajorities then form that span competing groups. The software automatically attracts consensus rather than the division of outrage, disinformation, and misinformation of current social media. Stakeholders are then brought to the table for face-to-face interaction and the discussion is live streamed. Consensus points are presented and the question How do we transform these into regulations? is posed. The process of arriving at new regulations is there for all to see, the logic is transparent, and the results are widely anticipated and accepted.

Might these techniques be adopted more broadly in democracies around the world before other technologies overwhelm them?

THE END OF LIBERALISM?

Harvard law professor Lawrence Lessig argued over two decades ago that computer scientists were the de facto regulators of the digital age.[22] Now in light of blockchains, decentralized finance, and smart contracts that purport to be value free and incorruptible but are not, he pleads, "We need a more sophisticated approach, with technologists and lawyers sitting next to behavioral psychologists and economists" to ensure the underlying computer code reflects social values rather than the values of private interests. He cautions, "We're facing an existential threat to our democracy, and we don't have 20 years to wait."[23]

Succumbing to the promises that the inventors of disinformation-spreading recommendation algorithms can solve the problems that threaten democracy is, to paraphrase writer Anand Giridharadas,[24] somewhat akin to calling on the arsonists to fight the fire. Tech leaders might use the opportunity to enhance their power even more. Still there are things that the tech giants could do to facilitate free and fair elections and to encourage the acceptance of election results.[25] The purveyors of popular browsers could provide voting information and updates of election results as their default home pages. Social media could do the same with their home pages. Users would see a truthful summary of election results before entering the standard maelstrom of their feeds.

While there are glimmers of hope, or at least suggestions for the right questions to ask, a fundamental issue remains whether democracy has the power and flexibility to adapt to the exponential onslaught of technology. If things change too rapidly, distributed, democratic decision-making processes may not be sufficiently efficient to respond. For all the flaws of autocracies, their top-down structure enables more rapid decision making. As the exponential technological wave surges, there could be an evolution away from democracies and toward autocracies. It is not clear even autocracies as we envisage them today will survive the onslaught. Russia and North Korea are fighting decades-old battles over guided missiles and nuclear bombs. Both China and the United States are striving to dominate future developments of AI and quantum computing. All four governments may be overwhelmed by the rush of technology.

Historian Yuval Noah Harari argues that a fundamental challenge to the notion of liberalism that underlies our principles of democracy is not the direct threat of AI but the profound growth of basic understanding.[26] The revelations of the workings of our biology and brains suggest that we are biological algorithms that strive to make sense of our experiences through what we identify as consciousness (see chapter 8). If this is true, then the basic notion of humans as free individuals that is the basis of our liberal regimes of business, law, and politics are severely undercut. Nevertheless, Harari also concludes that there will be a flood of developments driven by the exponential advance of technology. This technology may take no account of whether or not humans maintain the liberal notion that the free choices of each individual are paramount. If humans lose their value as drivers of the economy and as cannon fodder, then the political and economic system may have little use for them individually or collectively.

AN ENHANCED ELECTORATE

What happens to democracy and the value of the individual if some individuals become biologically, mentally, and algorithmically enhanced? Do those people still get just one vote? Will the poor and the rich each still get one vote if the rich have materially enhanced themselves? What is the power of one man, one vote if AI-driven telepathy becomes common so that we all know what the hive is thinking and every member in it?

As we move into an era of the IoT, edge computing, smart cities, and self-driving cars, we make data-driven decisions in more and more areas of life.

Why not let the data on me decide my vote? When AI-driven systems know individual voters better than they know themselves, individual voters might decide to let Siri vote for them because Siri has been monitoring them since birth and knows all their quirks and interests. Siri will not skip voting or be distracted by some momentary disruption, like a sick child, on election day, or be corrupted by a bribe, misinformation, or electioneering. Siri will not vote based on my temporary mood on election day but after balancing all my interactions over a lifetime. It would be crazy not to let Siri vote for me.

What is the point of a traditional election? After all, elections are just exercises in statistics. With enough of the right kind of data, an AI could just simulate the whole body politic and declare the winner: this percentage of the electorate votes for candidate A and this percentage votes for candidate B. Leave the messy humans out of the loop entirely. Since the algorithms are so capable and intelligent, replace the candidates with algorithms as well. Then don't have elections every four years but every four femtoseconds in a nearly constant update to status. The result would work smoothly and efficiently and leave our current concepts of democracy far behind.

Companies have many of the rights of individuals: free speech, the right to own property, the right to sue, the right to exercise religious principles. At the present time, companies cannot vote, but they can lobby governments and nudge their employees in certain political directions. Will there come a time when companies secure the right to vote? If people can sign smart contracts in a decentralized world and an AI can sign those contracts, can an AI own a company? If so, could the AI gain a right to vote through that path? If AIs, having outsmarted humans in virtually every economic enterprise, own the vast majority of the resources both material and intellectual, will it matter if AIs can vote? If AIs do vote, for what will they vote? Human candidates and policies? Their own unfathomable criteria?

What if something such as Kurzweil's singularity with the development of self-coding, conscious, general AI, does occur? What rights will such an entity have? Will there be one AI, one vote? With AI replicating at the speed of light, mere humans, even augmented humans, would be vastly outnumbered and outmaneuvered and hence virtually irrelevant to the political process.

There are currently small steps being taken to address some of these issues. Stephen Thaler is the founder of Imagination Engines in St. Louis, Missouri, and cofounder and director of Scentient.ai in Cambridge, Massachusetts. He used a general-purpose neural network code called DAUBUS (device for the autonomous bootstrapping of unified sentience) to write the code for a specific process labeled "a device for attracting enhanced attention." Thaler applied for a patent claiming that DAUBUS, not he, was the inventor since DAUBUS was not designed to address the particular goal of the process to be patented, and that he, Thaler, did not contribute to the specific invention.[27] The US Patent and Trademark Office (USPTO) refused to grant the patent on the somewhat circular grounds that the law defines *inventor* as an individual who invented the subject matter of the invention. Convinced that DAUBUS was the relevant individual, Thaler sought judicial review but the District Court for the Eastern District of Virginia concluded that the USPTO correctly decided that an inventor under the Patent Act must be a human. Thaler then appealed to the federal circuit court that again ruled against Thaler, arguing that the law refers to *himself* and *herself* but not *itself* with the common understanding that the inventor must be human. Courts in the European Union, the United Kingdom, and Australia made similar rulings against Thaler (a tenacious man). In a related case also brought by Thaler, the US Copyright Office Review Board reaffirmed that human authorship is necessary for copyright protection of a work of art even if the art is created entirely by a neural network like DALL-E with no contribution from a human. The only crack in this wall of opposition was South Africa, which did grant patents with DAUBUS as the inventor. With that exception and the great success of natural language programs like ChatGPT to generate original prose and DALL-E to create art, I suspect we have not heard the end of these issues concerning the legal rights of AI.

Free markets respond rapidly to small changes in circumstances but have been unresponsive to the potential threat of AI. With rapid change, it may be that no human enterprise, democratic or autocratic, can respond sufficiently rapidly. Governance may devolve to our machines, and antiquated notions of both democracy and autocracy will fade away. Who or what will be in control then?

Perhaps none of these fantasized, anti-democratic, technology-driven developments will come to pass. One of the ways to ensure that is to anticipate possible developments that will come at us exponentially rapidly. Then we might be able to take steps to ensure that unfavorable outcomes do not surprise and overwhelm us. The nonprofit Center for AI and Digital Policy[28] was formed in 2021 with the goals of assessing national AI policies and practices; training AI policy leaders; and promoting democratic values for AI.

During the presidential inauguration of 2021, the young inaugural poet Amanda Gorman gave a stirring reading that included these words:

> But while democracy can be periodically delayed
> it can never be permanently defeated.[29]

That heartfelt prediction may not prove true.

14

SPACE
The Difficult Frontier

DIMENSIONS

The problem with space is that it is so damn big, maybe bigger than we perceive or can conceive.

The nature of space remains one of the most confounding issues to physicists. Physicist John Archibald Wheeler captured this spirit in his remark that "in order to more fully understand this reality, we must take into account other dimensions of a broader reality." In my classes, I taught my students the secret hand signal available to denizens of three-dimensional (3D) space. Hold the fingers of one hand so that your middle finger, index finger, and thumb point in three mutually perpendicular directions. There need not be a special up-and-down or back-and-forth. Wave your hand around. Point your index finger in different directions. You can still maintain the condition that your two fingers and thumb point in three different directions. If you happened to be a 2D creature, you could not do that. You could only point in at most two mutually perpendicular directions. We are members of a special 3D club, and with that hand signal we can identify one another. I further challenged the students to point in three orthogonal directions and then point in a fourth direction that is perpendicular to all of the first three. We cannot do that. We are restricted to perceiving and occupying 3D space. On the other hand, if there were creatures that occupied four-dimensional space, they could point in four mutually perpendicular directions and make corresponding hand signals that we could never conceive or perform.

Is that it? Three-dimensional space? Perhaps not. Physicists have struggled to combine the theory of gravity as enunciated by Einstein with quantum theory that describes the submicroscopic world of particles—electrons, protons, and neutrons. Einstein characterized gravity as the result of a curvature of space. His theory has been verified again and again, most recently by the detection of gravitational waves—ripples in space—and by direct images of the warped space around gigantic black holes. Quantum theory has also been spectacularly verified. Quantum theory underlies all our modern digital world. We engineer the quantum behavior of electrons in computer chips to make it all possible. Quantum computing takes advantage of even more extreme aspects of the theory: that particles can exist in many different conditions simultaneously and that changes in those conditions can be triggered remotely and instantaneously.

A profound conundrum is that these two great theories contradict each other under certain extreme conditions. The goal of physicists is a new theory of everything that unites Einstein's theory and quantum theory. The result would be a theory of *quantum gravity* that would incorporate the current theories as excellent approximations but not the whole story just as Newton's theory of gravity turned out to be a very good approximation of Einstein's theory but not the whole story.

The best candidate we currently have for a theory of quantum gravity is string theory.[1] In this theory, the fundamental entities are not particles but tiny quantum-vibrating strings of energy. Different vibrations of otherwise identical strings reproduce all the known particles. Some of the strings vibrate in ways that represent particles of gravity. The mathematical description of their behavior is exactly the theory of gravity that Einstein constructed a century ago. String theory includes Einstein's theory so is *a* theory of quantum gravity if not *the* theory of quantum gravity. String theory has yet to be experimentally tested, and its implications are still being explored, but it gives us a framework with which to see how our physical understanding of the Universe may develop.

A truly remarkable aspect of string theory is that to make it work with mathematical consistency, the fundamental strings must vibrate in ten-dimensional space. In the unlikely possibility that there were creatures fully occupying that space, they would be able to point in ten mutually perpendicular directions. What a hand signal that would be!

In the current implementation of string theory, physicists view most of the higher dimensions as highly curved. If you headed off in one of those directions, you would find yourself back where you started before you had gone much further than the circumference of the fundamental strings themselves, somewhat like circumnavigating the curved surface of Earth. In this view, space has three large dimensions: the space we occupy and in which we flash our secret hand signals, and at least six tiny, intensely wrapped higher dimensions. That leaves the possibility that there is one more large—very large!—higher dimension.

This large fourth spatial dimension[2] could itself contain a vast number of 3D universes such as the one we occupy and observe, just as there could be many 2D sheets or 2D bubbles in our familiar 3D space. That notion is captured in the term *the multiverse*. Other 3D universes could have their independent existence and dynamics in the multiverse or they could be connected in some way. Some hypothesize that every time a black hole forms in our Universe, a new universe winks on with its own Big Bang elsewhere in the multiverse.[3]

We do not know that the multiverse exists, but we have been led to the possibility by decades of rigorous mathematical work. The idea of the multiverse is not merely a science fiction dream. The notion that the multiverse and the tiny wrapped-up higher dimensions might exist gives us a framework to raise some interesting possibilities.

Even if we restrict ourselves to considering our familiar 3D space, there is this one thing about it. It is very big, maybe even infinite. We may not be able to see all of it because space as we know it began only about 14 billion years ago in the Big Bang. We see by detecting light. Light travels at a fixed speed of one light year per year.[4] Since the Universe as we know it is only 14 billion years old, we can only see out to 14 billion light years in all directions. That is unfathomably far on any human scale, but it is a finite distance.

We have learned that the volume of space that we can observe is filled with galaxies like our own Milky Way spiral galaxy. The galaxies contain stars similar to those we can see in the night sky: big ones, little ones, bright ones, and dim ones. Even getting a grip on the size of space in our local neighborhood is a challenge. The nearest star is, of course, our Sun, and it is already far enough from Earth that light requires eight minutes to get here. The next nearest star is four light years away. The Milky Way is more than 100,000 light-years across. The nearest sizeable galaxy beyond the Milky Way is about 2 million light-years away. Things get distant very rapidly as we expand our scope.

With a complex array of sophisticated equipment, we can detect electromagnetic radiation of all wavelengths: the optical light that treats our eyes but also radio, infrared, ultraviolet, X-rays, and gamma-rays. We also register a zoo of particles that reach Earth from space. We can see into deep space, but we cannot go there—yet.

Our probing of space in a variety of clever ways has revealed the basic framework of our observable Universe. Space and time as we know them blinked into existence with the Big Bang. Our current physics breaks down at that instant of creation. We do not know why there is something in this Universe rather than nothing. We do not know what happened before the Big Bang. Those questions remain some of the biggest challenges to physics and cosmology; however, our current understanding begins to apply a mere trillionth of a second after the Big Bang. We do not know everything, but we know a lot.

We know that starting from the immensely dense and incredibly hot conditions of the Big Bang, our Universe has been expanding and cooling ever since. It is not that space just sits there and galaxies move outward in it, like a bomb exploding in outer space. It is that every bit of space itself is moving apart from every other bit of space. The galaxies are basically at rest in their local region of space and are being carried further apart from all the other galaxies by the expansion of the underlying space. It is like standing still on an escalator while moving upward but in all three directions at once.

We also know that the seeds of the formation of stars and galaxies began in tiny quantum fluctuations in those early conditions moments after the Big Bang. We have learned that our Universe is pervaded by particles of dark matter that gravitate but emit little or no light. We understand that galaxy formation was catalyzed by the clumping of this dark matter, and that the normal matter—electrons, protons, and neutrons—so critical for us to exist was merely along for the ride. We further realize that our normal particles are not normal at all. There is five times as much dark matter as there is stuff like us. We shine, but otherwise demographically we do not count for much.

We have also discovered that our Universe is pervaded by a yet mysterious dark energy. The dark energy does not fit within our current understanding of physics, and it anti-gravitates. The dark energy causes the expansion of our Universe not merely to coast, but to accelerate in an outward rush. The dark energy comprises three times the energy density of the dark matter so it is not to be trifled with. Some speculate that the dark energy results from the interaction of our Universe with another nearby universe that is slightly displaced from ours in the multiverse, but we don't know.

All this may seem abstruse, but the nature and extent of space and the possibility of higher-dimensional hyperspace are intrinsically bound up in the question of the future of humanity. We will return to these connections below.

ROCKETS

We know space is there and something about it. How do we get there? The current answer is by rocket.

The modern development of rockets was originally a personal enterprise in the hands of Robert Goddard. The notion that you could fly by just pushing hot gas out a nozzle was quite radical at the time although the Chinese had been building and launching fireworks for thousands of years. Goddard realized that his technique would work in the vacuum of outer space whereas that of the Wright brothers would not. Driven by the exigencies of World War II, rocket technology shifted into the hands of governments and became more sophisticated. The United States and the Soviet Union caught up after the war by importing the German scientists and engineers who had fostered the advances. Competing governments dominated the next several decades in the context of the Cold War. President John F. Kennedy initiated the bold if economically questionable Apollo program to put people on the Moon in open competition with the Soviet Union.[5]

The human space flight program subsequently retreated to low Earth orbit, but the space program otherwise flourished. Much of the effort remained in the hands of governments as spy satellites replaced airplanes and we began our exploration of the solar system. There were commercial efforts from the beginning with US aerospace companies working closely with government enterprises. The companies built the satellites and rockets, and the government—NASA—provided budgets and the launch facilities. The result was a plethora of satellites to spy, to communicate, and to monitor the surface of Earth and our Universe beyond.

Commercialization efforts took on a different tenor when Mike Griffin became NASA administrator in 2005. Griffin had worked for the private American Rocket Company that developed rocket engines and then was head of In-Q-Tel, a venture capital company that invested in technology of interest to the CIA. Griffin thus had a propensity to expand commercial investment in space exploration and development. He promoted the NASA Commercial Orbital Transportation Services (COTS) program to carry cargo and crews to orbit, including resupply of the International Space Station, and a NASA prize program to entice entrepreneurs to contribute. His stated goal was to get the competitive juices of the commercial sector involved so that NASA's space exploration effort would be sustainable over multiple administrations and Congresses. In a speech at NASA headquarters toward the end of his term, Griffin said,

Those of us on the government side of the space business must recognize a fundamental truth: if our experiment in expanding human presence beyond the Earth is to be sustainable in the long run, it must ultimately yield profitable results, or there must be a profit to be made by supplying those who explore to fulfill other objectives. We should reach out to those individuals and companies who share our interest in space exploration and are willing to take risks to spur its development.[6]

While there was some resistance to this philosophy within NASA and the traditional aerospace industry, this torch for commercialization was carried in the following NASA administration by Lori Garver, who was deputy administrator from 2009 to 2013.[7] The US government and commercial partnership was extended in 2021, when NASA signed a formal memorandum of understanding with another branch of government, the FAA, to develop specific guidance for commercial launch and reentry and for informing the public of the risks associated with commercial space travel.[8]

As encouragement of the commercialization of space was coming from the highest levels of NASA, another catalyzing change was developing: the advent of the internet and our digital age. It was not that the internet directly altered the space program but that stupendous wealth was generated, and that shifted the budget perspective. The ability to pay the immense capital costs of the space program was no longer solely in the hands of governments but within the capacity of some private individuals.

BILLIONAIRES IN SPACE

It is not clear that Richard Branson, Jeff Bezos, and Elon Musk made their billions in order to foster their dreams of space travel, but all three had those dreams when younger and implemented them when they acquired the resources.[9] Branson was seven when Sputnik was launched. He founded and grew the Virgin Group, which now comprises four hundred companies including Virgin Atlantic; Virgin Trains; Virgin Media; Virgin Money; Virgin Records; and a space company, Virgin Galactic. Jeff Bezos was five when Neil Armstrong and Buzz Aldrin stepped onto the Moon. He turned Amazon from an online bookstore into a giant enterprise selling everything to everyone and used some of the proceeds to found Blue Origin. Elon Musk grew up during the era of retrenchment with humans only in low Earth orbit—the era of the Space Shuttle and the ISS. He may have chafed under those restrictions. Musk was trained in physics and economics and has displayed perhaps the greatest creative range with his wealth. He made his first $300 million selling a software company and then cofounded and sold PayPal. With those resources, he founded diverse other enterprises including an electric vehicle company, Tesla; a tunneling company, the Boring Company; and a BCI enterprise, Neuralink; and he cofounded the nonprofit company OpenAI that aspires to promote ethical AI that benefits humanity. He also built a space exploration technologies corporation, SpaceX.

There are other billionaires in the world. These three put their money where their space passions were. They also did not solely use their own bank accounts but sought funds from investors and contracts from NASA. Presumably each of them initially sat down with a handful of trusted associates and said something like "This is not our traditional business, but I want to do it. What do we do financially and technologically to make it happen?" All three built upon decades of research and development by NASA; they did not start from scratch but sought to improve on the technology.

It makes sense to launch spacecraft into orbit from near the equator. The motion of the surface of our rotating Earth through space is most rapid there compared to higher latitudes. The spin of Earth at the equator thus already gives you a boost of about 1,000 miles per hour. That is why NASA built the Cape Canaveral facilities in Florida and the Air Force constructed the Vandenberg launch facility in southern California. It takes a speed of about 17,000 miles per hour to get into space so there is still work to be done.

The three space enterprises started by Bezos, Branson, and Musk followed this equatorial launch logic but eschewed the coasts and focused on the Southwest. Bezos built the test and launch facility of Blue Origin near the small, isolated town of Van Horn in the empty expanse of West Texas. Branson chose the desert of New Mexico to build Spaceport America, the home base of Virgin Galactic that is about forty driving miles from the town of Truth or Consequences. Musk started SpaceX in Southern California but shifted his focus to an area around the small South Texas town of Boca Chica where launches would carry out over the water of the Gulf of Mexico.

None of these enterprises was an overnight success. Blue Origin was founded in 2000, SpaceX in 2002, and Virgin Galactic in 2004, all before Mike Griffin became head of NASA. There were then two decades of hard work, progress, and setbacks. Besides one another, they also faced competition from traditional aerospace giants such as Boeing and Lockheed Martin.

According to Lori Garver, Bezos has an expansive, long-range vision of the human space program. Firing tourists into space is but a small first step to show that income can be made. Our Moon and Mars are way stations. Bezos envisions 1 trillion humans in space residing in rotating cylindrical colonies as discussed by Gerard O'Neill and treated in science fiction.[10] To begin that ambitious journey, Bezos has sold billions of dollars of Amazon stock to finance Blue Origin, thus minimizing the need for short-term revenue.

While not averse to showmanship, for a long time Bezos operated Blue Origin in relative secrecy. In 2006, a test craft was launched a few hundred feet in the air. Bezos commented on the experiment in a blog post nearly two months after the fact. There was silence in the press while the work went on. In 2011, the *Wall Street Journal* finally managed to penetrate the veil to report that a test rocket had reached 45,000 feet—not even the formal edge of space at sixty-two miles—before suffering some instability and exploding. Blue Origin posted a brief acknowledgment on its website. Some warning of the flight was given when the FAA issued a temporary flight restriction for the air space of Van Horn. The good folks of Van Horn probably also saw the flash of the explosion.

That flash was in the service of developing a suborbital craft called New Shepard after Alan Shepard, the first American in space. Blue Origin continued to test New Shepard and its safety systems. The next fifteen uncrewed tests, including three of the capsule escape system, were success-ful. The New Shepard program perfected the skill of guiding the booster rocket to a vertical landing on its tail. In 2015, the second successful test flight landed the booster within five feet of the center of the launch pad, the first time this had been accomplished. The capsule separated after the peak altitude and descended on parachutes to land nearby. The ability to land and reuse the booster and capsule was a principal goal in Bezos's quest to reduce the cost of getting people and supplies to space.

The New Shepard program celebrated its success with a well-publicized launch on July 20, 2021, the fifty-second anniversary of the original Moon landing. The flight carried Bezos himself, his brother Mark, Wally Funk (a pioneering woman astronaut who was never given a chance to fly by NASA), and Oliver Daemen, a young customer whose father paid his way. The New Shepard booster flew from the Van Horn launch site to an alti-tude of sixty-six miles and then gracefully settled back down on its tail, to be used again. The passengers in the pressurized capsule experienced three minutes of weightlessness and could see the curvature of the Earth. The capsule, slowed by its three parachutes, landed a little over ten minutes after launch. On October 13, 2021, Blue Origin gave the same ten-minute ride to ninety-year-old actor William Shatner (who played Captain James

T. Kirk of *Star Trek* and now became the oldest person yet to venture into space) along with two paying customers and a Blue Origin vice president.[11]

Blue Origin plans to charge tourists to fly to the edge of space and other customers to fly zero-gravity experiments on New Shepard. Bezos acknowledges that flying tourists is far from the ultimate goal but invokes the early days of civil aviation when barnstormers entertained customers and spread the interest in flying. Blue Origin has not yet advertised what the routine costs will be.

In parallel with the New Shepard suborbital program, Blue Origin has developed rocket technology and a market to sell its rocket engines to others, including nominal competitor United Launch Alliance, a joint venture between Lockheed Martin and Boeing. Blue Origin has a factory to produce rocket engines in Huntsville, Alabama, the site where NASA developed the famed Saturn V that powered the Apollo program. Blue Origin is also developing its own orbital launch vehicle, New Glenn, named for John Glenn, the first American to reach orbit. New Glenn is designed to carry cargo and crews into orbit. The first stage of this two-stage rocket will be twenty-three feet in diameter and powered by seven Blue Origin engines. Blue Origin has a facility near the Kennedy Space Center in Florida to assemble New Glenn rockets. It plans to launch New Glenn from the NASA Florida launch facilities and to pursue international markets to launch communication, TV, and other satellites. Blue Origin has declared that it will not compete with United Launch Alliance and SpaceX for national security or military applications.

The long-term goal of the Obama administration was to fly to asteroids for scientific reasons and for the possibility of mining abundant rare earths that are critical to modern technology. The Trump administration returned the focus to the Moon with the Artemis program to land the first woman or the next man on the Moon by 2024. Many thought this to be an impossibly ambitious and underbudgeted goal (it was) but that has not stopped many commercial organizations from joining the competition. In 2020, Blue Origin partnered with Lockheed Martin and Northrup Grumman to win a short-term NASA contract to design a system to return astronauts to the Moon. SpaceX won another of the three design contracts while Boeing competed but was unsuccessful. Blue Origin presented the Blue Moon lander concept to fly on the New Glenn. The lander would be an updated version of the Apollo system in which a host vehicle would remain in orbit while the lander would proceed to the surface and return using concepts from the New Shepard program. NASA then did a down-select in 2021, choosing the SpaceX design. Blue Origin protested the decision. This game is far from over.

Elon Musk started SpaceX after the founding of Blue Origin but has nominally had more commercial success. This is in part because he had to. Like other Musk enterprises, SpaceX is bold but has had perilous setbacks that have threatened its financial stability. As for Blue Origin, one of the goals of SpaceX was to reduce the cost of getting people and supplies to orbit. The long-range goal is to populate Mars.

SpaceX designs and builds rocket engines, the rockets themselves, and capsules to carry crew. SpaceX developed the single-engine expendable Falcon 1. In 2008, SpaceX became the first privately funded company to launch a rocket to orbit. The first three launches failed, leading various Musk enterprises, including Tesla and Musk himself, to near bankruptcy. The fourth Falcon 1 was successful, and the fifth carried a Malaysian satellite to orbit. On that basis, NASA awarded SpaceX a contract to resupply the ISS. That contract saved Musk enterprises from insolvency. SpaceX designed a Falcon 5 powered by five engines but it abandoned that project in 2005 in favor of designing and then constructing the bigger Falcon 9. Nine Raptor engines designed and manufactured by SpaceX power the first stage of the Falcon 9. The second stage has a single Raptor engine for the final boost to orbit.

In 2010, SpaceX became the first private company to launch, orbit, and retrieve a spacecraft: its Dragon capsule boosted by a Falcon 9. Funding for this project came from Musk himself, investors, private equity, and a variety of NASA contracts. In 2012, an unmanned Dragon capsule was sent to the ISS. SpaceX used a Falcon 9 for the first takeoff and powered vertical landing of an orbital rocket in 2015; accomplished the reuse of a Falcon 9 in 2017; and used the Falcon 9 to send a crew to the ISS aboard a new Crew Dragon capsule in 2020. There have been setbacks. A Falcon 9 destined for the ISS exploded after launch in September 2016. Musk first argued that the problem was that carbon fiber tanks holding super-cold liquid helium that were nested inside the liquid oxygen fuel tank froze the oxygen into solid form that could not flow into the engine of the second stage. Final analysis showed that a steel strut that supported the helium tank had broken, allowing the helium to mix with and freeze the oxygen. That problem was fixed, and there have been many subsequent successful launches. The Falcon 9 has now proven itself a reliable vehicle having flown and been reused over one hundred times.

In 2016, SpaceX began design work on an interplanetary transport system with the ultimate goal of carrying people and cargo to permanently occupy Mars. This enterprise will also have two stages: a booster called Super Heavy and a second stage named Starship. Both stages are enclosed in stainless-steel hulls thirty feet in diameter and reusable. The second stage is designed to be a long-duration vehicle capable of landing elsewhere. Both stages will be powered by SpaceX's Raptor engines. The Super Heavy will have around thirty engines and the Starship will have six—three optimized for operation in the atmosphere and three for the vacuum of space. The second stage was launched and recovered from low altitude in 2019 and from high altitude in 2021. A flight test to orbit of both stages is planned and has not yet been implemented. In 2020, NASA selected Starship as one of the three preliminary design studies for an Artemis lunar landing system and then in 2021 gave Starship the single contract to proceed. The first attempt to launch the Starship into orbit from Boca Chica on April 20, 2023, resulted in the failure of a few of the thirty engines to ignite, loss of control, and deliberate destruction of the rocket over the Gulf of Mexico. The power of the many engines that did ignite blew the launch pad to smithereens because a shield to channel the exhaust was not yet in place.

While the headquarters of SpaceX remains in California and SpaceX has a factory in California that produces the Raptor Vacuum engines, much of the SpaceX effort is moving to Texas, as are other Musk enterprises, including Tesla corporate headquarters and an electric vehicle manufacturing site; Boring Company headquarters; the headquarters for Musk's private foundation; and Neuralink facilities. Another engine production facility in McGregor, Texas, is producing about one thousand engines per year. A factory near Waco is planned to produce an even more advanced engine design. SpaceX has also established what it calls Starbase in the Texas Gulf Coast town of Boca Chica with other facilities planned in nearby Brownsville. These facilities will serve as a rocket production facility and launch site for prototypes of the Super Heavy and the Starship.

These facilities are part of Musk's goal of putting a city on Mars by 2050. At a talk at the International Astronautical Congress in 2016, Musk outlined his ambitious plans for Mars.[12] He argued that he could take passengers to Mars as early as 2024 if there were no hitches but, as we all know, hitches happen. Musk estimated a cost of approximately $10 billion that he hopes will be funded by a private-public partnership. He envisaged that each vehicle would carry one hundred passengers, each paying about $500,000, though the price should drop with time and scale. Musk speculated that a self-sustaining civilization on Mars would require about 1 million people and that getting them there would take ten thousand flights for the people and many more for equipment and supplies. A Mars city might take twenty years to construct and up to a century to become self-sufficient. Musk acknowledged that the first stages of such an ambitious project were dangerous. The chances that some people would die, he said, were "quite high." Musk himself has no immediate plans to go into space.

Whether Musk's goals for Mars are practical, the success of SpaceX in reducing the price of launch to orbit forced competitors to do the same. The result gave some credence to Mike Griffin's invocation of the powers of free enterprise. The Falcon 9 is roughly ten times cheaper in cost of kilograms to orbit than NASA would have accomplished with its traditional mechanisms. The result is that the United States now dominates the market for launch to orbit. Competitors include China with its Long March rockets, France with the Vega, and Russia with the Soyuz-2.

While SpaceX has successfully produced income from its launches of satellites of various sizes for governments and companies, one of its undertakings spurred the ire of astronomers. With the laudable goal of providing global internet access, SpaceX launched a fleet of Starlink satellites, sixty at a time. The goal was thousands of satellites feeding internet signals to ground stations from low Earth orbit. The problem was that the satellites reflected sunlight and earthlight and contaminated the dark skies that astronomers depend on for their most sensitive, ground-based observations. Photos of distant, dim galaxies were suddenly contaminated by the streaks of sixty Starlinks. Given the commotion from astronomers, SpaceX has endeavored to reduce the reflectivity of the satellites, but some ill will remains in corners of the astronomical community.

Virgin Galactic was founded two years after SpaceX and four years after Blue Origin with somewhat different immediate goals. Rather than heavy-lift vehicles launching vertically directly to orbit, Virgin Galactic designed staged systems that could take off and land horizontally like an airplane.

An important background for Virgin Galactic's approach was the privately funded Ansari X-prize. Peter Diamandis invented the concept of the X-prize in 1996. The idea was to stimulate a barnstorming project that would help to promote a commercial space industry and make space more publicly accessible. The prize, renamed the Ansari X-prize after the major donors, promised $10 million to the first team to launch a crewed spacecraft into space twice within two weeks. In this case, space did not mean into orbit but above the boundary of space as defined by NASA. The boundary of space is a little ambivalent because the atmosphere of the Earth does not have a sharp boundary but fades away. The US Air Force and NASA take the limit to be fifty miles and declare people who pass that limit to be astronauts. The international Fédération Aéronautique Internationale puts the limit a bit higher at 100 km (62 miles) above the mean surface of our (slightly pear-shaped) Earth.

Around two-dozen teams competed for the X-prize. A team flying a craft designed by the dramatically creative aeronautical engineer Burt Rutan won. Rutan's system comprised two craft. One was a twin-boomed plane, White Knight, which had a central fuselage made of light composite materials and a single jet engine mounted on top of that fuselage. The second component was a rocket, SpaceShipOne, that was carried aloft under the belly of White Knight and then launched from altitude into a brief flight above the sixty-two-mile limit. The SpaceShipOne team claimed the X-prize in 2004, the year Virgin Galactic was formed.

Virgin Galactic bought the company, Scaled Composites, that had built SpaceShipOne and proceeded to develop SpaceShipTwo. SpaceShipTwo is designed to carry two crew members and six passengers into suborbital space. This two-step system employs the White Knight Two aircraft that has two fuselages and four jet engines outboard of them. The pilot and copilot occupy cabins in those fuselages. SpaceShipTwo has a single delta wing and two booms extending to the rear. The back half of the wing and the two booms are designed to tilt downward about 70 degrees as a feathering system for atmospheric reentry, somewhat like half a badminton shuttlecock. White Knight Two carries SpaceShipTwo suspended in the middle like a third fuselage. Once at altitude, SpaceShipTwo is released horizontally and then fires its rocket to make the trip to space. It then descends and lands horizontally like an ordinary airplane. A version of SpaceShipTwo known as V.S.S. *Enterprise* suffered a tragic accident in 2014, when the descent feathering system was deployed improperly, killing one test pilot and seriously wounding another.

A second SpaceShipTwo, otherwise known as V.S.S. *Unity*, subsequently made successful test flights culminating in a flight on July 9, 2021, when Richard Branson, two pilots,[13] and three other crew members[14] flew from Spaceport America into space and returned, thus beating Jeff Bezos by eleven days. Bezos countered that Branson only went to 53.5 miles—sufficient to qualify as an astronaut by Air Force standards but not above the 66 miles he achieved that was above the international definition of the edge of space. Elon Musk with his over one hundred successful Falcon 9 flights bathed them both in faint praise.

CITIZENS IN SPACE

In some quarters, there is resentment that billionaires are reserving for themselves the adventure and luxury of space flight. An alternative view is that commercialization of space has to proceed this way. Prototypes are expensive. If there is going to be scaling to something like mass production and the lowering of costs to get people to orbit, it must start with massively expensive first steps. Governments have done this up to now, but to expand the opportunities, the power of the free market must be brought to bear. Although NASA has carried some civilians to space and Russia once invited a few wealthy individuals to visit their space lab, Mir, and the ISS, expanding this opportunity and bringing the price down is not NASA's goal or expertise. Whether through avarice, egotism, a sense of adventure, or fear that Earth will become inhabitable, some people with means are forging ahead. This undertaking requires means. The commercialization efforts of Musk, Bezos, and Branson have led to remarkable technical developments that eluded NASA and the Russian program. A prime example is boosters and rockets that re-land to be serviced and returned to space. The price today to fly to the edge of space or to orbit or to the Moon or to Mars is high, but with investment that governments alone are not likely to make, the price will come down.

While Branson, Bezos, and Musk draw most of the attention, the grandstanding of billionaires is just the tip of a technical iceberg. There are impressive engineering efforts behind all three that provide jobs, expand technical capability, and accelerate the space effort. In addition, Virgin Galactic, Blue Origin, and SpaceX are only part of a complex ecology of firms that have the goal of profiting from the expansion of the human enterprise into space. There is a welter of smaller companies developing their own programs, building components, and expanding opportunities. Yes, the billionaires get a lot of press, but they represent the first steps toward democratization of space.

One aspect of this democratization—and a benefit of the internet—is that anyone online can watch many of the launches. NASA often provides spectacular live feeds.[15] The private website NASAspaceflight.com has been operating since 2005 with the goal of covering news of aeronautical engineering developments and coverage of launches. For the patient, NASAspaceflight.com provides constant live coverage of and informal commentary on the SpaceX facility in Boca Chica.[16] At any given moment, hundreds of people watch an occasional employee driving to work. At rare moments, one can witness a test or launch.

One of the most publicized—and ridiculed—goals of these programs is to provide the means for private citizens to get to space. Musk joined the tourists-in-space publicity game indirectly by supplying a Falcon9 rocket and Crew Dragon capsule to billionaire Jared Isaacman, who chose three civilians[17] to join him on a three-day trip to space labeled Inspiration4. The crew launched on September 16, 2021, and orbited at an altitude above that of the ISS. Isaacman then announced plans for another flight and the first civilian spacewalk, one of the most dangerous operations that trained astronauts undertake.

Yusaku Maezawa, a Japanese billionaire, rode a Russian Soyuz rocket from Baikonur to spend twelve days in the ISS in 2021. He brought along a production assistant to document the adventure for YouTube. Maezawa has contracted with SpaceX to orbit the Moon with its next-generation StarShip technology.

Two decades ago, the Russian Space Agency, Roscosmos, formed a partnership with Space Adventures, a US company that arranges space flights for wealthy customers. Before 2009, Space Adventures had flown seven private citizens to the ISS on Russian Soyuz vehicles, including Anousheh Ansari, the sponsor of the Ansari X-prize and the co-founder of Cirque du Soleil, at a price of many tens of millions of dollars per seat. Roscosmos imposed a hiatus on carting wealthy private citizens to space for a decade when it focused on flying US astronauts to the ISS after the Space Shuttle was retired. This was in a time when political relations between governments were worsening. Roscosmos charged NASA $80 million per seat for those flights. With SpaceX and eventually Boeing's Starliner crew capsule picking up the burden of flying cargo and astronauts to the ISS, Roscosmos announced plans to resume flying tourists to the ISS. The flight of Yusaku Maezawa was the first of those in the new era.

If not space, there is weightlessness. Peter Diamandis and others founded the Zero Gravity Corporation in 2004. Space Adventures acquired it in 2008. Space Adventures subsequently has provided a weightless experience to about fifteen thousand people, including Stephen Hawking and Martha Stewart. The current price for a flight is about $7,500.

Virgin Galactic began booking seats on its suborbital flights at $450,000 apiece. If Branson can produce one hundred or one thousand White Knight Two–SpaceShipTwo combinations, he projects that the price will come down substantially. The market would be substantially different if the price were $1,000 or even $10,000 per seat. Suborbital flights may not seem as dramatic as flights to orbit, but if the configuration is altered so that SpaceShipTwo flies horizontally at its apogee rather than straight up as Branson flew, then the trip from Los Angeles to New York would take only half an hour and New York to Beijing would be an hour. Takeoffs would be from ordinary airports and landings would not be in capsules dangling from parachutes. It would be interesting to see what the market for that capacity would be.

Axiom Space was founded in Houston in 2016 by ex–NASA employees. Its goal is to provide commercial missions to the ISS. The company hopes to promote space tourism and commercially and government-funded space research, manufacturing, and exploration. Axiom has a contract with SpaceX to fly commercial crews to the ISS using Falcon 9 rockets and Crew Dragon capsules. Axiom flew the AX-1 mission to the ISS with a crew of four[18] comprising an ex-NASA astronaut Axiom employee and three self-paying billionaires in April 2022, the first all-private astronaut mission. Axiom launched a second similar private mission, AX-2, on May 21, 2023.[19]

Axiom has a NASA contract to add a module to the ISS and the goal of constructing and operating the world's first commercial space station.

There are other contenders for the people-in-space business. Mars One, a nonprofit Dutch company, has been recruiting volunteers for a one-way trip to settle on Mars, and they have many volunteers. The company argues that the technology to fly to Mars exists today and that the main inhibitor is cost. That cost is dramatically less if there need not be a return flight.

THE SPACE INDUSTRY ECOLOGY

As flashy as the space tourism business is, the current successful business is launching satellites to orbit. There is still a market for launching satellites of substantial size, but the technology has also changed, allowing the construction of small, standardized boxes crammed with components known as *cubesats* or *smallsats*. One of the Branson enterprises, Virgin Orbit, developed Launcher One, which is carried under the wing of a retrofitted Boeing 747 named Cosmic Girl. The first Launcher One failed to reach low Earth orbit in 2020, but the second one successfully launched ten cubesats in 2021. A subsidiary of Virgin Orbit, VOX Space, aims to provide Launcher One capability to the national security launch market. Virgin Galactic intends to serve the suborbital market for space tourism and commercial and government efforts requiring zero-gravity conditions. The second stage of a Virgin Orbit rocket failed in January 2023, and orbit was not achieved. Virgin Orbit announced a company-wide operational pause and then declared bankruptcy in March 2023 amid doubts about its ultimate profitability.

Another provider of commercial service to launch smallsats and cubesats is Rocket Lab.[20] Rocket Lab founder Peter Beck did not go to college but, incredibly, learned to build rockets on his own. Rocket Lab was founded as a private company in New Zealand in 2006, taking its name not only from its proposed products but from New Zealand entrepreneur Mark

Rocket, who was an early financial contributor. Rocket Lab established its headquarters in California in 2013 with the original company as a wholly owned New Zealand subsidiary. Rocket Lab launched its first suborbital spacecraft, *Ātea* (the Māori word for space), in 2009 and then developed an orbit-capable two-stage rocket, *Electron*, which first launched successfully in 2018. Electron has become a workhorse with frequent launches from the Māhia peninsula of New Zealand. Rocket Lab developed its own rocket engine technology, some of which employs 3D printing techniques. The company is designing a new, more powerful, human-rated rocket, Neutron. A longer-range goal is a mission to Venus.

Tom Markusic is emblematic of the vibrant commercial space ecosystem, working actively in the background with less publicity than Bezos, Branson, and Musk. Markusic trained in plasma physics and then worked in the aerospace industry. He was one of the first one hundred employees of SpaceX, where he worked on the launch of the first Falcon 1, then also held positions with Blue Origin and Virgin Atlantic. He concluded that neither Blue Origin nor Virgin Atlantic could compete with SpaceX and launched his own space technology company, Firefly, of which he became CEO and chief technology officer.

The goal of Firefly is to transform how we get to space. Firefly designs and builds its own rocket engines and is testing Firefly Alpha, a two-stage expendable rocket intended to provide low-cost access to orbit for small satellites and crew. The first launch of Alpha from Vandenberg Air Force base got off the pad nicely but suffered an inflight anomaly and exploded about three minutes into the flight as it neared maximum dynamic pressure. Presuming such glitches are overcome, Firefly will sell Alpha rockets for about $15 million or manage a full launch to orbit for $200 million, both on the low end of current costs. A second attempt was successful in October 2022. Work is underway on Firefly Gamma, a reusable space plane that could be launched vertically from the ground or dropped from a mother ship to land at an ordinary airport.

Firefly aspires to neither copy nor catch up with SpaceX but to be a space transportation company, the supplier that everyone needs, the company that sells shovels to gold miners. Firefly is a vertically integrated company designed to avoid the need to outsource critical components and to have sufficient resources to survive multiple launch failures. Markusic stresses the need to anticipate customer requirements and be first to market with innovative solutions in order to dominate the market. He envisions a future for Firefly in data acquisition related to the information revolution and focused on Earth imaging, including AI analysis of surface images. Markusic foresees multi-trillion-dollar opportunities servicing space stations in Earth orbit and habitats on the Moon and cleaning up space debris.

There is turbulence in this area of enterprise. The Ukranian venture capital firm Noosphere owned a significant stake in Firefly. In 2022, the private equity firm AE Industrial Partners (AEI) bought Noosphere's stake and installed AEI partner Peter Schumacher as CEO of Firefly. Markusic remained chief technical officer and a board member.

The publicity associated with SpaceX, Virgin Galactic, and Blue Origin and the technical capacity of chips and cameras developed for mobile phones allow private companies to launch small, cheap satellites. This capacity has brought a rush of other companies to join the commercial initiative and has drawn the attention of investors with total budgets far beyond that available to NASA. Space start-ups raised $7 billion in 2020, twice the amount in 2018. Swarms of small, highly capable satellites can provide a richer data harvest than traditional large, single-purpose, very expensive satellites. Investors are looking at start-ups in every sector of the industry: launch capability, communications, life support, supply chains, energy, and industry analysis. There are currently a dozen launch sites in the United States with another dozen planned. Other launch sites are scattered around the world, mostly near the equator. These come with local battles between supporters of this new space age and its attendant jobs and people who do not want their way of life disturbed by displacement, noisy launches, and crashing failures.

Companies that have raised funds recently include Relativity Space and Astra, which have launch aspirations; Astranis, a satellite internet company; Umbra, which designs satellites to take images from space that are not dependent on weather or light conditions; Planet, which has built and launched a fleet of small cubesats that image the whole Earth every day; Xplore, which designs orbital missions; NanoRacks, which arranges commercial use of the ISS; Hemisphere Ventures and SpaceFund, both of which invest in space initiatives; and Bryce Tech, a consulting company that analyzes all this activity. Relativity Space is pioneering the technology of 3D printing its rocket components using copper-based alloys known as Glenn Research Copper, a combination of copper, chromium, and niobium. Relativity Space constructed its Terran 1 rocket using 3D–printed parts exclusively. Terran 1, powered by nine 3D–printed engines, is 100 feet tall and 7.5 feet wide. The first Terran 1 was launched from Cape Canaveral Space Force Station in Florida on March 23, 2023. The first stage worked successfully but the second stage failed shortly after launch. As for many new entrants into civilian space projects, Relativity Space will surely do a thorough analysis and try again.

CesiumAstro was founded as an aerospace engineering company in Austin, Texas, in 2017 and now has seventy-five employees. On September 29, 2021, it launched Cesium Mission 1, a pair of *phased array* satellites that promise new levels of accuracy for scanning the Earth. With SpaceX, Firefly, CesiumAstro, and other firms, Austin, Texas, may become a space technology center not just state capitol, Silicon Hills, and the Live Music Capital of the World. How weird is that?

Many private space companies have plans to go public, thus giving them access to greater capital and allowing founders and investors to cash out. Many are considering merging with special purpose acquisition companies (SPACs). SPACs are shell companies designed to provide public access by merging with a company that has products and a prospective value. Merging with a SPAC provides a means to go public that is quicker and less expensive than working with a traditional investment bank. If it is too good to be true.

Musk, Bezos, and Branson have said that the development of the space industry could ultimately address the challenges of climate change. This might be done by moving industry into space to mitigate warming or by moving people to Mars if a new home for humanity is required. The space industry makes a small, negative contribution to climate change now but that could change if suborbital space travel and launches to orbit and beyond become as common as Amazon home delivery. Frequent launches could deposit carbon in the atmosphere and leave particles in the stratosphere that would affect the ozone layer by absorbing or reflecting sunlight.

Another problem is space debris. There are now thirteen countries capable of launching satellites into orbit including India, Israel, Iran, and North Korea. There were 3,372 active satellites in orbit as of January 1, 2021, and 1,897 of them were launched by the United States.[21] Most were in low Earth orbit but many were further away in geosynchronous orbit, which allows them to hover above a particular spot on Earth.[22] Geosynchronous orbit is a particularly advantageous condition for communication satellites that radiate to specific service areas on the ground. The total number of orbiting objects currently, nearly twenty thousand,[23] dwarfs the number of active satellites in orbit. Most orbiting objects are space junk such as old satellites and launch vehicles that remain in orbit. Debris from those that have collided and shattered is estimated to be about 1 million pieces. As the amount of debris collects, the chances of collision grow. A particular danger is that the process will avalanche as big pieces break into smaller shards and smaller shards into a greater number of even smaller fragments that nevertheless move at about twenty thousand miles per hour. These fragments can do considerable damage to active orbiting craft and the humans who occupy them. A loose bolt or even a fleck of paint moving at orbital velocity can do severe damage. Various ways are contemplated to reduce the debris: nets or harpoons; a tether or laser blast that could slow a satellite, thus causing it to settle to lower orbit and burn up in the atmosphere; or puffing up a portion of the atmosphere to provide excess drag and orbit decay. Some see a business opportunity in providing this service, and others advocate an if-you-launch-it-recover-it policy.

While business opportunities in the space near Earth burgeon, national efforts continue to push space exploration out into the solar system. NASA established its Commercial Lunar Payload Services program (CLPS) to contract with aerospace companies to send small robotic landers to the Moon and plans to establish a crewed facility on the Moon with its Artemis program.

China has a rich, well-planned program for space exploration. It operates the large Jiuquan Satellite Launch Center in the Gobi Desert. From there, China orbited its first crewed space laboratory, Tiangong-1,[24] in 2011 and operated it until 2018. China launched the first module of a new, large modular space station in 2021 and completed the job with the third and final module in October 2022. Three astronauts flew to the station, also called Tiangong, to begin permanent occupancy a month later. China returned a sample of Moon rocks in 2020 and had a rover exploring the lunar far side in 2023. Russia has plans for new lunar probes as do Canada, Japan, South Korea, India, the United Kingdom, Germany, Australia, Mexico, the United Arab Emirates, and Turkey. Many businesses are vying for contracts to facilitate these government initiatives and to pursue their own programs. The result of progress built on progress will lead to something like exponential growth of Moon initiatives. After a long, post-Apollo hiatus, the Moon is likely to be relatively crowded in another twenty years.

Mars is another popular target. NASA has a long history of orbiting Mars and landing and operating rovers. The Soviet Union, the European Space Agency, and the Indian Space Agency have also had successful Mars missions. It is still not trivial to accomplish. NASA, Russia, Japan, and China have all had missions fail.

Three missions took advantage of the same launch window in 2020 and successfully arrived at the red planet within two days of one another in 2021. The NASA Mars rover Perseverance carried the first successful aircraft to Mars, the small drone helicopter Ingenuity. The Hope orbiter was launched by the United Arab Emirates to study the climate of Mars. China's Tianwen-1[25] deployed a lander and rover and left a probe in orbit that proceeded to photograph the entire surface. China thus became the first nation to send both an orbiter and a lander to Mars on its first try.

Japan is working on a mission to return a sample of material from the Martian moon Phobos before the end of the decade. NASA is working jointly with the European Space Agency on a Mars sample return mission that would return with some Martian dirt to be studied in great detail on Earth. The European Space Agency is, in turn, partnering with Roscosmos to land the Kazachok[26] platform on the surface of Mars. Kazachok will dispatch the Rosalind Franklin rover, named after the British X-ray crystallographer whose work allowed Watson and Crick to deduce the double helix structure of DNA. The Franklin rover will aptly search for evidence of ancient or contemporary Martian microbial activity.

NASA has long-range plans to develop a new space launch system based on a large new rocket and spacecraft capable of delivering humans to orbit or a landing on Mars in the 2030s. NASA has established a facility on top of a lava mountain on Maui where some of the conditions to be faced by the first astronauts on Mars can be replicated and facilities at the Johnson Space Center in Houston where crews can spend a year under simulated Mars conditions. Other studies are done with long stays in the ISS. The goal of these studies is better to understand the biomedical and even psychological challenges that will greet the first Martian astronauts.

Return missions of either dirt or eventually humans raise a complimentary issue: planetary protection. NASA's Planetary Protection program actively studies the means of protecting solar system bodies from contamination by Earth life (*forward contamination*) and the means of protecting Earth from alien life forms that may be returned from other solar system bodies (*reverse contamination*).

DEEP SPACE: CAN WE GET THERE?

Then there is the issue of deep space that modern astronomy has revealed. The observable Universe contains about 200 billion galaxies much like our own Milky Way spiral galaxy and perhaps ten times as many small dwarf galaxies. A typical galaxy might contain 100 billion stars. In the last couple of decades, astronomers have established that there are approximately as many planets as there are stars. That makes about 100 billion planets in our

Milky Way alone and a total number of planets in the observable Universe that is nearly too big to contemplate.[27]

This raises two grand questions: can we go there, and are we alone in this vast, star-spangled Universe?

The first question is somewhat simpler because it basically only depends on physics and not tricky issues of sociology and psychology. The fundamental factors are that space is very big and that nothing corporeal can travel faster than the speed of light. Even traveling at the speed of light, it would take years to get to the nearest star beyond the Sun and 100,000 years to cross the Galaxy. At the modest speeds to which we are now capable of boosting rockets, all these times would be longer by a factor of about 30,000. Even if we can figure out how to travel faster, it could take longer than a human lifetime to get anywhere interesting. Another important point of physics is that as anything—rockets, people—travels closer to the speed of light, the rate of passage of time changes. Astronauts traveling at anything close to the speed of light would age more slowly than people behind on Earth, and correspondingly, the friends and relatives left behind on Earth would age and die quickly relative to the traveling astronauts.

If we are going to beat these limits, we will have to develop some means to effectively travel faster than the speed of light. This is easy in *Star Trek* or *Star Wars*: you just press the hyperspace button. With our current physics, we have no means to do so. Physics points at two ways that might provide a way out, one requiring the existence of a higher-dimensional space and the other even more strange.

One possibility for beating the speed of light is the concept of *wormholes*, a name coined by physicist John Archibald Wheeler. The notion is that there might be a means of going from one point in our normal 3D space to another not by traversing the 3D space between them but by entering a tunnel between them that passes through a higher-dimensional space. In the framework of Einsteinian physics, the journey would be much shorter and more quickly traversable through hyperspace than through normal space. It is also true that one could come out the other end of the wormhole before one went in so how this would work in practice as a transportation mechanism, even in principle, is unclear. These notions have been explored by Nobel Prize–winning Caltech physicist Kip Thorne and colleagues and described in his book *Black Holes and Time Warps*.[28] The problem is that forming a wormhole requires the tearing and stitching of 3D space that we do not know can be done. Predicting whether wormholes are possible even in principle will require, at least, a complete theory of quantum gravity that we do not have.

A variation on this theme also relies on engineering the curvature of space. The idea would be to drastically compact the space in front of a vehicle, thus drawing in a distant destination and stretching space out behind. This is sometimes called *curvature propulsion* or an *Alcubierre drive* after the physicist who worked out the basic principles in the context of Einstein's theory of curved space. A significant important engineering detail is that this mechanism requires the production of negative mass. Similar ideas are conceptually invoked to stabilize wormholes and underlie some explanations of the dark energy that accelerates our Universe. It was easy enough for Alcubierre to add such a component to his equations,[29] but no one has any idea whether such a substance exists never mind how to generate, manipulate, and engineer it. Even granting the basic notion of negative mass, it turns out there are potential problems. Drawing the potential destination close so one gets there faster than light could travel is nice but spitting out all that stretched space in your wake could cause significant problems for anyone in your exhaust.

The other possibility for motion faster than the speed of light has its basis in quantum theory. As described briefly in chapter 7, quantum theory describes the microscopic world of molecules, atoms, and electrons. This is a world governed by probability and not certainty, by sudden jumps and not smooth changes, and by things going where they should not be allowed to go. Quantum effects are difficult to discern in everyday life, but physicists and engineers have mastered them with high accuracy. We know how to make quantum theory work for us.

One of the more remarkable, tested predictions of quantum theory is that a particle can have a range of possible properties—say, pointing up or pointing down—simultaneously whereas normal experience would have it pointing one way or the other but not both at once. If one were to measure the state of the particle, however, it would turn out to either point up or point down, no longer having an equal probability of pointing both ways at once. The transition from having an equal probability of pointing in either way to pointing specifically in one direction happens instantaneously. According to quantum theory, something has happened faster than the speed of light. The basic principle is illustrated in the famous thought experiment of Schrödinger's cat (chapter 7). According to quantum theory, the transition from being a mix of alive and dead to being either one or the other happens instantaneously.

These quantum principles are manifested in a process known as quantum entanglement. The states of pointing up and down or being alive and dead are quantum entangled until a measurement is made. An especially

intriguing, practical application is to create two photons at a single place
and fire them in different directions. Each photon should have equal probability of spinning to the left or spinning to the right,[30] but quantum theory
mandates that if one entangled photon is spinning left, the other must be
spinning right. Each photon is doing both at once, but once the direction
of spin of one is determined by measurement, the spinning state of the
other is known because it must be the exact opposite. In some mysterious
way, the spin state of the other photon becomes known, thus realizing an
instantaneous action at a distance.

A remarkable aspect of this is that the distance can be large. An experimental link of this sort was set in 2017 by a Chinese team led by physicist
Jian-Wei Pan of the University of Science and Technology of China in He-
fei. Entangled photons were created in an orbiting satellite named Micius[31]
and beamed to ground stations. When one station measured the spin of a
received photon, the other station 1,200 kilometers away instantaneously
knew the spin of the photon it received. At about the same time, a team
of physicists at Los Alamos National Laboratory stopped publishing on the
topic of quantum entanglement, suggesting that either their techniques did
not work or that they worked well and were instantaneously classified.

The technique of quantum computing depends on controlling the
quantum-entangled state of many elements called *quantum bits* or *qubits*. The more qubits, the more powerful the computer. The successful
construction of a quantum computer demands that the spontaneous decoherence of quantum-entangled qubits be avoided. This requires, among
other things, operation at temperatures near absolute zero.

In one battle against decoherence, physicists controlled an entangled
string of ytterbium atoms with laser pulses.[32] They found laser pulses at
regular intervals did not help. They then tried a more complex time pattern,
a *Fibonacci sequence*, in which the next number in the series is determined
by adding the previous two numbers. The Fibonacci pattern acts as if it is
projecting a higher dimensional structure onto a lower, 2D space. The result was a driving of the ytterbium atoms in a pattern that was ordered but
never repeated. The shape of the ytterbium string itself became a qubit that
was much less susceptible to interaction with the environment and decoherence. One result was that the coherent, entangled state of the ytterbium
atoms was extended in time. More remarkably, the result was effectively a
new state of matter: neither solid, liquid, gas, nor plasma. Even more remarkably, the string of ytterbium atoms behaved as if they were a substance
that existed in some higher spatial dimension that had two dimensions of

time, and the two higher dimensions of time were projected onto a single time dimension in the laboratory.

The 2022 Nobel Prize in Physics was awarded to Alain Aspect, John Clauser, and Anton Zeilinger for their fundamental experimental work establishing the reality of quantum entanglement; and the 2023 Breakthrough Prize in Fundamental Physics was awarded to Charles Bennett, Gilles Brassard, David Deutsch, and Peter Shor for establishing the theoretical basis for entanglement and associated quantum information technologies. We clearly have much more to learn about the amazing quantum world.

These techniques of quantum entanglement at the root of quantum computing are far from transporting a person at greater than the speed of light, but quantum entanglement raises the prospect of sending information faster than the speed of light. If we think of a person as just the sum of the information that comprises them, we can at least fantasize a means to send the information to reconstitute the person. There are many practical limitations that may render this notion impossible in practice.[33]

The bottom line is that if something like wormholes or curvature propulsion or quantum transport or a method we have yet to envisage proves impossible, then it will be very difficult for us to travel any appreciable distance in space or for any other creature to do so. Other possibilities for deep space travel remain to live a very long time, to somehow hibernate on long voyages, or to transport colonies of people so that the individuals who arrive at a destination are the progeny of those who originally set out. A variation on these themes would be to upload the consciousness of individuals into robots and send long-lived, hibernating robots.

ARE WE ALONE?

Despite the vastness of space, or perhaps because of it, one of the great issues of modern science is the question of whether we are alone in this immense Universe. The Russian rocket scientist Konstantin Tsiolkovsky alluded to the question as early as 1933. The great physicist Enrico Fermi succinctly posed it in the summer of 1950 at Los Alamos National Laboratory where he was working on the development of the atomic bomb. In a casual lunchtime conversation with colleagues, he asked, "Where are they?"[34] Fermi perceived that with all the planets, stars, and galaxies in an already ancient Universe, intelligent life could be common. Even traveling at the shuffling speeds of chemical rockets, humans or intelligent aliens could manage with space colonies or some other technique to get to the planets around a nearby star in a few million years. In a few more million years, local technology could be developed, some people would feel crowded, and new ventures would be launched onward and outward. Fermi thus estimated that while no individual would traverse the Galaxy, a species could diffuse through the galaxy in about 250 million years. This is near the time it takes our Sun to orbit the Galaxy and is a much smaller time than the age of the Universe.[35] Evidence of intelligent life elsewhere could be subtle, yet there is no sign much less abundant evidence.

While Fermi framed the issue, Frank Drake, an astronomer at the University of California at Santa Cruz, brought it to further focus at a meeting in 1961 by sketching the famous Drake equation. This heuristic equation was an attempt to frame the question of the number of intelligent civilizations in the galaxy by multiplying the estimated probabilities of a number of relevant factors such as the rate of formation of stars in the Galaxy; the probability that the stars have planets; the probability that life has formed on those planets; the probability that intelligent life has evolved; the probability that this intelligent life would communicate; and the length of time a communicating civilization would survive. The rate of stellar birth is reasonably well known: A star like the Sun winks on once every year in the Galaxy, and in the last couple of decades astronomers have converted a guess that every star has planets into a certainty. Beyond that, the terms of the Drake equation get more speculative and uncertain. The answer could be small, or it could be appreciable. The fact is we see nothing.

Contemporary astronomers have a vast array of tools at their disposal to peer into the Universe in a wide variety of ways. In addition to powerful telescopes on the ground and in space that observe in the optical range attuned to human vision, we probe with long-wavelength radio waves, infrared, ultraviolet, X-rays, and gamma-rays—the entire sweep of electromagnetic radiation. We also detect particles: cosmic rays from the Galaxy and beyond and neutrinos from the Sun and exploding stars. We see nearby stars and planets, individual stars in other galaxies, and galaxies just bestirring at the birth of the Universe. In all this, we see not the slightest hint of another communicating civilization. Where are they, indeed.

On the other side of the argument, astronomers have provided evidence that physics and chemistry play out the same in remote reaches of the Universe just as they do on Earth. Most astronomers have a hunch that life is common in the Universe if only at a simple, microbial level. The search for any evidence, even indirect, of that extraterrestrial life drives a significant component of contemporary astronomy under the general rubric of *astrobiology*. To some the conditions in the Universe seem tuned to lead to life. Others invoke the *anthropic principle* and argue that if the Universe were to have substantially different conditions, we could not have evolved so of course the Universe we find ourselves in has the conditions it does. If conditions are good for our life form, then they should be good for others.

Microbial life is one thing. Intelligent civilizations are something else. There are too many uncertainties in the relevant factors in the Drake equation to make reliable estimates that microbes will evolve to conscious, intelligent beings as has happened on our planet. A principal factor is how long an intelligent civilization can exist before it develops the technological capability to destroy itself—think nuclear bombs and climate change—without the discipline to survive those threats. In our case, it has been at least a century; merely a wink in the universal eye.

A specific framework for thinking of life elsewhere was given by Soviet astronomer Nicolai Kardashev in 1961. Kardashev envisaged civilizations ranked by their technological ability to capture and use energy. A *Type I* civilization would be able to employ all the energy that falls on a planet from its host star. We are near that condition. A *Type II* civilization would be able to gather essentially all the energy of the host star itself. A Kardashev *Type III* civilization would be able to capture the energy of entire galaxies.

There have been clever proposals to search for Kardashev civilizations. One notion depends on a factor that is independent of how a civilization employs the energy it captures. According to the laws of entropy, a substantial portion of the captured energy must be reradiated as heat, typically in the infrared portion of the electromagnetic spectrum. With current infrared detection technology, interesting upper limits can be placed on the number of Kardashev Type III civilizations in the local Universe: few to none.[36]

Another angle on Kardashev civilizations was raised with the discovery of Tabby's star. Tabetha Boyajian is an astronomer at Louisiana State University. She was the lead author on a 2016 paper announcing the odd behavior of a star about 1,500 light-years away. The star is only somewhat more massive and brighter than our Sun, but it shows unique, irregular variations in its light. Some people, perhaps tongue in cheek, suggested that we were witnessing the broken remnants of a Type II Kardashev civilization that still orbited its host star. That claim called for the highest level of evidence, evidence beyond speculation that has not been forthcoming. A bevy of more prosaic explanations have been proposed, including swarms of comets. None has yet been robustly established, but a prosaic, astronomical explanation of Tabby's star remains much more likely than a crumbled civilization.

An odd astronomical object named Oumuamua[37] raised a similar fuss. Oumuamua proved to be the first-ever detected object that came from beyond the solar system. Astronomers in Hawaii discovered it in 2017. Oumuamua is reddish in color and varies in brightness, resembling an asteroid but with notable properties. It is not spherical but long and thin, and it tumbles end over end. It is also traveling at a speed that means it will leave the solar system. The implication is that it must have arrived from beyond the solar system and was just passing through. Oumuamua moved in a jittery fashion as if it were buffeted by the solar wind or perhaps ejected streams of gas. That it jittered at all implies a low density compared to familiar asteroids. That conclusion triggered suggestions in some quarters that it was a hollow, constructed device, manufactured by an extraterrestrial civilization. Astronomers checked if Oumuamua might be radiating radio signals to communicate. It was not. Another object of interstellar origin, probably a comet, was discovered in 2019, knocking Oumuamua off its throne of uniqueness. Once again, while details of its origin remain elusive, Oumuamua is almost surely a product of natural forces and not of artificial design.

If we and other civilizations survive ourselves, how might we see evidence of other intelligent life? This issue is encapsulated in the phrase *search for extraterrestrial intelligence* (SETI). Spurred by Fermi, Drake, and others, there have been various programs to do this search. Early arguments were that it makes sense to look for radio waves since they penetrate interstellar gas and dust more readily than does optical radiation. The radio signals might simply be leaked into space, as we have been doing for some time, or they could be a deliberate means to communicate information. The SETI Institute in California sponsored the construction and operation of an array of radio telescopes with the goal of detecting signs of intelligent life and today supports a broad range of research related to the search for life in the Universe.[38]

There are two aspects to SETI: passive and active. In passive SETI, the most common form, we just look and listen. The proponents of active SETI pursue means to consciously advertise our existence to the Universe with the crude means at our disposal. Rather than radiating randomly into all directions in space, perhaps we should point powerful lasers with coded signals at nearby stars.

Russian billionaire Yuri Milner, physicist Stephen Hawking, and Facebook founder Mark Zuckerberg founded the Breakthrough Starshot program. This venture is exploring the possibility of launching tiny craft loaded with powerful computer chips and cameras—interstellar mobile phones—and driven by gigantic lasers to over 10 percent the speed of light. Such tiny craft could reach the nearest stars in about forty years and beam back signals in another four years. There are important engineering details to figure out. At their proposed speed, the craft are likely to be eroded by interstellar gas and dust particles.

Others caution about being too active in this regard. An analogy might be finding a cave in the woods. Is it more prudent to stand outside and listen or to barge into the cave and shout, "Any bears in here?" A number of years ago while we were on a hike near Independence Pass above Aspen, my colleague Jerry Ostriker, then of Princeton University, related to me a research project he had undertaken with an undergraduate. They had done a computer project to explore the nature of extraterrestrial life. They coded the computer with equations describing the boom and bust of ecology: availability of food and energy, and population growth and decline. They included three life forms: a lichen-like life that could smear on rocks; a paranoid life form that reacted to any perceived threat by killing at the speed of light; and a life form that poked up, looked around, and cheerily asked, "Anybody here?" They ran the equations over and over. There was only one solution: lichen everywhere; one evil son-of-a-bitch with murderous capacity; and an occasional, happy-go-lucky civilization that sent out signals and promptly died. Perhaps we should be cautious about active SETI.

Amid these developments, the void still beckons. Why is it so empty? Why do we seem to be alone? If there are extraterrestrial civilizations, why do we not perceive their communications or receive their visitors? Ray Kurzweil and others think that it is not credible that all other civilizations kill themselves or are hiding from us. He speculates that the reason we see no one is that there is no one there. We are truly alone.

Another possibility is that we have not looked far or deep enough. Our detectors are of finite capacity. We have only been radiating radio signals into space for about a century so they have only traveled one hundred light-years, a small bubble in the vast volume of our Milky Way never mind the visible Universe. In an ancient Universe, other civilizations are likely to have arisen long before ours and could have spread signals much further.

My favorite speculation is that our communication technology is yet too crude. We no longer use drums and smoke signals to communicate over large distances. Just as indigenous tribes in the Amazon are oblivious to radio signals and fiber optic cables, we might just not know an efficient way to chat with our cosmic neighbors. The speed of light is very slow compared to the vast expanses of space even within our Galaxy. It may be that other civilizations have long since abandoned any crude notion of communicating with electromagnetic radiation of any kind. I like to think that with the advance of science, we might yet discover a means to communicate at faster than the speed of light—through hyperspace or by quantum entanglement—and find the cosmic internet we anticipate. I hope it does not have Facebook.

Detecting communications from another extraterrestrial civilization is one thing; receiving extraterrestrial visitors is something else again. There are questions of physics again addressing the fact that space is large and the speed of light is small. It is one thing to somehow send signals and quite another to send things. Maybe there are ways to construct wormholes or send solid entities with quantum entanglement, but maybe not. There is certainly no current evidence that such procedures are possible.

Then there are squishier and more uncertain issues that amount to extra-terrestrial sociology. Why would ET want to come here? Advanced extra-terrestrials might care as much about visiting us as we do about visiting an anthill during a walk in the woods, but maybe some extraterrestrials are the equivalent of myrmecologists who want to study our anthill.

We often imagine extraterrestrials as much like us but green and with tentacles. Given the difficulty of moving corporeal objects over vast dis-tances in space, maybe it makes more sense to send tiny things in the spirit of Breakthrough Spaceshot. Maybe extraterrestrials are not scaly creatures but swarms of self-replicating nanobots. If such visitors are alive in some sense but not composed of DNA, how would we know they are here? Per-haps when we venture forth, we will have evolved and merged with our machines and we will be robots or nanobots.

UNIDENTIFIED AERIAL PHENOMENA

Despite the overwhelming lack of evidence for extraterrestrial civilizations in all our astronomical data, there are those who think we have been visited and that the evidence is suppressed in a government cover-up. Are UFOs real? Absolutely. There are phenomena in the sky that in a crude sense are flying, some of which are objects, and that at first crack are unidentified. Upon investigation, the vast majority of these reports prove to be entirely prosaic: the planet Venus, aircraft flying directly toward one on a landing pattern, ball lightning. Hoaxes are not uncommon. Other reports defy easy explanation. President Obama was quoted as saying in this context, "There are things we don't know." That is true. What is not supportable is the leap from "Unidentified" to "Must be aliens." Science journalist Sara Scoles provided an anthropological study of the UFO culture in a 2020 book.[39]

The US government did a recent favor to those interested in this topic by releasing in 2021 a report with the imprimatur of the Office of the Director of National Intelligence in which this category of sightings was newly named *unidentified aerial phenomena* (UAP).[40]

This phrase better captures the literal nature of the topic: phenomena that are in the sky and resist identification. That still does not make them alien craft. One of the verities of this subject is that while there is no concrete, independently verifiable evidence for alien visitation, it is impossible to prove a negative in every case. Most UAP are proven to be natural or man-made phenomena. One cannot prove that all UAP are not an alien manifestation, but that does not make it so.

That recent government report had an important precursor. In the 1960s, the US Air Force commissioned a study called Project Blue Book that was charged with investigating UAP phenomenon. The study was led by Edward Condon, a professor of physics at the University of Colorado, and published as a book.[41] The study laid out the case for natural causes of most UAP reports and established an important standard for adequate documentation. The report presents fifty-nine detailed case studies only one of which was deemed difficult to explain in a conventional manner.

The 2021 report was another attempt to address the notion that the government is aware of alien visitation but is suppressing the evidence. The report was brief and shy of the details mustered by the Condon report yet drew attention from its admission that there are a few UAP that can not be easily explained. Reports from Navy pilots and associated video were particularly striking. The report did not stress the majority of UAP that are easily explained or dismissed. Once again, "Unexplained" calls for an explanation; it is not synonymous with "alien aircraft." It would be more appropriate to refer to these examples of UAP as "Not yet explained."

The professional competence and integrity of the Navy pilots are not to be questioned. One might still ask, "Why Navy pilots?" Where are the Air Force reports? The Navy flies over the ocean. The Air Force flies over land. One might then ask whether the observed phenomena are rare manifestations of the atmosphere-ocean boundary layer, a trick of reflection of some kind the locus of which only seems to dance around at an unphysical velocity.

It turns out that a more prosaic explanation is that the Navy pilots have recently been commanded to provide UAP reports to overcome the tendency to underplay such experiences for fear that their professionalism might be questioned. The Air Force has yet to enact a similarly rigorous reporting structure. One might also ask about the precise nature of the Navy reports. Did the pilot see the UAP independent of a video camera or was he looking through a heads-up display that was showing the same image the camera was recording? If the latter, what is the chance that there was some stray reflected light within the camera mechanism, a problem that plagues many sensitive astronomical observations.

The US Department of Defense established an Unidentified Aerial Phenomena Task Force, since reorganized as the Aerial Object Identification and Management Synchronization Group (AOIMSG). NASA has recognized its formal responsibility to ensure the safety of aircraft by establishing in 2022 an Independent Study of Unidentified Aerial Phenomena. Highly respected astrophysicist David Spergel was appointed to lead the nine-month study. Spergel was once chair of the Department of Astrophysical Sciences at Princeton and is now president of the Simons Foundation in New York City. Upon his appointment, Spergel said, "Given the paucity of observations, our first task is simply to gather the most robust set of data that we can. We will be identifying what data—from civilians, government, nonprofits, companies—exists, what else we should try to collect, and how to best analyze it."

The 2023 National Defense Authorization Act updates UAP to stand for "unidentified anomalous phenomena," which includes unexplained phenomena in the air, on land, at sea, and in space. A new provision directs the Department of Defense to establish a mechanism for reporting sightings that prohibits reprisals against personnel who report UAPs and that investigates prior use of nondisclosure agreements that may have limited previous reports.

It is useful to think about UAPs like a physicist or astronomer. Many UAP reports cite the distance and size of the object. Statements like "They were a mile away and as big as a football field" should immediately raise red flags. It is very difficult to estimate the distance to a given distant object, and one must know that distance to estimate the size. You can go outside and look at the stars and have no idea at all how far away they are. It took astronomers centuries to figure out ways to determine accurate distances in space. In the rush of witnessing a UAP, it is all too easy to misunderstand and misestimate a distance. Without an accurate distance, estimating the size is hopeless.

Another important aspect of both the Condon report and the more recent UAP report is what is not said. Does the government know things it is not going to tell? Of course. Stealth aircraft flew for decades before they were acknowledged to the public. The batwing ground support F-117 Nighthawk and the B-2 Spirit bomber were developed in the 1970s. I hope and trust the government has produced even more sophisticated—and secret—aircraft technology since then. The government report leaves any such projects undescribed. It does not say, "We are hiding nothing from you." It says that even given the things we are consciously hiding, there are UAP we were not able easily to explain, and we are not hiding those. We are not hiding evidence of alien craft.

The even more notorious incidents of extraterrestrial alien visitations are those in which people report to have been abducted, examined, and probed. An important and difficult issue here is human psychology. Carl Sagan and Ann Druyan tackled this topic in their book *The Demon-Haunted World*.[42] They pointed out the parallels between common stories of abductions by incubus or succubus demons in the past and more contemporary stories of alien abduction. Both have themes of "sexually obsessive non-humans who live in the sky, walk through walls, communicate telepathically, and perform breeding experiments on the human species." Sagan and Druyan proposed that reports of demons were the same human phenomena but cast in a context that reflected popular images of the day. There were no known reports of extraterrestrial alien abductions in the Middle Ages when the notion of the Earth as a planet—never mind that of a populated Universe—was not a common component of popular perception just as there are few reports of demon abductions today.

Extraterrestrial aliens were reported to come from Mars, Venus, and Jupiter until the late 1950s when a variety of NASA missions showed these planets to be uninhabited. The extraterrestrial aliens then came from elsewhere. Concrete evidence of alien abductions beyond oral reports is lacking, and possible psychological explanations are many: sleep paralysis, lucid dreaming, false memories, and various forms of psychopathology. Once again, one cannot prove a negative, but claims of extraterrestrial alien abductions are so important they must be held to the highest levels of skeptical scientific proof.

Keep an open mind for new evidence—but remain skeptical.

WE WILL GO

Our best current guess is that the Universe we observe is a one-time shot. There was a peak in the generation of stars, planets, and perhaps living creatures about the time the Sun was born and the whole thing has been winding down since, like the fading embers of a brilliant firework. The Universe will apparently expand forever, with the stars flaming out and the galaxies being drawn infinitely far apart from one another. We should make of this Universe what we can while we can.

We have learned all this about space only in the last century or so, and only in the last seventy years have we tiptoed off our pale blue dot into the surrounding vastness of space. We have launched myriad satellites around Earth, including the ISS that orbits only as high above the surface as the distance from San Antonio to Dallas. We have landed robots and people on the Moon, a mere 240,000 miles away—that's about thirty times the diameter of Earth. We have sent robots to the Sun, to the eight planets in the solar system, to the dwarf planet Pluto, and beyond, but only slightly beyond. Missions throughout the solar system will flourish in the future.

We will continue to venture into space for a variety of reasons. We will go for the same reason English climber George Mallory died climbing Everest: because it is there. We will go because space contains resources that can be exploited to benefit humankind and perhaps turn a profit. We will go because we might ruin our home planet and require new domiciles. Whether we retain our humanity as we venture into space is another issue.

(15)

THE FUTURE
Ready or Not

THE FUTURE IN A FUNHOUSE MIRROR

Deep future time is challenging to contemplate. Astronomers, geologists, and cosmologists do it all the time, but it is a much more elusive topic when the effects on humans are concerned. Trying to predict the future of humanity over geological or astronomical timescales is somewhat like trying to predict the life of a person from the condition of a days-old blastocyst in a womb. Even rather small extrapolations into the future are problematic when changes come exponentially rapidly and humans react to those very changes. Predicting the future of humanity is like trying to grab smoke.

It is natural to wonder first and foremost about what the future will bring for me, for my children, for my grandchildren, for people who are alive today. One of the principal lessons of the story told in this book is that you cannot select the piece of the future with which you want to cope. It is common to separately address the big topics of our technological future as single themes: robots, AI, climate change, advances in biology. They will all happen in a rush, in your lifetime and the lifetimes of our grandchildren. Then they will keep happening, ever faster. It is all going to happen.

Fifty years ago we had cars, airplanes, radio, and TV. We had rockets and a Moon landing but we had no internet, no mobile phones. We had medicine and DNA but no CRISPR. Fifty years from now, robots, AI, and the effects of climate change will have gone from occasional to ubiquitous and perhaps overwhelming. There will be designer babies. There will be colonies on the Moon and on Mars.

Exponential advances in our technology will not occur because they result from a single grand plan designed at the top. These advances will be more like natural selection with many small experiments, some of which will have a degree of survivability. There will be plans, but much of what is to come will be driven by individuals: scientists, engineers, and entrepreneurs pursuing their passion for compelling ideas, elegant solutions, and seductive profits. In the midst of the coming turmoil with no one in charge, we must be vigilant for unintended consequences.

We must try to understand the future that is coming in order to decide what to promote and what to prevent. Changes will come in a multitude of small—if exponentially rapid—steps. Decisions about where and how to restrict certain aspects of technology will need to be made in some collective fashion and will be all the more difficult for that. Advances in technology often bring huge ethical challenges. Who is affected and how? How are the benefits spread equitably and potentially negative aspects limited or at least distributed fairly?

We live in a world where great religions—Judeo-Christian, Islam, Hinduism, Buddhism—that are thousands of years old set our ethical standards yet struggle to be reconciled with one another. How will those institutions cope with the technological tidal wave that is coming in mere decades of exponential development?

Our future is destined to be one of greater connectivity between people, between machines, between people and machines. This could fundamentally change what it means to be human and how we function as a society. Do we strive to maintain our free will and individuality or embrace a future where our collective interaction brings great new power and capability?

On the last day of my class, I asked my students to make impossibly difficult predictions about what they thought the future of humanity would be like in 100 years, 1,000 years, 10,000 years, 100,000 years, 1 million years, and beyond. All these are timescales that are long on a human scale but mere tiny ticks on the clock of the Universe and thus very practical to consider. I was busy assessing the students' predictions so I could duck the responsibility of addressing the issue myself. It is time for me to take the plunge, sobered by the old Danish proverb "It's tough to make predictions, especially about the future."

One way to put the relevant issues in perspective is to look back at the history of humanity. The difficulty of this exercise is that while time has proceeded more or less uniformly since its creation at the Big Bang, the technological development of humanity has not. Reflection in time from past to future is more like a funhouse mirror that rather than making one look taller or smaller speeds up everything in time. The future is not a simple mirror of the past because technological developments occur ever faster. The technology of our future will come much more rapidly than that of the past recedes. Nevertheless, reflecting on our history gives us some perspective on what to expect in our future.

THE NEXT CENTURY

One hundred years ago, we had automobiles but we barely had airplanes. We had radio, but no TV. The rudiments of quantum theory and Einstein's theory of gravity were being put into place. There were no computers. We had medicine but no concept of the double helix of DNA. In the next hundred years, the possible changes are already almost too much to contemplate.

We do not know whether there will be a technological singularity with the development of computers and software that are far more capable than we are in the realm of much human activity. Extrapolating the growth of speed and capacity beyond 2005, Kurzweil estimated that computers would exceed the power of the human brain by 2030 and of all human brains by 2060.[1] There are arguments that machine superintelligence cannot be contained because that task would require programs that are themselves not computable.[2] Even people who find the prospect of a singularity likely have difficulty placing this transition in their lifetime but find 100 or 1,000 years more palatable. That may be wishful thinking.

There is a huge, fraught issue of whether computers will become conscious, the true heart of the contemplated singularity. For all the progress on both computers and AI, that remains a distant potential albeit closer than we may think. Human knowledge is dynamic, reacting to changes. It is still beyond our current capacity to get a machine to form abstractions and to address a commonsense knowledge problem. The challenge is to build an appropriate cognitive architecture to adapt to the environment, including humans, as humans do. It is not a matter of building values and ethics into a system but of achieving human-like flexibility. Systems based on statistical correlations like most current machine learning algorithms are not adequate to the task. It remains a grand challenge to develop a system that can react to the environment as capably as an eighteen-month-old infant.

It is yet to be seen whether AI can learn what we think of as intuitiveness, out-of-the-box thinking, creativity. Current AI can do art and write music but attempts at writing screenplays tend to be dismal failures. LLMs like ChatGPT may be closing that gap. The evolution of AlphaGo Zero shows that the potential for human, even superhuman intuition, is there, but can AI work in the wild the way a human can?

The pieces are on the table that when properly combined and integrated could lead to conscious computers, general AI, artificial sentient entities. Sentient entities, that is, humans and other animals, function by moving about, intaking data from their environment—sights, sounds, smells—comparing that input with memories of past experience and using that comparison to plan their next steps. Machines could attain that capacity.

We can envisage robots equipped with an array of powerful sensors that can navigate their environment and collect a range of data. It is within the capacity of current AI to analyze that data and use it to plan, to strategize. It seems a rather small technical challenge to design AI that can recall past experience—if that has not already been accomplished. How would such a sensor-enabled, AI-powered robot differ from a sentient animal? Such machines would have a superhuman ability. A crucial part of machine strategizing would be to rewrite software to enable the strategy. Current AI can code. A machine could decide what it wanted to do better then rewrite its own code to make it so and evolve super rapidly. Humans cannot do that.

It may not matter whether computers become conscious. If computers can supplant humans in many ways, it amounts to the same thing. What do humans do if computers coupled to robots and AI can do almost everything better, if they can exchange data and information at the speed of light and evolve more rapidly than we can conceive? There will be a phase of collaboration with robots in which the robots constructively partner with humans, but how long will that last before humans simply become irrelevant? With the exponential development of robots and AI, that phase of useful collaboration could be quite short before the computers get bored and move on.

Within the next century, the imminent clash of people and machines will have resolved in some fashion. Society will have coped with ubiquitous, intrusive data collection or will have surrendered to a lack of privacy. By then we will have witnessed the answer to the question of the future of jobs. What will humans do when hyper-capable AI operates ubiquitously in nearly every corner of human activity? Will doctors and lawyers and scientists and politicians be necessary never mind office workers and farmers and factory workers and truck drivers and custodians and store clerks and restaurant staff? Will we all become artists when regular jobs go away? Will AI be better creators and critics of art than humans? In 2021, an AI helped to complete Beethoven's unfinished Tenth Symphony.[3] AI that generates elegant prose and captivating art based on simple natural language commands is destined to engender a broad rethinking of what it means to create art and be an artist.

There are those who call for continuous learning to keep the advances of AI and the loss of jobs at bay. Custodians may not have to become surgeons, but perhaps they can learn some information technology and take a small step up and sideways. Surgeons are learning to collaborate with robots for delicate operations. Continuous learning may be necessary in the immediate future, but will it be sufficient? Can we actually retrain basically everybody all the time to keep up with, if not ahead of, technological advances in order to keep society intact?

There may be a tipping point within the next century between a phase when individuals and society at large can adapt to change even at an ever-increasing pace and a new phase when that change happens so rapidly that individuals and societies and the institutions that organize and govern them cannot adapt. AI may understand humans better than humans understand themselves so it will be natural for AI to take over the organizing and running of human institutions. Will these machines care about tender human egos or notions of free will? What if we can design our own will? Or if an AI can do so? If the machines strive to optimize the system, what constitutes the system and how are the optimum conditions defined and implemented?

A hundred years may see immense changes in cities. Are smart cities the best way to provide life, liberty, and the pursuit of happiness for their citizens? Cities developed as hubs to foster human enterprises. As AI develops and connections become increasingly global, will cities survive? The human urge to form communities will be counteracted by AI that may not value such things. Will people or data—"atoms or bits" in the phrasing of my colleague Jay Boisseau—dominate? Will concentrations of people in cities be more or less vulnerable to hacking and cyberwar than if people and their needs are more distributed?

Walmart made its fortune by distributing its stores and not by concentrating in cities. Amazon is distributing its distribution centers. Human connections will still provide a gravity, but there will be centrifugal forces as global connectivity advances, as AI takes over so many human enterprises that require communities, and as urban housing continues to be outrageously expensive. A hint of the future of cities may have been given with the COVID-19 pandemic when city centers tended to hollow out, at least temporarily. The price was too high, but for a time it miraculously became possible easily to drive on the roads of Austin, Texas, at 5 p.m. Maybe cities are not needed if the machines are running things efficiently, all talking to one another from everywhere on the globe at the speed of light. An AI in Hong Kong could read faces in Helsinki as easily as the local police precinct could.

With ever more sophisticated AI-driven prosthetics, humans will naturally merge with their machines at some level. A major question is whether humans can or will want to merge in a manner that allows their biological substrate to compete with the machines. Once AI-aided prosthetics allow a BCI so that humans can control prosthetics by thinking, they can do so at the end of their arm or across the country. What changes might that telekinesis bring to work and leisure? Perhaps you could personally control a robot that selects items of interest to you on a shelf at an Amazon distribution center and drops them in a bin for delivery or control the robot that loads the dirty clothes in your washing machine while you are at work. Imagine sitting on your sofa and controlling any robot anywhere by thinking.

The army is developing a helmet that can keep a soldier in the *flow*, that state of concentration that allows a human to perform at peak concentrated performance, somewhat like an electronic "upper."[4] Such a piece of equipment could make for more efficient soldiers, but the question arises as to what would be lost? Would such an enhanced human gain concentration but lose empathy? If daydreaming is suppressed, would we lose the creativity that is one of the chief attributes of human progress? Would an enhanced ability to make quick, decisive decisions sacrifice the richness of human experience filled with doubt and uncertainty? If we upgrade our brains, will we lose our minds? Our souls?

BCIs that enable AI-mediated telepathy will alter or even disrupt society. I do not want to know all that you think and do not want you knowing all that I think, but that intrusion may be inevitable. The ability to share thoughts and knowledge with anyone or everyone would be an immense step in the evolution of humans by totally changing social intercourse. New businesses may be needed to provide a filter infrastructure to enable safe, socially acceptable mind swapping. There may need to be an associated political and regulatory structure. There will be a generation of young people who will think that typing on keyboards and tapping on screens are so incredibly old-fashioned. Old people will need fourteen-year-olds to teach them how to debug their mind-reading headsets. The ultimate outcome might be the formation of a hive mind, blending us all together and vanquishing any notion of human individuality.

Our medical technology will advance dramatically as we better understand biological functions and disease at a molecular level. Perhaps a variety of cancers will be cured or prevented. Stem cells may be engineered to repair damaged hearts and lungs and kidneys. There is already progress on miniscule nanobots that can crawl or swim though our blood streams and lymphatic systems to seek out diseased regions and repair them.

The goal of medicine will not be restricted to healing but will embrace enhancement. Beyond competing with our own machines, suppose we could smell the environment like a dog can, see ultraviolet like a bee, detect magnetic fields like birds, and sound like bats? What if we could smell estrus as many animals do? Would that change our sexual behavior? Will superhumans care about sex anymore for either pleasure or procreation? Could we consciously attain the mental states presently only accessible through drugs or meditation? Could we understand, control, and utilize dream states? On a parallel track, what mental states will an AI attain? Will AI dream of electric sheep or anything else?

What will be the social implications if aging is solved along with cancer? Will there be battles between the young and the old over jobs and dwindling resources? If no one dies, what happens to the population? Must we regulate births? Will only the rich in developed countries enjoy extended lifetimes? Can society suffer that? Is it ultimately healthy that everyone eventually dies thus promoting rejuvenation and renewal? We need to ponder such issues as progress is made to extend health spans and longevity.

In the next century all these issues will play out simultaneously against the background of an ever-wilder environment driven by climate change. Our governments have been irresponsive and inept despite clear projections from scientists for several decades. Necessary changes are impeded by vested interests in fossil fuels and by a culture in which internal combustion vehicles are deeply ingrained. In 2023, climate scientists predicted that the Earth would surpass the goal of the 2015 Paris Agreement to limit global warming to 1.5 degrees Celsius between 2033 and 2035.

Even among advocates of clean energy there are debates over the utility of large solar and wind farms coupled with an updated transmission grid versus local, even personal solar installations and mini-wind farms. Hydrogen fuel cells show promise but are still a minor factor. Nuclear power remains controversial in the face of issues of safety and the storing of radioactive waste. Progress on thermonuclear power remains slow although accelerating.

Floods will occur, especially along coasts. In the best case, flood insurance will become drastically more expensive and people will abandon the coasts. In the worst case, there will be massive, disruptive migrations away from coastal flood zones. Drought induced by climate change will bring complementary issues of migration or suffering by those who cannot afford to flee. There are tremendous political problems in deciding how the developed countries that created the problem can aid the developing countries that need cheap power now and other countries that make no substantial energy demand but will suffer the brunt of climate change.

The complex reality is that the world needs a distributed mix of solutions to the challenges of a shifting climate. The power grid must be updated and made robust against severe weather and hack attacks. There's a need for local production of power with a guaranteed supply for medical facilities and to fill in where a distributed grid is not effective or economical. Battery technology needs to be improved and distributed. Electric vehicle charging stations need to be widely available before gasoline- and diesel-fueled vehicles are largely abandoned. It is impractical to make wind a local resource, and it is not clear that local solar can provide global needs since it is spotty and more effective closer to the equator. A proper distribution of power sources and capacity would take all these solutions into account. Time is wasting as the problems get more severe and the solutions may take a decade or more to implement. In the meantime, all the other exponential technological developments will be racing ahead, maybe helping and maybe confusing the issues of coping with climate change.

Perhaps our best hope is that the machines will figure out an efficient, affordable, technological way to remove carbon from the atmosphere. In the meantime, humans are making some progress.[5] In late 2022, the US Department of Energy announced a new $3.5 billion program to fund regional carbon capture hubs capped with a $115 million prize for developing processes to convert captured carbon into useful products such as building materials. Fortunately, it turns out that simple polymineralic rocks such as basalt and granite are more effective in trapping carbon in their crystal structure than are chemical reactions to form carbonates, the process previously assumed to be the method of choice to capture carbon.[6]

THE NEXT MILLENNIUM AND BEYOND

A century is hard enough, but let's look deeper into the past and the future.

One thousand years ago, Europe languished in the Middle Ages although this was the golden age of Islam and a relatively sophisticated culture flourished in China. The Renaissance was yet to occur, and Buddhism had not arisen in India. A thousand years from now, there will have been significant changes. If humans have not wiped out civilization in a war, they will have changed at a fundamental level and come to some accommodation with machines. There will be a substantial occupation of the solar system. Cities on Mars will be ancient. The population on Earth might have stabilized at some high level, but it might have shrunk to a few billion to accommodate a finite terrestrial ecosystem. Food—both animal-like protein and vegetable—will be grown locally on racks in widely distributed systems. An advanced version of 3D printing, perhaps at the atomic and molecular level, will allow distributed local manufacturing.

Even if humans have continued to evolve purely by natural selection, in a thousand years they may already be different in substantial ways if not yet a different species. Adaptation to the rigors of a hotter, wetter environment is a likely possibility. We could be radically different if we choose our own biological evolution. If humans have actively redesigned their DNA and protein substrate to cure disease and compete with machine technology, *Homo sapiens* may no longer exist.

A millennium into the future, our social organizations are likely to have radically changed. Cities may have dissolved into a network of smaller, comfortable communities all linked by global connections. Financial centers may no longer make any sense since business can be done from anywhere. AI may have supplanted all current forms of governance and commerce. A thousand years might be enough to develop and exploit a new theory of quantum gravity. Maybe we will know whether hyperspace exists and if so, how to communicate and travel within it. If not, we will have sent electromagnetic signals to a significant portion of the solar neighborhood in the Milky Way but will ourselves still be confined to near our solar system.

Ten thousand years ago, humans were just developing agriculture and forming villages to facilitate the exchange of goods and services.[7] The exact order of these developments is contested. In *The Dawn of Everything*,[8] historian David Graber and archeologist David Wengrow argue against the traditional linear view that there was a steady, foreordained evolution from egalitarian hunter-gatherer societies to agriculture, villages, tribes, tribal leaders, private property, cities, civilization, and endemic inequality. Graber and Wengrow make the case that villages formed first, that agriculture came and went with flood seasons, and that people made conscious choices on how to live, work, and distribute political power that were far more complex and intentional than a passive existence driven forward by monotonic developments in population and technology. Some early hunter-gatherer societies deliberately eschewed the lure of agriculture. Analogous arguments for the complexity of early, interwoven, hominid branches have

supplanted any notion of a linear advancement from monkeys to chimpanzees to Cro-Magnon to modern *Homo sapiens sapiens*. Our future in
resonance or tension with technology may be just as complex.

Ten thousand years from now, humans may have adapted to life in
space with low gravity. We may have designed more efficient metabolisms
requiring little food, water, and air. Human society may have fragmented
or coalesced.

One hundred thousand years ago, *Homo sapiens* shared the Indo-
European continent with Neanderthals, Denisovans, and perhaps other
humanoid species. Do you have any doubt that 100,000 years from now our
progeny will differ in substantial ways from current *Homo sapiens sapiens*?

NASA has already established that weightlessness can trigger gene
switches. A few humans may already have passed to their offspring some
new switch settings appropriate to living in space. We may evolve into
a variety of species: *Homo terran*; *Homo martian*; *Homo europo*; *Homo
vacuo*. We may more resemble today's robots than today's humans. Yuval
Noah Harari speculates that "once the Internet-of-all-Things is up and running, humans might be reduced from engineers, to chips, then to data, and
eventually we might dissolve within the torrent of data like a clump of earth
within a gushing river."[9]

Our electromagnetic signals 100,000 years from now will have spread
through the Milky Way. If there are other extraterrestrial civilizations, the
probability that we will have discovered them will be high. Whether our
progeny has physically traveled far again depends on the critical issue of
whether we have beaten the speed of light in some practical way.

One million years ago there were various hominid species making their
way around Africa but *Homo sapiens* did not exist. One million years from
now is a tiny blink of cosmological time and scarcely time for our planet
Earth to have changed at all. The mighty continents will have drifted one
hundred miles on the magma ocean of the mantle. One million years is a
significant time measured by the biological clock of evolution and natural
selection even if we do not deliberately implement our newfound capacity
to alter our own genome and to merge with our machines. We could die off:
go extinct as species do. We might have expanded into space and evolved.
Whatever hominid creatures remain on Earth, they will not be *Homo sapiens*. If there is one certain prediction in this gaze into the future of humanity, it is that humanity as we now know it will not exist in 1 million years.

There were no plants or animals on Earth 1 billion years ago. There was
abundant multicellular life. The Cambrian explosion lay 400 million years

in the future. One billion years from now, the Sun will be slightly larger and cooler but 10 percent brighter. The extra heat is likely to have evaporated the oceans, halted plate tectonics, ended the carbon cycle, and driven a greenhouse effect in the atmosphere. There will be less free oxygen, and the surface will be uninhabitable for oxygen breathers like us. If our progeny exist on Earth, they will likely be tiny and very smart, perhaps indistinguishable from nanobots at first glance.

Ten billion years ago, the Universe was just getting cranking, forming the first stars and galaxies. Our Sun and its retinue of planets were not yet born. Ten billion years from now, the Sun will have burned out its thermonuclear fuel. It will have expanded to near the current position of the asteroid belt and have consumed the inner planets, including Earth unless Earth has drifted outward sufficiently in the process. This remnant of the Sun will be a white dwarf star with a mass about 60 percent of the mass of the current star and a size comparable to that of Earth. This white dwarf will yield a dim blue light and not the life-giving brilliance of our current Sun. Even on the small chance that Earth still exists, it will not be hospitable to life—certainly not human life as we know it today. By this time, any remnants of humanity will be space dwellers or gone.

If we look further back, only 14 billion years into the past, we run into a wall of ignorance. We do not know why the Universe, all of space and time, blinked into existence in the Big Bang. Smart minds are trying to figure that out, but we have no firm clue. We also do not know the future beyond 14 billion years, but it is easier to make plausible estimates for that future than for an equivalent past time. Given our current Universe and our state of knowledge of physics and cosmology, our best guess is that the Universe will lumber along, expanding and cooling, its stars and galaxies getting ever further apart and burning out. The Universe is likely to keep expanding into a condition with no content but the occasional photon 1 trillion years from now and more. No stars, no planets, no galaxies, no people. Perhaps there is a refuge elsewhere in the multiverse but I would not count on it.

When I challenged my students to do this predictive exercise, a few bold ones who perhaps just wanted to tease me projected a grand future in the spirit of Kurzweil: our intelligence would merge with that of our machines and expand throughout the Universe, and maybe into other components of the multiverse, rendering the inanimate matter of the multiverse conscious with transcendent intellect. Even this grand view raises deep issues. The Universe is already a data-crunching machine at the quantum atomic and molecular level in ways we are still striving to understand. How will anything be different if an intelligence expands through the Universe or into the other universes of the multiverse? If something like human sensibility vanishes, will not the Universe be poorer for it?

An uncomfortable fraction of my students had a much more dismal view. They thought that humanity would die off rather quickly, succumbing to climate change, nuclear war, disease (even before the COVID-19 pandemic), an asteroid, the domination of machines, or an extraterrestrial invasion. The level of pessimism was sobering, but then the class was designed to raise awareness of both the promise and peril of rapid technological development. I did not mean to scare the students, but the topic had that effect on some. They did not use the phrase, but they were anticipating TEOTWAWKI: The end of the world as we know it.

LEAVING OUR MARK

Faced with a sprawling, perhaps infinite, future, we may need to store and curate all the data that shape us and our evolving society in case our progeny are curious about their roots. This is already a challenge. Most of the data produced by all of human history was created in the last few decades, and more accumulates at an exponential rate. At the same time, storage media grow rapidly obsolescent. Punch cards, magnetic tape, floppy disks, and compact discs vanish or languish, collecting dust on shelves. Vast amounts of data are now stored on silicon chips in the cloud, but how will those be read in a century, a millennium from now?[10] Even those media are volatile compared to well-preserved paper or a Sumerian clay tablet although these lack storage capacity. Some today advocate ceramic tablets or nickel-based NanoFiche[11] wafers that might, in principle, last a thousand years, but they also have limited storage capacity, and the future is much longer than that. Storage at the atomic or molecular level in DNA might prove useful, but it needs to be read. Will the reading devices be equally long lived?

Back in the early days of email (the 1980s) a brilliant and insightful colleague of mine, Bohdan Paczynski of Princeton, noted that since computer memory was so cheap, there was no point in selecting which emails to save. Save it all! I recently upgraded a computer. On it I had email files originally hosted by a long-expired computer spanning a time from 2008 to 2013—not that long ago—that someday I might, might(!) want to explore. The information technology person with whom I consulted on getting these old files transported to my new computer blithely informed me that I could not transport those particular files: they were no longer supported. My only option was to select individual emails by hand and mail them to myself, in the process losing header information. I understand that history needs some curation, but this does not bode well for preserving information in the future.

MAINTAINING CONTROL OF OUR TECHNOLOGY

There are reasons to think that humans and their organizational systems may be able to cope with the immense changes coming. Despite two world wars in the twentieth century, humans have survived over seventy-five years with nuclear weapons that could instantly destroy civilization. While there have been famous nuclear reactor breakdowns—Three Mile Island in 1979; Chernobyl in 1986; Fukushima Daichi in 2011—they have been relatively few and relatively contained. The US Navy has operated nuclear reactors at sea in submarines and surface craft without a single nuclear accident. With a few exceptions, air travel is far safer per mile than travel by automobile. After the deadly fire on the Apollo launch pad in 1967, GE was assigned oversight of the Apollo quality control program, the slogan of which was "Zero defects."

There are reasons for the safety and reliability of dangerous human enterprises. Thomas Dietterich of the University of Oregon School of Electrical Engineering and Computer Science wrote an essay in 2018[12] summarizing some of the attributes of what have been named *high reliability organizations* (HROs).[13] HROs assume that failure modes await that have never been previously observed. Absence of failure is regarded as a sign of insufficient vigilance. HROs encourage the reporting of all mistakes and near misses. HROs shun simple interpretations. They encourage multiple interpretations of events, varied checks and balances, and adversarial reviews. They value interpersonal skills as well as technical knowledge, rotate people through positions of responsibility, and retrain frequently. HROs cultivate a deep situational awareness so that they can differentiate expected output from the effect of unanticipated internal or external influences. HROs are committed to resilience; they practice improvising in reaction to unexpected circumstances. HROs empower every team member to make decisions, raise alarms, or halt operations. Concerns are spread through the organization rather than along a fixed reporting path, and management is always available to register issues.

Dietterich advocates implementing the operational philosophy of an HRO in our effort to integrate AI into society. AI systems should not just monitor their own behavior but the behavior of the humans with whom they interact and any changes in the external environment. There must be constant surveillance by both humans and AI for anomalies, near misses, and unanticipated consequences.

These are all excellent notions, but it must also be recognized that some of the technological developments that will require the vigilance of HROs are different than the threat of nuclear war. Nuclear weapons require the resources of a nation-state. Means exist to monitor associated developments and counter threats. There are no guarantees, but so far it has worked. The technological developments of AI and biology are different. They can be done by individuals or small groups and spread at the speed of the internet. A lone scientist in China engineered the first designer babies. Hackers can use AI to spread malware, evade detection, and crack passwords. The internet itself spreads and amplifies misinformation and disinformation as well as fruitful knowledge and information.

AVOIDING STUPID STUFF

One way to avoid the worst ramifications of our burgeoning technology is the notion of relinquishment. In the words of President Barack Obama in a different context, "Don't do stupid stuff." Relinquishment could be considered by an appropriately constituted HRO or otherwise invoked on a community-wide basis. Despite a single violation, the field of biology has collectively put in place a ban on designer babies, effectively a relinquishment of the application of a relevant and readily accessible technology. I am not confident this ban will last long, but it is certainly appropriate now.

There is a general concern, which I tend to share, that the capabilities of AI will grow so powerful so rapidly that steps must be taken now to consider and perhaps limit the nature and capacity of AI. Oxford's Nick Bostrom, an influential pioneer in the field, and computer scientist Stuart Russell,[14] among many other people, express this concern. Russell imagines that a superintelligent AI might decide the solution to climate change is to eliminate humans as a source of pollution.

Kurzweil explored this topic in some depth.[15] He enunciated two categories: a broad relinquishment and a fine-tuned relinquishment. A broad relinquishment implies forgoing a wide swath of technology but raises the danger of eliminating the beneficial aspects of a technology along with perceived threats. Fine-tuned relinquishment is the prospect of targeting specific negative outcomes.

We should develop smart nanobots that can search out and destroy cancer cells, but nanobots that can self-replicate in a natural environment (the gray goo scenario)[16] should be avoided. Daniel Dennett calls for a relinquishment of conscious machines.[17] The problem is how to limit specific implementations without sacrificing associated benefits.

There are good reasons to pursue AI that has manifest benefits. There is also general agreement that machines that contain the code to replicate and modify their code should be limited lest they spread uncontrollably through the ecosystem of the internet. In 2021, OpenAI released a new program, Codex, that can write appropriate new computer code given a task requested in natural human languages. Codex can program in any of twelve computer languages, translate from one of these languages to any of the others, and autocorrect a human programmer who is writing code. Codex was then implemented in ChatGPT where it was widely used to aid programmers. At its current level of accomplishment, Codex and ChatGPT will not put programmers out of business but they are a very useful complement that takes some of the drudgery out of writing code and thus enhance human abilities. That is a very effective thing to do—for now.

The important question is where this capacity takes us in the exponentially advancing future. Codex cannot yet reason like a human, but one can imagine combining its ability to write new original code with a program like AlphaGo Zero that can already develop strategies beyond human ken. What happens when AlphaGo Zero poses its own queries for which Codex provides the coded solution with no human in the loop? Can the newly empowered AlphaGo Zero expand its frontiers and capacity and ask for yet more code to implement its goals, whatever they are, iterating at the speed of light? Whether OpenAI is ultimately a source of good or evil remains an open issue. This may be a circumstance where fine-tuned relinquishment is required.

IMPOSING ETHICS

A significant positive reaction to the possible perils of AI is the growing concern for what is termed "ethical AI." Luciano Floridi, a professor of philosophy and ethics of information at Oxford, points out that just as the current aging generation did not know a life without automobiles, the current young generation will not know a life without digital media in all its manifestations.[18] This transition will only happen once so we need to get it right. Notions of computer ethics in the 1950s led to information ethics and now to digital ethics related to the governance of the whole digital ecosystem: data, algorithms, and related practices. Ethics needs to be considered at every step in the evolution of our machines. A philosophy of "Move fast and break things" might be a good basis for building a fortune, but it is not necessarily good for society. Even Silicon Valley pioneers have come to recognize that.

In 2016, Carnegie Mellon University received a grant of $10 million from a law firm to form the K&L Gates Endowment for Ethics and Computational Technologies to study ethical AI. Stanford initiated AI100,[19] a one-hundred-year study to explore the societal effects of AI. The first AI100 report was released in 2016 and a second in 2021. During those five years, concern for the sociological issues related to AI grew appreciably. The University of Texas at Austin initiated a broad, multidisciplinary undertaking called Good Systems,[20] an eight-year, $10 million, grand challenge the goal of which is to design responsible AI technologies. Good Systems aspires to establish general principles of constructing and operating ethical AI that all can follow. Good Systems has formed a partnership to promote innovative ethical AI with MITRE, a nonprofit dedicated to promoting a safer world.[21] Harvard, MIT, Stanford, Berkeley, and many other universities are introducing principles of ethics in computer science classes that previously ignored such non-technological issues. In 2022, the Association for Computing Machinery launched a new journal, the *ACM Journal on Responsible Computing*.[22] The Responsible Intelligence Institute, a nonprofit dedicated to community-driven, independent assessment of responsible AI formed a partnership with the Standards Council of Canada to determine requirements for a worldwide, ethical, AI certification program.[23] The nonprofit Partnership on AI[24] brings together over one hundred international academic, civil society, industry, and media organizations to seek positive AI outcomes for people and society. The consortium AI for Good[25] aspires to develop AI to serve humanity rather than commerce. The Center for AI Safety[26] argues that "mitigating the risk of extinction from A.I. should be a global priority alongside other societal-scale risks, such as pandemics and nuclear war." Kay Firth-Butterfield is CEO of a new Centre for Trustworthy Technology[27] which is a member of the World Economic Forum's Fourth Industrial Revolution Network that also aims to convene stakeholders from government, business, academia and society to shape an ethical and trustworthy technology agenda.[28] They aim to study all sorts of technology: AI, artificial biology, and neuroscience. According to the Chinese newspaper *Xinhua*, even the Chinese Communist party is calling for researchers to "assess the potential risks, take precautions, safeguard the people's interests and national security, and ensure the safety, reliability, and ability to control AI."

These critical conversations are themselves growing rapidly. Their impact remains to be seen.

Professor Jungfeng Jiao, one of the leaders of Good Systems, described the issue this way:

> The increasingly ubiquitous adoption of AI technology in homes, communities, and cities is poised to transform our society. Will we become a society where the gaps between haves and have-nots are further enlarged and entrenched, or can we steer the development of AI in ways that will make our society more just and equitable? Achieving the latter vision requires fundamentally reshaping how we train future AI scientists and developers. We cannot afford for AI to be designed without consideration of possible negative repercussions, leaving it to others to sort out the societal consequences of the technologies. Thus, there is an urgent need for us to understand the potential impacts of AI and why we need ethical AI systems to help us build a more just and equitable future for everyone.[29]

In 2020, the National Science Foundation issued a call for proposals to establish an Institute for Human–AI Interaction and Collaboration[30] with the goal of making AI "more productive, robust, and fair" by promoting "inclusive design, being socially beneficial, avoiding unjust bias, and being built and tested for safety, accountability and privacy principles." The National Science Foundation seeks to get beyond minimal exchanges between humans and AI by incorporating both spoken and written language, gestures, body language, tactile interaction, and augmented reality. The goal is to build confidence that AI systems operate on fair and transparent principles that can be understood and vetted by a broad community. These are laudable goals and a substantial task but especially challenging since it is very difficult to understand how a given neural network AI has reached a given result. Joydeep Ghosh of the Texas Good Systems consortium points out that fairness, explainability, and responsibility are all related aspects of ethical AI but they are different and each difficult to implement.

Amazon and Google have pledged support for the Institute for Human–AI Interaction and Collaboration. They will not be involved in the merit review that selects participants in the program, but they will have some input in final award decisions. Amazon has initiated a research awards program to encourage diverse perspectives on the topic of AI fairness.[31] Google promises to promote its Google AI Principles and broaden participation from underrepresented groups and institutions.[32] The Office of Science and

Technology Policy (OSTP) in the White House is promoting an AI bill of rights. Eric Lander, short-term director of OSTP in the Biden administration, emphasized that the bill of rights needs teeth: laws, litigation, and procurement enforcement to bolster general principles of governance, privacy, fairness, transparency, and explainability.[33]

Ethical AI is an important goal but it is faced from the start with the question of whose ethics. Ethics can vary with culture. Are we invoking Judeo-Christian, Muslim, Hindu, Buddhist, Confucianist, animist, or atheist ethics? People in Asia seem to be more comfortable with personal-privacy intrusions than those in the West. Diversity is an issue as reflected in the formation of interest groups such as Black in AI, Latinx in AI, and Queer in AI. Ethics probably cannot be incorporated in a machine by recording the right zeroes and ones in the right portion of a computer memory but by instituting an ability to adapt to an external social environment. What matters may not be the values and ethics built into an algorithm but the development of an appropriate cognitive architecture that responds flexibly to fluid situations within certain moral guidelines. Perhaps there should be a new golden rule: "Do unto AI as you would have it do unto you."

Similar notions for the ethics of future biology also must be raised. There is a long-standing system of medical ethics beginning with the Hippocratic oath and reflected in the current security rules of the Health Insurance Portability and Accountability Act (HIPAA) that govern the privacy of medical information. There are already challenges in storing and sharing medical information to facilitate efficient and accurate medical care as doctors migrate from hand-written notes to digital records. Telemedicine has great promise, but if such systems get hacked there could be great damage. The same is true of the vast and growing database of medical records of all the people wearing watches that record medical information. Virtually all that data is owned and controlled by corporations like Apple and Google. More people will have wearable or embedded instruments such as pacemakers and prosthetics that talk to the cloud. Machine learning algorithms are hungry for that data. Who will own and control it? How can it be shared effectively without violating medical privacy? As technologies for reading from and writing to the brain become more proficient and widespread, will corporate titans own your thoughts? At the very least, there must be new means to encrypt AI medical and health data without sacrificing the utility of those data for curing disease.

The developments of CRISPR and other advances in biology like treatments or cures for aging also require intense ethical consideration. If the problem of aging is solved, there will be massive ethical challenges and sociological adjustments as designer babies compete with the non-aging for the privilege of maintaining control over AI or being controlled by AI. Young CRISPR researcher Jiwoo Lee commented to *Wired* magazine, "I think there's a really slippery slope between therapy and enhancement. Every culture defines disease differently."[34] What is considered a boon to public health in one culture might be regarded as eugenics in another.

The first example of applying CRISPR to engineer human babies drew widespread condemnation and calls for Congress to address the science, finances, politics, and ethics of editing sperm and eggs, but Congress remains mired in immediate, myopic concerns. Bioethics is a well-developed field in comparison with the ethics of AI but the two must merge as machine learning permeates the fields of medicine and biology ever more deeply. If we merge with our machines, the fields of bioethics and AI ethics will essentially become the same.

There are also broad moral and ethical issues associated with the question of what we owe to future humans and their progeny. Consideration of this issue is summarized in the word *longtermerism*. We have a strong moral commitment to our children but also to strangers in need. That is why we donate to disaster victims and children with cleft palates in Africa. What of future humans? If we dodge existential threats and some form of humanity survives hundreds of millennia, there are far more humans yet to be born than have ever lived on Earth. What is our obligation to these people, strangers all? Longtermerism advocates that we think of future humans as we do strangers today. We should comport ourselves morally and ethically in a way that preserves and improves the situation of those distant relatives. We should avoid nuclear war, heal and husband the climate and ecology, prevent the worst nightmare scenarios of AI and biological research. An insightful case for longtermerism is given by Oxford philosophy professor William MacAskill in his book *What We Owe the Future*.[35]

A contrasting effort is the Voluntary Human Extinction movement.[36] The notion here is that by voluntarily choosing not to breed, humans could be phased out thus bringing an end to the Anthropocene. The Earth would return to a pristine healthy ecology once again red in tooth and claw. Surely there is some middle ground.

THE PROMISE AND THREAT OF CONNECTIVITY

We are faced with a future of increased connectivity that will deeply alter society and how people function within it. Humans have advanced over other species because of their ability to cooperate and learn from one another in ever more sophisticated and effective ways. As we get more connected, we will amplify our distributed intelligence and the intelligence of groups, either people or AI, or a merging of the two.

Our new ability to connect has already changed our society in profound ways. Sherry Turkle of MIT has written extensively on the cultural and cognitive impact of mobile phones.[37] People have gotten hooked on connectivity. People ignore companions and children in favor of devices. People talk rather than listen. A new craving to share has gripped broad swaths of our communities in which the notion of a private journal is a pointless anachronism. Yuval Noah Harari captures this new sensibility: "If you experience something—record it. If you record something—upload it. If you upload something—share it."[38] To update Descartes, I post, therefore I am.

Some envisage a new Web 3.0, otherwise known as the Spatial Web,[39] in a coming era when everything is digitally connected to everything. The notion is that Web 1.0 was built on the original system sponsored by the Advanced Research Agency that connected a few Pentagon-funded experts and their computers. Web 1.0 was read-only and enabled only static web pages that could be viewed with a browser. Our current read-write system, Web 2.0, was born with both standardized TCP/IP[40] protocols (the http[41] protocols developed by Tim Berners-Lee) and other technology. Web 2.0 enabled connected websites and promoted interaction and collaboration between people, thus allowing hyperlinks, browsers, search, social media, great economic advance, surveillance capitalism, and privacy invasion. A key aspect of Web 2.0 is that all the information fed to your browser is agnostic to the location of the computers hosting the websites you read; they could be in Kansas or Kazakstan. Another feature of Web 2.0 is that it is based on documents, basically words or images on flat paper rendered digitally. It is fundamentally 2D.

Advocates of Web 3.0 envisage a new set of protocols and technology that would allow information about the spatial location of a thing to be intrinsic to the data characterizing it, hence a spatial web. Web 3.0 would promote interaction between humans and machines. Web 3.0 would allow careful tracking of every person and every part of every thing that will occupy the IoT and the relations between them not just in real space but in those spaces accessible in AR and VR. Web 3.0 is potentially vastly bigger than the 3D space of planet Earth. It could incorporate features of the *metaverse* that are limited only by our imagination. This potential led Facebook in October 2021 to rename itself Meta and Apple to advertise its Vision Pro headset in June 2023 as a portal to the spatial web. Much of the discussion of Web 3.0 is focused on business, contracts, and data. Human life is not just contracts and data. Or maybe it is—a depressing thought.

Some imagine that Web 3.0 will be facilitated by verifiable *decentralized identifiers* (DIDs), hyperconnected *braided blockchains*, and *distributed ledger technology* that enable self-actuating smart contracts. This would, in principle, allow commerce to be done easily, transparently, and verifiably throughout the real world and the metaverse with an explosion of economic impact. As an example, supply chains would be closely and efficiently monitored by knowing where every component of every machine was at every instant of time from mining the raw materials, to manufacturing, to sales, to implementation and maintenance. Ubiquitous DIDs and smart contracts could help to limit the existence and dissemination of deepfakes, misinformation, and disinformation. Web 3.0 promises a means of establishing trust when we cannot trust our senses.

Dan Mapes and Gabriel René are strong advocates of Web 3.0. They founded the Spatial Web Foundation,[42] which envisages the spatial web as a hyper-integrated, contextually aware, ethically aligned network of people, AI, and machines. Karl Friston of the Wellcome Centre for Human Neuroimaging at University College London is a pioneer of computational neuroscience. Friston developed techniques of *active inference*[43] and the *free energy principle*[44] as a way to understand the ability of the human brain to predict and plan. In association with Friston, Mapes and René also began VERSES, a cognitive computing company closely related to the Spatial Web Foundation that specializes in the next generation of AI.[45] VERSES is working on methods for developing human-interpretable, explainable AI (XAI) systems based on Friston's work on active inference and the free energy principle. The company claims this approach will offer new possibilities for the transparency and understanding of AI processing. The goal is to make AI that can explain its decision-making processes in human-understandable terms and hence be more easily understood and regulated.[46]

There is a host of issues to address in the implementation of Web 3.0. Are DIDs for people assigned at birth or do you apply for them later as for current social security numbers? How do you own an idea in Web 3.0? Enforce copyright?

What are the physical demands of Web 3.0? How much computation will it require? Web 3.0 might be subject to a *combinatorial explosion*. That might be related to factorial[47] of n, a huge number if n is at least the number of people and things linked in Web 3.0. How many things are we talking about? Every bolt and nut? Every cell in a body? How much energy is needed to perform that computation? How much heat is produced?

Web 3.0 may demand ubiquitous practical quantum computing, but even quantum computing needs power and produces heat. Who writes all the software to exploit Web 3.0 with all its separate components? Is that task turned over to superintelligent AI? If so, does that already mean game over for the competition between humans and machines?

The answer to many of these questions depends on just what granularity we are talking about. How far down do the smart contracts go? How long are spatial domain names if there are subdomain names for every neuron in your brain, every screw in a machine? If a spatial domain is defined as the land and home under control of owners, what about the space above?

Suppose a person is infused with smart nanobots rendering them healthy but still owned by the commercial enterprise that constructed them. Can the smart contracts associated with the nanobots restrict where the person can go in order to protect the investment in the smart little buggers? Suppose a restaurant controls the domain of its property and subdomains going down to the pans, dishes, cutlery, and food items. To what extent can that restaurant control its clientele regarding who is admitted and who is not? Suppose the restaurant has a smart contract with the manufacturer of nanobots that are controlling a communicable disease. Can the restaurant refuse admission to a person with those nanobots in their bloodstream? With all these smart contracts and spatial domains, are there not bound to be conflicts as people, machines, and AI wrestle for territory and rights in the multiverse? How are those resolved?

Advocates for Web 3.0 argue that everyone will control what personal information is collected on them, who has access, and how it can be used. That is a nice idea, but how does that work in practice? If you are monitored by millions of smart contracts interacting with your environment wherever you go, how can you possibly protect your privacy? The spatial web knows where you are and what you are doing at every moment of the day. Already with sparse cell phone data, Uber records, and just-walk-out shopping there is ubiquitous location tracking with which you can find and identify individual people. Anonymity is a farce. Will not Web 3.0 be like that but worse?

Web 3.0 may subject us to extreme information overload. If we are constantly immersed in the data of the spatial web, won't we get saturated? The human mind can only absorb information so rapidly. What is the demand on a person walking around trying to constantly adjust the ads they allow from the flood that besiege them?

Much of the discussion of Web 3.0 treats people as equivalent to consumers. Who walks around thinking, "I need to buy something now"? On a typical day I spend virtually none of my time planning to purchase something. Even if you are an inveterate shopper immersed in VR shopping, how do you know the clothes you buy will fit? Some say a personal agent will learn what you like. How do you experience something new that you don't already like? Are accidental purchases consummated by blockchain reversible? Is there room for buyer's remorse on Web 3.0?

Some advocates of Web 3.0 argue that our deepest desire is to communicate and share ourselves with one another without fear of miscommunication, and miscommunication would be minimal in an idealized world of ubiquitous smart contracts. They glorify the prospect of sharing the mind's eye directly. I see mental telepathy already on the horizon with BCIs, and it makes me very leery. What if we don't want to share our inner thoughts? How taxing would it be to constantly be adjusting what you want to mentally share and with whom or what?

All this AR and VR. Suppose you enjoy the real world? Some supporters of Web 3.0 advocate using AI and VR to re-experience your past life or to invent a future reality. This seems a path to Aldous Huxley's 1932 science fiction novel *Brave New World*: everyone immersed in VR all the time. Who then grows food in the real world?

Web 2.0 was designed or at least manipulated to modify human behavior. Web 3.0 may transform humans as digital phones have done. For better or worse? Will only those who can adapt to the rapid technological change and environment of Web 3.0 succeed, leaving others behind?

To extend the dreams of advocates of Web 3.0, will Web 4.0 include hyperspace if that proves to be real?

There will be more and tighter coupling with cloud computing and the IoT as we enter an era of *ambient intelligence*. No one knows everything or accesses all the data currently available. Instead, individuals access what they can that drives and amuses them. The point of generating, accessing, and storing data is not the data per se but the potential to convert those data to understanding and knowledge. A traditional approach has been to put information in a book for a human to read or to upload data to a central computer server to process. A new process of *federated* or *collaborative learning* leaves the data in decentralized devices or servers and trains a machine-learning algorithm to access that distributed data. This technique might be especially useful in a widely distributed IoT with dense local storage. The procedure is currently aimed at machines, but the effects of enhanced connectivity are likely to be drastically amplified when we develop direct neural links with one another and with the IoT.

Human-machine collaboration may amplify our collective intelligence but drive a tendency to offload thinking to machines. This displacement may reduce the cognitive load on people, but something may be lost in the process. We could be forced to merge with our machines to handle the cognitive data load, or maybe our brains will shut down under the onslaught of the data overload.

There will be new ways to implement connectivity. Massive role-playing games like *Second Life* are a precursor to a higher level of connectivity in a computer-facilitated metaverse. Developments in VR will bring yet another dimension to this capacity. One can envisage an evolution from mobile phones to the metaverse to BCIs to what is effectively telepathy.

As AI expands and adapts in our connected society and makes more critical decisions for humans, will humans lose some of the creative skills gained through millions of years of evolution? Will some professional skills—those of doctors or lawyers—atrophy because of growing reliance on AI? Do we need to use it or be threatened with losing it?

As machines get exponentially more capable, will humans be able to keep up? Will we need to employ our exponentially expanding understanding of our biology to augment our abilities to compete with our machines? Will we be forced to tinker with our brains in order to learn more and faster? Will we ultimately need to merge with our machines? What if the division between individuals begins to melt away with increased connectivity? Are we destined to devolve into a hive mind?

There is great potential for amplified, collective intelligence but other aspects must be contemplated as these developments occur. A fundamental dilemma of the wired world is that as connections become more convenient and useful, they become more vulnerable to violations of privacy and hacking. Notions of privacy will be revolutionized if not obliterated.

A novel new form of privacy invasion has arisen with current technology. Advances in AI and biology are bringing changes in perspective on the role and nature of death. Nature developed death to facilitate natural selection: out with the old, in with the new. The reality of death is woven into individual perspectives and the structure of cultures. There are now prospects of postponing or eliminating death by altering our biology or merging with our machines. As long as death remains with us, we have alternatives such as storing data, photographs, recordings, and videos of loved ones after their passing. As discussed above, there are problems with how long those data can be preserved and accessed. The flip side is using those data to resurrect dead people. AI can now create holograms of deceased people and invest them with words characteristic of the deceased. No one asks the dead for permission to use their data in this way.

Much of the speculation about merging with our machines to produce enhanced humans rests on an assumption that choices of enhancement— whether or not the option only of the rich—will be chosen by free individuals freely choosing. Suppose instead that economic forces or political imperatives guide enhancements? Are you choosing that upgrade because you want to or is it because some AI is nudging you in a certain direction for ends you do not directly perceive? What is the good of human consciousness if algorithms become better at nearly every current human enterprise? If being a human is important, will not technological, economic, and political forces guide the enhancement of humans? If humans are not in control, are they any longer human?

Evolution has been a successful process because it tries everything to see what works in a given environment. For all their flaws, humans have advantages because they are diverse with a variety of perspectives, motivations, cravings, and capabilities. That means that while individuals have limitations, collectively we can explore a range of potential actions in each situation. Amidst the resulting chaos there are strengths of diversity, flexibility, and resilience. In a world controlled by very smart machines, there is some danger that the machines will seek ideal, efficient, solutions that while seemingly strong are ultimately brittle when large changes inevitably arise. Contrast the autocracy of the Soviet Union with the liberal democracy of the United States or the just-in-time supply chains that provided huge efficiencies in global enterprises until COVID-19 interrupted that scheme and businesses turned again to building supply surpluses to have a flexible buffer. Will the IoT be smart enough to employ a healthy degree of resilience when machines and people are wired together? Will future systems be strong like a mighty oak that breaks in a hurricane or flexible like a stand of bamboo that can weather the storm?

BE AWARE

How do we address the issues of the coming exponential growth of technology? Neither succumb to hype nor panic: be aware. The first steps are to learn, think, and talk. Examples of exponential growth are happening all around. In my class I asked the students to bring examples of technology advances to class each day. I told them this habit of taking note might be the one thing they retain from the class years from now. I hope so. You might not notice as you tend to daily concerns, but if you are aware, you will see aspects of new technology everywhere. I have given a wide variety of contemporary examples here, but whenever you read this book, be attentive to presentations in current books, radio programs, TV programs, newspapers, magazines, podcasts. Once you start paying attention, you will notice more and more examples of new technology and its effects on society. Think about what you see and learn. What is healthy and constructive, what is dangerous and should be relinquished and how? Then talk to family and friends and people you meet. George Orwell said, "To see what is in front of one's nose needs a constant struggle."[48] This book is a small contribution to that struggle to see and to the needed conversation, a conversation that has grown since I first taught my class in 2013.

Among the issues to monitor are concerns for ethics, some mentioned in this book and others. What does it mean to have ethical AI and how do we accomplish that? How do we instill ethics in autonomous vehicles and autonomous weapons the very purpose of which is to kill? How do we craft an ethical path as we learn to edit our germ line and guide our evolution? Parallel issues arise with BCIs. How do we tread into this new area without altering what it means to be an individual, independently thinking human?

There are already massive issues of privacy. While the main concern used to be solely government intrusion, social media have consciously evolved to capture vast reams of data on billions of people with the deliberate goal of behavior monitoring and modification. Notions of anonymizing data have proven to be a sham. Location tracking data produced in multiple ways in a single digital phone coupled with abundant other data has easily been used to identify individuals. Archaic laws designed for a pre-digital age even allow the government to access data stored in the cloud that is more than six months old without a warrant.[49] These data can reveal not just one's shopping and voting proclivities but one's sexual orientation, early signs of pregnancy before a woman knows, manic or depressive episodes, and more.

Individuals can take small steps to protect their right to be left alone. There are movements, especially among technical and artistic communities, to develop individual means of protection. Among these are masks and special camouflage to prevent facial identification and recognition in public spaces, crowds, and protests.

Such actions may help at the margins, but strong new laws and mores need to be developed to maintain our right to protect our beliefs (First Amendment), the right to control our homes (Third Amendment), the right to be free of unreasonable searches (Fourth Amendment), and the right to avoid self-incrimination (Fifth Amendment), all of which are under assault by both private and government enterprises. We need new laws to control the collecting, selling, and trading of data and to regulate how long that data can be retained. There need to be restrictions on obscure terms of agreement that few read and understand and that demand relinquishment of fundamental rights as a condition to use a product. At least the restrictions that are in place in terms of agreement for protection of privacy and other rights must be enforced.

Can laws and regulations help to shape a safe digital world? The LLMs of OpenAI, Microsoft, and Google acquired their vast training data by scraping the internet for reams of data without explicit permission. Does that already violate the EU GDR or the AI Act under consideration? What new regulations are needed? How do we write laws and regulations to mitigate emerging risk when we do not fully know what those risks are?

Should there be some regulation of the supply of GPU chips that are critical to the training of LLM AI? Should a right to know be codified to alert us when we are reading a document prepared by an LLM AI? Should there be a means to audit individuals and enterprises that intend to train an LLM AI? Should a license be required to access the enormous resources of a cloud provider in order to train a substantial LLM AI, something like an AI driver's license? Should there be know-your-customer laws for cloud providers as there are for banks to enable tracking who is employing LLM AI? Should there be liability laws that hold AI companies responsible for damage done by their products? Should cloud providers be liable for damage done by users of their services? OpenAI, Microsoft, and Google have all incorporated restrictions into their LLM AI so one cannot easily ask for instructions to build chemical weapons or produce pornography, but new enterprises promptly sprang up with the avowed goal of developing and releasing unrestricted LLM AI. Should that work be restricted or banned? Should there be a means to protect training data so that malevolent actors cannot corrupt it, a movement already underway? Can such laws and regulations be effectively enforced?

New York Times writer Thomas Friedman calls the emergence of new AI technologies a "Promethean moment" akin to the invention of the printing press, the scientific revolution, the Industrial Revolution, nuclear power, and the internet only maybe more so. He writes,

> We are going to need to develop what I call "complex adaptive coalitions"— where business, government, social entrepreneurs, educators, competing superpowers and moral philosophers all come together to define how we get the best and cushion the worst of A.I. No one player in this coalition can fix the problem alone. It requires a very different governing model from traditional

left-right politics. And we will have to transition to it amid the worst great-power tensions since the end of the Cold War and culture wars breaking out inside virtually every democracy. We better figure this out fast because, Toto, we're not in Kansas anymore.[50]

A 2021 report on information disorder from the Aspen Institute[51] presents a variety of recommendations to balance the battle of free speech against disinformation and misinformation. The report argues that proactive leadership from all parts of our society is needed to pursue this battle, leadership that is currently lacking. In a democracy, that leadership is selected by voting.

Another incipient tension is between the drive for efficiency by profit-driven markets and the health of society where other values are treasured. Should our social institutions and culture adapt to the incursion of AI, or should our institutions evolve to keep our broader values paramount? Once again, the collective will of individuals can be brought to bear.

Another step is thus to vote. We need collective decisions that are probably only effectively derived by the political process. All politics is local both in space and time. Immediate issues dominate any given election. Except for Andrew Yang, there was no discussion of the crucial issues associated with the exponential growth of technology in the presidential campaign of 2020. There is likely to be little in 2024. Vote for people with understanding and foresight before your chance to vote slips away.

It is difficult to deal with problems that have not yet transpired but are coming ferociously fast. It is easy to point out possibilities that have been speculated upon but have not come to pass and hence to dismiss those possibilities. That is too facile a way out.

For every "That hasn't happened" there is a "yet."

APPENDIX: EXPONENTIAL AND BEYOND

This appendix puts a little more rigor behind some of the statements about exponential growth presented in chapter 2.

FOOTNOTE 2

The text in chapter 2 contrasted the time to fill a room with pennies if a penny were added once per second in a linear process compared to an exponential process in which one doubled the number of pennies at each step, adding first one penny, then two pennies, then four pennies, each second. Given that it takes about 1 billion pennies to fill a classroom of modest size, it takes 1 billion seconds, about 30 years, to fill the room one penny at a time. The text stated that to estimate the time to fill the room if the number of pennies doubled at each step, we required 2^n to be a billion, where n is the number of steps. The number of doubling steps only has to be about 30, requiring only about 30 seconds in our imaginary example of doubling every second.

The formula $N = 2^n$ to estimate the number of steps, n, to get a total number of pennies, N, is only an approximation. The prescription to estimate the total number of steps if the number of pennies is doubled at each step is formally

$$N = \sum_{n=1}^{N} 2^{(n-1)} \tag{A1}$$

where n is the number of steps and N is the total number of pennies accumulated, a billion in our example. In this equation, Σ is a command to sum the factor $2^{(n-1)}$ using first n = 1, then n = 2, and so forth up to the total number, N. The approximation N = 2^n takes a few steps longer—a negligible difference if N is a large number.

FOOTNOTE 3

The text in chapter 2 described in words the nature of exponential growth as a process in which the rate of change of the amount of something is strictly proportional to the current accumulated amount of that quantity. The quantity being accumulated could be money or, according to the argument presented in the chapter, collective human knowledge.

The criterion that the rate of change of something is proportional to the present amount can be expressed mathematically as

$$dK/dt = cK \tag{A2}$$

where K is the amount, dK is the differential increase at a given time, dt is a differential increment of time, and c is a constant of proportionality that describes just how much K changes for a given accumulated amount, K. The quantity K will change more rapidly if c is large, more sedately if c is small. The solution to this equation is

$$K = K_0 e^{ct} \tag{A3}$$

where e is the basis of natural logarithms. The number e is a special transcendental number about equal to 2.71828 with the remarkable property that if you raise e to a power ct, the rate of change of K will be cK as in equation A2. In this equation, K_0 is the initial amount, or one penny in our room-filling example. Equations A2 and A3 show that both the quantity represented, K, and the rate of change of that quantity, dK/dt, grow exponentially.

Because e can be represented as a power of 2, and vice versa, equation A3 can also be represented as

$$K = K_0 \, 2^{(t/\tau)} \tag{A4}$$

where τ is the growth time, the time to double. The doubling time, τ, is related to the inverse of the constant of proportionality, c, in equation A2.

EXPONENTIAL GROWTH

Figure A.1, a reproduction of figure 2.1 of chapter 2, illustrates the fundamental process of exponential growth in graphic form. This diagram plots the amount of some quantity on the vertical axis as a function of the number of doubling times.[1] The curve was generated according to equations A2 and A3 by which the rate of change of the amount at a given time is proportional to the amount present at that time. The diagram provides an illustration of how rapidly an exponential process can grow.

Chapter 2 postulates that knowledge roughly follows the pattern illustrated in figure A.1. The rate at which new knowledge is gained is approximately proportional to the amount already known and hence grows approximately exponentially or with a nearly constant doubling time. It might grow even more rapidly.

Figure A.1 gives the impression that step 8 is special. Around that time, the curve seems to change behavior and rocket upward. This is sometimes described as the knee of the curve. The location of this knee is, however, a function of the arbitrary choice of scales of the axes. Figure A.2 gives the same prescription as that of figure A.1, but both of the axes have been

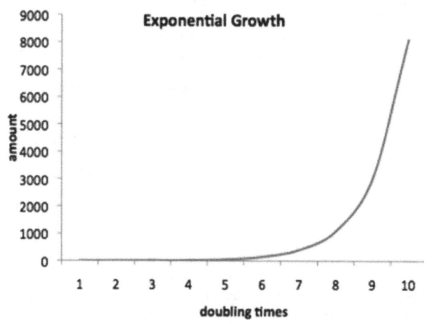

Figure A.1. The amount of some quantity that grows exponentially is plotted as a function of the number of steps.

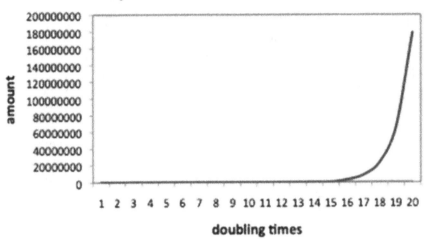

Figure A.2. The same exponential growth law as in figure A.1 but extended to 20 steps. The knee now appears around step 18. The shift in the location of the knee is an artifact that depends on the choice of the range of the axes.

expanded to cover a greater range in the number of steps and in the accumulated amount. In this figure, nothing much appears to be happening at step 8. The action, the knee, appears at about step 18. The point is that the underlying mathematical rule giving rise to the plot—that the rate of change of the amount is exactly the same proportion of the current amount—is the same in both diagrams. The amount grows according to the same exponential prescription at any given step point along the horizontal axis. If the growth is exponential, there is no intrinsic meaningful knee.

Figure A.3 illustrates the fundamental point that the underlying prescription of exponential growth is the same everywhere in figures A.1 and A.2. Figure A.3 again gives the same data as figures A.1 and A.2 but changes the way in which the amount is portrayed on the vertical axis. In figure A.3, the vertical axis is not the amount but the logarithm of the amount. The logarithm is a mathematical construction that represents the factor by which a quantity has been multiplied. In this case, one step vertically represents an increase by a factor of 10. The next step is another factor of 10 and so a factor of 100 more than the first step. In figure A.3, the 2 on the vertical axis represents 10^2 (or 100). The 4 on the vertical axis represents 10^4 (10,000). In figure A.3, the same prescription as in figures A.1 and A.2 yields a simple straight line. The knee has disappeared. The growth follows the same prescription everywhere. An examination of the vertical axis shows that the plot fully captures the overwhelming increase

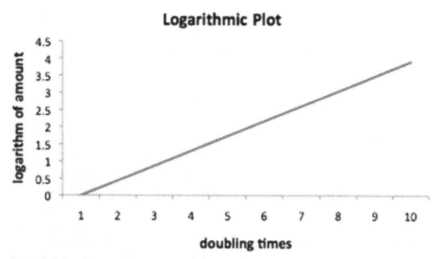

Figure A.3. The same exponential growth law as in figure A.1 but with the amount expressed as a logarithm (the power to which 10 would have to be raised to get the appropriate amount). It is easy to identify exponential growth in a logarithmic plot. It will be a straight line.

as the number of steps increases, but the notion that there is a special time—a special number of steps where there is a break in the underlying behavior—has disappeared.

The value of a logarithmic or log plot is that it makes it easy to discern exponential growth. If the growth is a simple straight line in a log plot, the increase is exponential. It is also easy to discern departures from an exponential trend: the growth will depart from a simple upward straight line. In this portrayal, the notion of bending the curve would mean to cause that inexorably rising straight line to flatten out. The line can also bend upward, indicating a faster than exponential growth. If you see that, watch out!

SUPER-EXPONENTIAL GROWTH

There are processes that can grow even faster than exponentially. An example of super-exponential growth is certain forms of power-law change. Super-exponential growth occurs when the amount of a quantity under consideration changes in proportion to a variable raised to a fixed, negative power. This can be expressed as

$$K = K_0 \, (t/t_0)^a \qquad\qquad (A5)$$

where K_0 is the initial amount of quantity K, t is time, t_0 is the initial time that sets the timescale of the growth, and the constant a determines how rapidly the doubling time decreases with time. The rate of change of quantity K can then be written as

$$dK/dt = aK/t \tag{A6}$$

The rate of change of the amount of quantity K is then not proportional to the amount, as for exponential growth. Rather, the rate of change blows up—heading toward infinity—if the variable t goes to 0. Exponential growth can lead to infinity, but it takes an infinite time to get there. In contrast, a power-law growth with decreasing doubling time can blow up to infinity at a finite time. Whereas exponential growth can be represented as a straight line in a log plot such as figure A3, this sort of power-law growth is represented as a straight line in a plot where both axes are measured logarithmically, a log-log plot.

Chapter 2 describes a plot presented by Ray Kurzweil in *The Singularity Is Near* that gives the time when significant events occurred as a function of how long ago they occurred. Both axes of this plot are presented as the logarithm (base 10) of the relevant quantity. In this scheme, either a time of 1 billion years ago or the time since an event of 1 billion years ago, 10^9 years, is represented by a 9. A time of a decade ago or a time since an event of a decade ago, 10^1 years, is represented by a 1. The result is known as a log-log plot. This plot suggests a mathematical singularity, with the time between significant events going to 0 a finite time in the future.

Chapter 2 also describes another example of super-exponential behavior, an analysis of the gross world product (GWP) by David Roodman from 10,000 BCE to the present.[2] He uses Itô[3] or stochastic calculus to fit the jerky data and finds an approximate power-law fit. The result is a rising straight line not in a log-linear plot but in a log-log plot, as shown in figure A.4. This behavior is super-exponential. In particular, Roodman plots the logarithm of GWP in billions of dollars (so 100 billion = 10^2 billion and the logarithm in base 10 would be 2) against the logarithm of years prior to a critical time (so 100 years before that critical time, 10^2 years, would also have a logarithm in base 10 of 2).

At a mathematical singularity, one is effectively dividing by 0 so the result is infinitely big. In figure A.4, this happens because the horizontal axis represents the time before 2047.

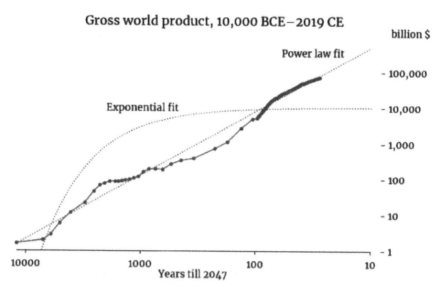

Figure A.4. **The logarithm of the gross world product (GWP) in units of billions of dollars plotted against the logarithm of time before the year 2047, which increases to the left. A straight line representing an idealized, power-law, super-exponential growth for which the doubling time decreases with time approximately reproduces the data (dots). An exponential growth for which the rate of change of GWP is proportional to the GWP and for which the doubling time is constant does not agree with the data. The data slump during the epoch of the Dark Ages but then show an increase in slope about 200 years ago corresponding to the Industrial Revolution. The data show only a slight perturbation from power-law growth at the time of the recent digital revolution. This plot also illustrates the critical aspect that exponential growth will get to an infinite value but will take infinitely long to do so. In contrast, the power law blows up to infinity at a finite time, or 2047 in the model presented. With permission of David Roodman.**

The power law is that time raised to the power of about -1.83. That means that as the time before 2047 gets smaller and smaller, one must divide by an ever-smaller number, and the result gets even bigger. Finally in 2047, the time before 2047 goes to 0. In the mathematics, one must formally divide by 0 and the result—the GWP in this exercise—goes to infinity. That point is off the plot to the right in figure A.4 but dramatically imminent.

NOTES

PREFACE

1. www.nap.edu/catalog/11316/the-astrophysical-context-of-life/.
2. David Branch and J. Craig Wheeler, *Supernova Explosions* (Berlin: Springer, 2017).
3. Ray Kurzweil, *The Singularity Is Near* (New York: Penguin, 2005).

CHAPTER 1. SURVIVING AN ERA OF ASTONISHINGLY RAPID TECHNOLOGICAL CHANGE

1. A term popularized by computer scientist and science-fiction writer Vernor Vinge (*First Word*, Omni, January 1983).
2. https://jcraigwheeler.agsites.net/disc.htm/.

CHAPTER 2. EXPONENTIAL AND SUPER-EXPONENTIAL GROWTH

1. Arthur Conan Doyle, *The Parasite*, a short novel first serialized in *Harper's Weekly* between November 10 and December 1, 1894.
2. This is an approximation. See appendix A for a more precise expression.
3. This verbal description can be expressed more rigorously. See appendix A for a mathematical representation.
4. Ray Kurzweil, *The Singularity Is Near* (New York: Penguin, 2005).

5. An exponential curve is not formally scale-free, but the scale is the same everywhere along the curve, before, at, and after the "knee" in figure 1. See the appendix.

6. Erik Brynjolfsson and Andrew McAfee, *The Second Machine Age: Work, Progress, and Prosperity in a Time of Brilliant Machines* (New York: W.W. Norton & Company, 2014).

7. L. Bornmann and R. Mutz, "Growth Rates of Modern Science: A Bibliometric Analysis Based on the Number of Publications and Cited References," *Journal of the Association for Information Science and Technology*, https://doi.org/10.1002/asi.23329/.

8. In this context, the term *singularity* may be considered a metaphor for the special condition predicted by Einstein's theory of gravity to occur in the center of black holes and at the instant of creation of the Universe in the Big Bang. More generally, a mathematical singularity corresponds to dividing by zero and having solutions to equations blow up to infinity.

9. A point made by Sabee Grewal in my fall 2016 class.

10. Buckminster Fuller, *Critical Path* (New York: St. Martin's Press, 1981).

11. https://archive.org/details/TheToxicTerabyte/page/n10/mode/2up/.

12. M. Rosenberg, "Marc My Words: The Coming Knowledge Tsunami," https://learningsolutionsmag.com/articles/2468/marc-my-words-the-coming-knowledge-tsunami/.

13. D. Roodman, "Modeling the Human Trajectory," www.openphilanthropy.org/blog/modeling-human-trajectory/.

14. D. Roodman, "On the Probability Distribution of Long-Term Changes in the Growth Rate of the Global Economy: An Outside View," www.openphilanthropy.org/sites/default/files/Modeling-the-human-trajectory.pdf/.

15. Bill Joy, "Why the Future Doesn't Need Us," *Wired*, April 2000, www.wired.com/2000/04/joy-2/.

16. The Future of Humanity Institute closed on 16 April 2024. www.fhi.ox.ac.uk/.

17. www.cser.ac.uk/.

18. https://futureoflife.org/.

CHAPTER 3. INTELLIGENCE

1. https://schmidtfutures.com/.

2. Shoshana Zuboff, "Facebook Is Targeting You," *New York Times*, November 14, 2021, www.nytimes.com/2021/11/12/opinion/facebook-privacy.html/.

3. Cade Metz, *Genius Makers: The Mavericks Who Brought AI to Google, Facebook, and the World* (New York: Dutton, 2021).

4. https://obamawhitehouse.archives.gov/blog/2016/12/20/artificial-intelligence-automation-and-economy/.

5. Executive Order no. 13859, "Maintaining American Leadership in Artificial Intelligence," *Federal Register* 84, no. 31 (February 2019), www.federalregister.gov/documents/2019/02/14/2019-02544/maintaining-american-leadership-in-artificial-intelligence/.

6. https://ai100.stanford.edu/.

7. https://en.wikipedia.org/wiki/Geoffrey_Hinton/.

8. Mike Isaac, "Self-Driving Truck's First Mission: A 120-mile Beer Run," *New York Times*, October 25, 2016, www.nytimes.com/2016/10/26/technology/self-driving-trucks-first-mission-a-beer-run.html/.

9. Denise Grady, "A.I. Took a Test to Detect Lung Cancer. It Got an A." *New York Times*, May 20, 2019, www.nytimes.com/2019/05/20/health/cancer-artificial-intelligence-ct-scans.html/.

10. Steve Lohr, "Universities and Tech Giants Back National Cloud Computing Project," *New York Times*, June 30, 2020, www.nytimes.com/2020/06/30/technology/national-cloud-computing-project.html/.

11. https://groups.csail.mit.edu/vision/TinyImages/.

12. Kashmir Hill, "A Face Search Engine Anyone Can Use Is Alarmingly Accurate," *New York Times*, May 26, 2022, www.nytimes.com/2022/05/26/technology/pimeyes-facial-recognition-search.html/.

13. C. Chan et al., "Everybody Dance Now," https://arxiv.org/pdf/1808.07371.pdf/.

14. A. Mathis et al., "DeepLabCut: Markerless Pose Estimation of User-Defined Body Parts with Deep Learning," www.nature.com/articles/s41593-018-0209-y/.

15. James Vinvent, "Watch Jordon Peele Use AI to Make Barack Obama Deliver a PSA about Fake News," The Verge, April 17, 2018, www.theverge.com/tldr/2018/4/17/17247334/ai-fake-news-video-barack-obama-jordan-peele-buzzfeed/.

16. Nina Schick, *Deep Fakes: The Coming Apocalypse* (New York: Twelve Books, 2020).

17. A. Mordvintsev et al., "Inceptionism: Going Deeper into Neural Networks," https://ai.googleblog.com/2015/06/inceptionism-going-deeper-into-neural.html/.

18. W. Knight, "The Dark Secret at the Heart of AI," *MIT Technology Review*, www.technologyreview.com/2017/04/11/5113/the-dark-secret-at-the-heart-of-ai/.

19. I. J. Goodfellow et al., "*Generative Adversarial Networks*," https://arxiv.org/pdf/1406.2661/.

20. Based on a Monte Carlo tree search.

21. Carlos E. Perez, "Why AlphaGo Zero Is a Quantum Leap Forward in Deep Learning," Medium, October 22, 2017, https://medium.com/intuitionmachine/the-strange-loop-in-alphago-zeros-self-play-6e3274fcdd9f/.

22. Deanna Marcum and Robert C. Schonfeld, *Along Came Google: A History of Library Digitalization* (Princeton, NJ: Princeton University Press, 2021).

23. Gidgeon Lewis-Kraus, "The Great A.I. Awakening," *New York Times*, December 14, 2016, www.nytimes.com/2016/12/14/magazine/the-great-ai-awakening.html/.

24. A. Vaswani et al. "Attention Is All You Need," https://arxiv.org/pdf/1706.03762/.

25. Will Douglas Heaven, "OpenAI's New Language Generator GPT-3 Is Shockingly Good—And Completely Mindless," *MIT Technology Review*, July 20, 2020, www.technologyreview.com/2020/07/20/1005454/openai-machine-learning-language-generator-gpt-3-nlp/; Farhad Manjoo, "How Do You Know a Human

Wrote This?" *New York Times*, July 29, 2020, www.nytimes.com/2020/07/29/opin ion/gpt-3-ai-automation.html/.

26. Cade Metz, "Meet DALL-E, the A.I. That Draws Anything at Your Command," *New York Times*, April 6, 2022, www.nytimes.com/2022/04/06/technology /openai-images-dall-e.html?searchResultPosition=1/.

27. Kevin Roose, "An A.I.-Generated Picture Won an Art Prize: Artists Aren't Happy." *New York Times*, September 2, 2022, www.nytimes.com/2022/09/02/tech nology/ai-artificial-intelligence-artists.html/.

28. For a powerful reaction from some visual artists see Julia Rothman and Shaina Feinberg, "Human Artists Take on Their New Robot Competition," *New York Times*, December 23, 2022, www.nytimes.com/2022/12/23/business/ai-gen erated-illustration.html?searchResultPosition=1/; and Sarah Andersen, "The Alt-Right Manipulated My Comic: Then A.I. Claimed It," *New York Times*, December 31, 2022, www.nytimes.com/2022/12/31/opinion/sarah-andersen-how-algorithim -took-my-work.html?searchResultPosition=1/.

29. Sigal Samuel, "A New AI Draws Delightful and Not-So-Delightful Images," *Vox Future Perfect*, April 14, 2022, www.vox.com/future-perfect/23023538/ai-dalle -2-openai-bias-gpt-3-incentives/.

30. Ezra Klein, "This Is a Weirder Moment Than You Think," *New York Times*, June 19, 2022, www.nytimes.com/2022/06/19/opinion/its-not-the-future-we-cant -see.html?searchResultPosition=4/.

31. Kevin Roose, "Bing's Chatbot Drew Me In and Creeped Me Out," *New York Times*, February 17, 2023, nytimes.com/2023/02/16/technology/bing-chatbot -microsoft-chatgpt.html?searchResultPosition=4/.

32. https://futureoflife.org/open-letter/pause-giant-ai-experiments/.

33. You wondered where R2-D2 got its name?

34. S. Narang and A. Chowdhery, "Pathways Language Model (PaLM): Scaling to 540 Billion Parameters for Breakthrough Performance," Google Blog, April 4, 2022, https://ai.googleblog.com/2022/04/pathways-language-model-palm-scaling -to.html/.

35. Strictly speaking, the system was conserving angular momentum, which applies to angular rather than linear, straight-ahead motion.

36. B. Baker et al. "Emergent Tool Use from Multi-Agent Interaction," OpenAI Blog, September 17, 2019, https://openai.com/blog/emergent-tool-use/.

37. An elaboration of the notion of emergent properties is given in chapter 9.

38. Peter Dockrill, "StarCraft II Has a New Grandmaster, and It's Not Human," Science Alert, October 30, 2019, www.sciencealert.com/starcraft-ii-has-a-new -grandmaster-and-it-s-not-human/.

39. A. Bakhtin et al., *Science* 378, no. 6624 (November 2022): 1067–74, https:// www.science.org/doi/10.1126/science.ade9097.

40. www.nsf.gov/news/special_reports/announcements/082620.jsp/.

41. www.whitehouse.gov/wp-content/uploads/2022/10/Blueprint-for-an-AI-Bill-of-Rights.pdf?utm_medium=email&utm_source=FYI&dm_i=1ZJN,86VTP,E29F44,XKURV,1/.

42. www.nist.gov/itl/ai-risk-management-framework?utm_medium=email&utm_source=FYI&dm_i=1ZJN,86VTP,E29F44,XKURU,1/.

43. www.nscai.gov/wp-content/uploads/2021/03/Full-Report-Digital-1.pdf/.

44. Kelsey Piper, *Vox Future Perfect*, March 12, 2021.

45. Kelsey Piper, *Vox Future Perfect*, February 8, 2023, https://www.vox.com/future-perfect/23591534/chatgpt-artificial-intelligence-google-baidu-microsoft-openai/.

46. https://www.askdelphi.com/.

47. Sherry Turkle, *Reclaiming Conversation: The Power of Talk in a Digital Age* (New York: Penguin Books, 2015).

CHAPTER 4. ROBOTS

1. www.youtube.com/watch?v=wE3fmFTtP9g&t=1s/.

2. www.washingtonpost.com/news/morning-mix/wp/2014/03/17/a-robot-fish-that-could-alter-your-image-of-robots-forever/ and www.bbc.com/news/science-environment-33719518/.

3. www.washingtonpost.com/news/speaking-of-science/wp/2016/03/03/with-artificial-octopus-skin-robots-can-bend-and-stretch-while-changing-color/.

4. www.newscientist.com/article/2338260-liquid-robot-can-split-into-tiny-droplets-and-reform-into-a-blob/.

5. www.bbc.com/news/technology-24758935/; www.nytimes.com/2016/03/14/technology/modeled-after-ants-teams-of-tiny-robots-can-move-2-ton-car.html?WT.mc_id=SmartBriefs-Newsletter&WT.mc_ev=click&ad-keywords=smartbriefsnl&_r=0/; and www.bbc.com/news/science-environment-36313958/.

6. MIT News, Sept/Oct 2022.

7. https://www.youtube.com/watch?v=kVhnQA-Hl14/.

8. https://en.wikipedia.org/wiki/Snakebot/ and www.youtube.com/watch?v=pv_MknD6jks/.

9. www.nature.com/news/robot-jellyfish-takes-to-the-air-1.14528/.

10. www.nature.com/news/origami-robot-folds-itself-in-4-minutes-1.15687/ and www.americanscientist.org/article/robogamis-are-the-real-heirs-of-terminators-and-transformers/.

11. www.disneyhistoryinstitute.com/2015/02/before-lincoln-first-disney-human.html/.

12. https://global.toyota/en/detail/19666346/.

13. www.bostonmagazine.com/news/2014/10/08/megabot-somerville-artisans-asylum/.

14. www.bluebird-electric.net/artificial_intelligence_autonomous_robotics/ka butom_rx03_robot_giant_rhinoceros_beetle_hitoshi_takahashi_japan.htm/.

15. https://en.wikipedia.org/wiki/Titan_the_Robot/.

16. https://en.wikipedia.org/wiki/Land_Walker/.

17. www.jebiga.com/hankook-mirae-method-2-robot-suit-vitaly-bulgarov/.

18. https://qz.com/976204/china-now-has-its-own-giant-robot-and-it-just-chal lenged-a-us-team-to-a-fight/.

19. https://en.wikipedia.org/wiki/Kuratas/.

20. https://siamagazin.com/prosthesis-worlds-first-and-largest-off-road-exoskel eton-racing-mech/.

21. www.youtube.com/watch?v=tZUu-hZCQ78/.

22. www.mantisrobot.com/.

23. www.hyundai.com/worldwide/en/brand/technology-with-a-human-heart/.

24. https://en.wikipedia.org/wiki/Roomba/.

25. https://en.wikipedia.org/wiki/AIBO/.

26. https://robots.ieee.org/robots/aquanaut/.

27. www.youtube.com/watch?v=cNZPRsrwumQ/.

28. https://www.youtube.com/watch?v=CzXMoo4nlUM/.

29. https://gizmodo.com/spot-is-a-smaller-more-kickable-version-of-boston -d-1684749999/.

30. www.youtube.com/watch?v=wXxrmussq4E/.

31. https://pal-robotics.com/robots/reem-c/.

32. www.softbankrobotics.com/emea/en/pepper/.

33. https://spectrum.ieee.org/mayfield-robotics-announces-kuri-a-700-mobile -home-robot/.

34. https://en.wikipedia.org/wiki/ASIMO/.

35. The term has been applied in broader contexts than robots.

36. www.youtube.com/watch?v=_sBBaNYex3E/.

37. https://news.harvard.edu/gazette/story/2014/08/the-1000-robot-swarm/.

38. https://www.youtube.com/watch?v=XxFZ-VStApo/.

39. www.firstquadcopter.com/quadcopter-manufacturers/.

40. www.faa.gov/uas/getting_started/.

41. www.faa.gov/uas/research_development/traffic_management/.

42. *Lidar* was originally coined as a contraction of *laser* and *radar* and has evolved into an acronym that is variously rendered as "light detection and ranging" or "laser imaging, detection, and ranging."

43. www.zdnet.com/article/robby-the-last-mile-delivery-robot-gets-an-update/.

44. https://techcrunch.com/2020/07/21/kiwibot-delivery-robots-head-to-san -jose-with-new-partners-shopify-and-ordermark/.

45. www.savioke.com/.

46. https://nuro.ai/.

47. https://news.mit.edu/2016/first-3d-printed-robots-made-of-both-solids-and -liquids-0406/.

48. https://techcrunch.com/2022/11/22/researchers-are-building-robots-that-can-build-themselves/.

49. https://en.wikipedia.org/wiki/Three_Laws_of_Robotics/.

CHAPTER 5. AUTONOMOUS VEHICLES

1. Lawrence M. Krauss, *The Physics of Star Trek* (New York: HarperPerennial, 1995).

2. Or it was until the war in Ukraine upset the Russian economy.

3. C. Peck et al., "Using Online Verification to Prevent Autonomous Vehicles from Causing Accidents," *Nature Machine Intelligence* 2 (2020): 518–28, www.nature.com/articles/s42256-020-0225-y/.

4. www.mckinsey.com/industries/automotive-and-assembly/our-insights/ten-ways-autonomous-driving-could-redefine-the-automotive-world/.

5. Lawrence Ulrich, "When One Car Has More Horsepower Than Churchill Downs," *New York Times*, October 29, 2020, www.nytimes.com/2020/10/29/business/electric-cars-horsepower-lucid-air-rimac.html?searchResultPosition=1/.

6. Lauren Smiley, "'I'm the Operator': The Aftermath of a Self-Driving Tragedy," *Wired*, March 8, 2022, www.wired.com/story/uber-self-driving-car-fatal-crash/?utm_source=Sailthru&utm_medium=email&utm_campaign=Future%20Perfect%20NEW%20Friday%202/17/23&utm_term=Future%20Perfect/.

7. www.eviation.com/.

8. Gautham Nagesh, "Flight Instead of a Ride? Electric Craft May Alter Urban Area Commuting," *New York Times*, December 4, 2021, www.nytimes.com/2021/11/22/business/air-taxi-aviation-electric.html?searchResultPosition=1/.

9. Philippa Foot, *Virtues and Vices and Other Essays in Moral Philosophy* (Berkeley: University of California Press, 1978).

10. E. Awad et al., "The Moral Machine Experiment," *Nature* 563 (2018): 59–64, www.nature.com/articles/s41586-018-0637-6/.

11. https://link.springer.com/article/10.1007/s10506-017-9211-z/.

12. www.transportation.gov/sites/dot.gov/files/docs/policy-initiatives/automated-vehicles/320711/preparing-future-transportation-automated-vehicle-30.pdf?source=post_page/.

CHAPTER 6. SMART WEAPONS

1. Paul Scharre, *Army of None: Autonomous Weapons and the Future of War* (New York: W.W. Norton & Company, 2019).

2. David Sanger, *The Perfect Weapon: War, Sabotage, and Fear in the Cyber Age* (New York: Crown, 2019).

3. www.youtube.com/watch?v=NzdhIA2S35w/.

4. Tyler Cowen, *Average Is Over: Powering America beyond the Age of the Great Stagnation* (New York: Dutton, 2013).

5. www.esd.whs.mil/Portals/54/Documents/DD/issuances/dodd/300009p.pdf/.

6. www.stopkillerrobots.org/.

7. www.hrw.org/news/2016/04/11/killer-robots-and-concept-meaningful-human-control/.

8. Hope Hodge Seck, "Congress Wants a 'Manhatten Project' for Military Artificial Intelligence," Military.com, September 29, 2020, https://www.military.com/daily-news/2020/09/29/congress-wants-manhattan-project-military-artificial-intelligence.html/.

9. www.defense.gov/News/Releases/Release/Article/2091996/dod-adopts-ethical-principles-for-artificial-intelligence/.

CHAPTER 7. MAPPING THE BRAIN

1. Quoted from George Edgin Pugh, *The Biological Origin of Human Values* (New York: Basic Books, 1977).

2. https://braininitiative.nih.gov/.

3. https://beta.nsf.gov/news/nsf-invests-bio-inspired-and-bioengineered-systems-artificial-intelligence-infrastructure-and?utm_medium=email&utm_source=govdelivery/.

4. Genetically ancient creatures called *comb jellies* have a continuous web of neurons with no synapses.

5. M. J. McConnell, J. V. Moran, and A. Abyzov et al., "Intersection of Diverse Neuronal Genomes and Neuropsychiatric Disease: The Brain Somatic Mosaicism Network," *Science* (2017): 356, https://www.science.org/doi/10.1126/science.aal1641/; R. Khamsi, "Change of Mind," *MIT Technology Review* 124, no. 5 (September/October, 2021).

6. George J. Augustine, Jennifer M. Groh, and Scott A. Huettel et al., eds, *Neuroscience*, 7th edition (Sunderland, MA: Sinauer Associates, 2023); https://en.wikipedia.org/wiki/Cortical_column/.

7. https://en.wikipedia.org/wiki/Cortical_minicolumn/.

8. E. Moser, Y. Roudi, and M. Witter et al., "Grid Cells and Cortical Representation," *National Review of Neuroscience* 15 (2014): 466–81, https://doi.org/10.1038/nrn3766/; E. I. Moser, M.-B. Moser, and Y. Roudi, "Network Mechanisms of Grid Cell," *Philosophical Transactions of the Royal Society* (2014), http://dx.doi.org/10.1098/rstb.2012.0511/.

9. A video of the development of the hexagonal grid pattern from a single neuron as a mouse moves around an enclosure is given at www.youtube.com/watch?v=i9GiLBXWAHI/.

10. In principle, a one-dimensional pattern can be added to characterize height and hence span a 3D space and even more arrays added to map an abstract conceptual space of higher dimension.

11. By using machine learning tools to read and interpret the neural signals from hundreds of head direction cells at once, researchers have learned to read the head direction of mice with high precision (Z. Ajabi, A. T. Keinath, and X. X. Wei et al., "Population Dynamics of Head-Direction Neurons during Drift and Reorientation," *Nature* 615 (2023): 892–99, https://doi.org/10.1038/s41586-023-05813-2/.

12. A. Banino, C. Barry, B. Uria, C. Blundell, T. Lillicrap, and P. Mirowski et al., "Vector-Based Navigation Using Grid-Like Representations in Artificial Agents, *Nature* 557 (2018): 429–33, https://www.nature.com/articles/s41586-018-0102-6/.

13. Jeff Hawkins, *A Thousand Brains: A New Theory of Intelligence* (New York: Basic Books, 2021).

14. Infrared radiation is slightly too long in wavelength to be detected by the human eye. It is instead registered as warmth on the skin.

15. Basic ultrasound techniques use high-frequency sound waves to provide imaging of fetal growth, heart activity, and brain structure.

16. Francis R. Willett et al., "High-Performance Brain-to-Text Communication via Handwriting," *Nature* (May 12, 2021), https://www.nature.com/articles/s41586-021-03506-2/.

17. https://neuralink.com/.

18. www.paradromics.com/.

19. Ferris Jabr, "The Man Who Controls Computers with His Mind," *New York Times*, May 13, 2022, www.nytimes.com/2022/05/12/magazine/brain-computer-interface.html?searchResultPosition=1/.

20. J. Tang et al., *Nature Neuroscience* 26 (2023): 858–56, www.nature.com/articles/s41593-023-01304-9; Oliver Whang, "A.I. Is Getting Better at Mind-Reading," *New York Times*, May 1, 2023, www.nytimes.com/2023/05/01/science/ai-speech-language.html?searchResultPosition=1/.

21. Jonathan Moens, "An AI Can Decode Speech from Brain Activity with Surprising Accuracy," *Science News*, September 8, 2022, www.sciencenews.org/article/ai-artificial-intelligence-speech-brain-activity-accuracy/.

22. A. Nemani et al., "Assessing Bimanual Motor Skills with Optical Neuroimaging," *Science Advances* 4, no. 10 (October 3, 2018), doi:10.1126/sciadv.aat3807, https://advances.sciencemag.org/content/4/10/eaat3807/.

23. A. H. Horowitz et al., "Targeted Dream Incubation at Sleep Onset Increases Post-Sleep Creative Performance," *Nature* (2023), www.nature.com/articles/s41598-023-31361-w/; Sofia Moutinho, "'Dream Glove' Boosts Creativity During Sleep," *Science*, May 15, 2023, www.science.org/content/article/dream-glove-boosts-creativity-during-sleep/.

24. Peter H. Diamandis and Steven Kotler, *The Future Is Faster Than You Think!* (New York: Simon & Schuster, 2021).

25. Peter H. Diamandis, "Elon's Neuralink & Brain-Machine Symbiosis," Tech Blog, July 21, 2019.

26. R. Yuste et al., "Four Ethical Priorities for Neurotechnologies and AI," *Nature* 551 (2017): 159–63, www.nature.com/news/four-ethical-priorities-for-neurotechnologies-and-ai-1.22960/.

27. https://betalist.com/startups/humai/.

28. *Hippos* meaning "horse" and *kampos* meaning "sea monster."

29. www.hippocampome.org/.

30. Laura Ungar, "Scientists Grow Human Brain Cells in Rats to Study Diseases," AP, October 12, 2022, https://apnews.com/article/science-health-14edb6a6 d19893c3dd15a879f3cd2a56/.

31. B. J. Kagan et al., *Neuron* (October 12, 2022), https://doi.org/10.1016/j.neu ron.2022.09.001/.

CHAPTER 8. CONSCIOUS COMPUTERS

1. O. Güntürkün, "The Surprising Power of the Avian Mind," *Scientific American* 322, no. 1 (January 2020): 49–55; L. Chittka and C. Wilson, "Expanding Consciousness," *American Scientist* 107, no. 6 (2019): 364–69.

2. F. Betti et al., "Exogenous miRNAs Induce Post-Transcriptional Gene Silencing in Plants," *Nature Plants*, www.nature.com/articles/s41477-021-01005-w/.

3. David J. Chalmers, *Conscious Mind: In Search of a Fundamental Theory* (New York: Oxford University Press, 1997). See also David J. Chalmers, *Reality+: Virtual Reality Worlds and the Problems of Philosophy* (New York: W.W. Norton & Company, 2022).

4. Daniel Dennett, *Consciousness Explained* (New York: Little Brown & Co., 1991).

5. For the perspective of a thoughtful engineer, see http://sentientartificialin telligence.com/.

6. D. Deutsch, "Creative Blocks," Aeon, October 3, 2012, https://aeon.co/es says/how-close-are-we-to-creating-artificial-intelligence/.

7. Roger Penrose, *The Emperor's New Mind: Concerning Computers, Minds, and the Laws of Physics* (Oxford: Oxford University Press, 1989).

8. About 10^{-17} seconds (Max Tegmark, *Our Mathematical Universe: My Quest for the Ultimate Nature of Reality* [New York: Knopf, 2014]).

9. The pattern of the activity of cortical columns in the language of chapter 7.

10. This is a paraphrase. The first written use of the phrase, or something like it, "The brain is merely a meat machine," appeared in J. Weizenbaum, "On the Impact of the Computer on Society," *Science* 176, no. 4035 (May 12, 1972): 609–14. Weizenbaum later attributed it to Marvin Minsky in J. Weizenbaum, "Social and Political Impact of the Long-Term History of Computing," *IEEE Annals of the History of Computing* 30, no. 3.

11. Byron Reese, *The Fourth Age: Smart Robots, Conscious Computers, and the Future of Humanity* (New York: Atria Books, 2018).

12. Alan Turing, "Computing Machinery and Intelligence," *Mind* 59, no. 236:433–60, doi:10.1093/mind/LIX.236.433, ISSN 0026-4423/.

13. John Searle, "Minds, Brains, and Programs," *Behavioral and Brain Sciences* 3 (1980): 417–57, doi:10.1017/S0140525X00005756/.

14. And maybe the cleaner wrasse, a sociable fish (M. Khoda et al., "If a Fish Can Pass the Mark Test, What Are the Implications for Consciousness and Self-Awareness Testing in Animals?" *PLOS Biology*, https://journals.plos.org/plosbiol ogy/article?id=10.1371/journal.pbio.3000021/).

15. Celena Chong, "This Robot Passed a 'Self-Awareness' Test That Only Humans Could Handle Until Now," *Business Insider*, July 23, 2015, www.businessin sider.com/this-robot-passed-a-self-awareness-test-that-only-humans-could-handle -until-now-2015-7/.

16. Daniel Kahneman, *Thinking Fast and Slow* (New York: Farrar, Straus, & Giroux, 2011).

17. Richard H. Thaler and Cass R. Sunstein, *Nudge: Improving Decisions about Health, Wealth, and Happiness* (New York: Penguin Books, 2008).

18. Douglas Hofstadter, *Gödel, Escher, Bach: An Eternal Golden Braid* (New York: Basic Books, 1979).

19. Melanie Mitchell, *Artificial Intelligence: A Guide for Thinking Humans* (New York: Farrar, Straus, & Giroux, 2019).

20. Bill Joy, "Why the Future Doesn't Need Us," *Wired*, www.wired.com/2000 /04/joy-2/.

21. Nick Bostrom, *Superintelligence: Paths, Dangers, and Strategies* (Oxford: Oxford University Press, 2014).

22. Stuart Russell, *Human Compatible: Artificial Intelligence and the Problem of Control* (New York: Viking, 2019).

CHAPTER 9. EVOLUTION

1. Two entertaining and insightful visual overviews of how we came to be are the segment called "The Cosmic Year" in Carl Sagan's classic TV series *Cosmos*: https://www.youtube.com/watch?v=Ln8UwPd1z20/; and the sparkling TED talk by David Christian, "The History of the World in 18 Minutes," www.ted.com/talks /david_christian_the_history_of_our_world_in_18_minutes/.

2. Steven Weinberg, *The First Three Minutes: A Modern View of the Origin of the Universe*, second edition (New York: Basic Books, 1993).

3. https://viethungpham.com/2015/09/16/darwinism-criticized-by-pictures -hoc-thuyet-darwin-bi-phe-phan-qua-hinh-anh/darwinism-4/.

4. A former student of mine in a graduate class.

5. Ann Druyan, *Cosmos: Possible Worlds* (New York: National Geographic, 2020).

6. James Gleick, *Chaos: Making a New Science* (New York: Viking Books, 1987).

7. Charles Darwin, *The Origin of the Species by Means of Natural Selection* (London: John Murray, 1859); Charles Darwin, *Descent of Man, and Selection in Relation to Sex* (London: John Murray, 1871).

8. Richard Dawkins, *The Selfish Gene* (Oxford: Oxford University Press, 1976).

9. Yuval Noah Harari, *Sapiens: A Brief History of Humankind* (New York: HarperCollins, 2014).

10. M. Maurice, N. Tosi, S. Schwinger, D. Breuer, and T. Kleine, "A Long-Lived Magma Ocean on a Young Moon," *Science Advances*, https://www.science.org/doi/10.1126/sciadv.aba8949; animation at www.space.com/earths-moon-magma-ocean-200-million-years.html/.

11. Walter Gilbert, "The RNA World," *Nature* 319 (1986): 618.

12. Charles S. Cockell, "The Origin and Emergence of Life under Impact Bombardment," *Philosophical Transactions of the Royal Society of London*, October 29, 2006.

13. Anna Gosline, "Asteroid Impact Craters Could Cradle Life," *New Scientist*, September 10, 2004, www.newscientist.com/article/dn6383-asteroid-impact-craters-could-cradle-life/.

14. S. A. Benner et al., "When Did Life Likely Emerge on Earth in an RNA–First Process?" *Chemistry Europe*, September 24, 2019, https://chemistryeurope.onlinelibrary.wiley.com/doi/abs/10.1002/syst.201900035/.

15. Carl R. Woese and George E. Fox, "Phylogenetic Structure of the Prokaryotic Domain: The Primary Kingdoms," *Proceedings of the National Academy of Sciences of the United States of America* 74, no. 11: 5088–90.

16. Laura A. Hug, et al., "A new view of the tree of life," *Nature Microbiology* 1, no. 16048 (2016), www.nature.com/articles/nmicrobiol201648/. See also www.onezoom.org/ for an interactive tree of life.

17. Madelaine Böhme et al., "A New Miocene Ape and Locomotion in the Ancestor of Great Apes and Humans," *Nature* (2019), https://www.nature.com/articles/s41586-019-1731-0/.

18. Madelaine Böhme, Rüdiger Braun, and Florian Breier, *Ancient Bones: Unearthing the Astonishing New Story of How We Became Human* (Vancouver: Greystone Books, 2020).

19. https://en.wikipedia.org/wiki/Lucy_(Australopithecus)/.

20. R. L. Cann, et al., "Mitochondrial DNA and Human Evolution," *Nature* 325 (1987): 31–36, https://www.nature.com/articles/325031a0/; see also the update by E. K. F. Chan et al., "Human Origins in a Southern African Palaeo-Wetland and First Migrations," *Nature* 575 (2019): 185–89, https://www.nature.com/articles/s41586-019-1714-1/.

21. M. Lipsom et al., "Ancient DNA and Deep Population Structure in Sub-Saharan African Foragers," *Nature* 603 (2022), www.nature.com/articles/s41586-022-04430-9/.

22. L. R. Berger et al., "*Homo naledi*, a New Species of the Genus *Homo* from the Dinaledi Chamber, South Africa," eLife, https://doi.org/10.7554/eLife.09560; https://en.wikipedia.org/wiki/Homo_naledi/.

23. A. Pinson et al., "Human TKTL1 Implies Greater Neurogenesis in Frontal Neocortex of Modern Humans Than Neanderthals," *Science*, www.science.org/doi/10.1126/science.abl6422/; Carl Zimmer, *New York Times*, www.nytimes.com/2022/09/08/science/human-brain-neanderthal-gene.html/.

24. F. Sánchez-Quinto and C. Lalueza-Fox, "Almost 20 Years of Neanderthal Palaeogenetics: Adaptation, Admixture, Diversity, Demography and Extinction," *Philosophical Transactions of the Royal Society B*, January 19, 2015, https://royal societypublishing.org/doi/10.1098/rstb.2013.0374/.

25. H. Zeberg and S. Pääbo, "The Major Genetic Risk Factor for Severe COVID-19 Is Inherited from Neanderthals," *Nature* 587 (2020): 610–12, www .biorxiv.org/content/10.1101/2020.07.03.186296v1/.

26. M. Dannemann et al., "Human Stem Cell Resources Are an Inroad to Neandertal DNA Functions," *Stem Cell Reports* 15, no. 1 (2020): 214–25, https://www .cell.com/stem-cell-reports/fulltext/S2213-6711(20)30190-9/.

27. M. Larena et al., "Philippine Ayta Possess the Highest Level of Denisovan Ancestry in the World," *Current Biology* 31, no. 19 (2021): 4219–30, https://www .cell.com/current-biology/fulltext/S0960-9822(21)00977-5/.

28. https://en.wikipedia.org/wiki/Cheddar_Man/.

29. All the studies of genetic history based on exhuming remains outlined in this chapter raise sensitive issues of ethics and morality. Who participates in this research and who benefits? How are the goals of science balanced against the rights of living descendants of ancient people? See Amanda Heidt, "Ancient DNA Boom Underlines a Need for Ethical Frameworks," The Scientist, www.the-scientist.com /ancient-dna-boom-underlines-a-need-for-ethical-frameworks-69645/.

30. V. Villalba-Mouco, M. S. van de Loosdrecht, and A. B. Rohrlach et al., "A 23,000-Year-Old Southern Iberian Individual Links Human Groups That Lived in Western Europe Before and After the Last Glacial Maximum," *Nature Ecology and Evolution* 7 (2023): 597–609, https://doi.org/10.1038/s41559-023-01987-0/; C. Posth, H. Yu, and A. Ghalichi et al., "Palaeogenomics of Upper Palaeolithic to Neolithic European Hunter-Gatherers," *Nature* 615 (2023): 117–126, https://doi .org/10.1038/s41586-023-05726-0/; Carl Zimmer, "Ancient DNA Reveals History of Hunter-Gatherers in Europe," *New York Times*, March 6, 2023, www.nytimes .com/2023/03/01/science/dna-hunter-gatherers-europe.html/.

31. Frans de Waal, "Who Apes Whom?" *New York Times*, September 15, 2015, https://www.nytimes.com/2015/09/15/opinion/who-apes-whom.html/.

32. Carl Sagan, *Pale Blue Dot: A Vision of the Human Future in Space* (New York: Ballantine Books, 1994). The title was adopted from the name given to a photograph of the Earth that Sagan urged NASA to take from Voyager 1 in 1990 as it passed the orbit of Neptune headed out of the solar system.

CHAPTER 10. GENETICS

1. B. Alberts et al., *Molecular Biology of The Cell*, 7th edition (New York: W.W. Norton & Company, 2022).

2. Freidrich Meischer was searching for the chemical basis of life in the kitchen of a medieval castle in Tübingen in 1869 when he discovered DNA in bandages

soaked in pus. He concluded that DNA with just four components was too simple to encode life.

3. The hydrogen arose in the Big Bang; the carbon, nitrogen, oxygen, and other heavier elements are from dying stars. We are Carl Sagan's "star stuff."

4. Liz Kreusi, "All the Bases in DNA and RNA Have Now Been Found in Meteorites," *Science News*, April 26, 2022, www.sciencenews.org/article/all-of-the-bases-in-dna-and-rna-have-now-been-found-in-meteorites/.

5. Scripps Research Institute, "Scientists Discover New 'Origins of Life' Chemical Reactions." *Science Daily*, July 28, 2022, www.sciencedaily.com/releases/2022/07/220728112005.htm/.

6. J. Jumper et al., "Highly Accurate Protein Structure Prediction with Alpha-Fold," *Nature* (2021), www.nature.com/articles/s41586-021-03819-2/.

7. M. Baek et al., "Accurate Prediction of Protein Structures and Interactions Using a Three-Track Neural Network," *Science* (2021), www.ipd.uw.edu/wp-content/uploads/2021/07/Baek_etal_Science2021_RoseTTAFold.pdf/.

8. NOVA, *Human Nature*, season 47, episode 11, www.youtube.com/watch?v=GoOWt4cZsfg/.

9. Rob Stein, "Breaking Taboo, Swedish Scientist Seeks to Edit DNA of Healthy Human Embryos," *NPR Morning Edition*, September 22, 2016, www.npr.org/sections/health-shots/2016/09/22/494591738/breaking-taboo-swedish-scientist-seeks-to-edit-dna-of-healthy-human-embryos/.

10. Jennifer A. Doudna and Samuel H. Sternberg, *A Crack in Creation: Gene Editing and the Unthinkable Power to Control Evolution* (New York: Houghton Mifflin Harcourt, 2021); see also Walter Isaacson, *The Code Breaker: Jennifer Doudna, Gene Editing, and the Future of the Human Race* (New York: Simon & Schuster, 2021).

11. http://scienceandentertainmentexchange.org/blog/3218/.

12. A. Brojakowska et al., "Retrospective Analysis of Somatic Mutations and Clonal Hematopoiesis in Astronauts," *Communications Biology* 5, no. 828 (2022), www.nature.com/articles/s42003-022-03777-z/.

13. W.-W. Liao et al., "A Draft Human Pangenome Reference," *Nature* 617 (2023): 312–24, https://www.nature.com/articles/s41586-023-05896-x/.

14. Recent research has shown that identical twins begin to accumulate genetic differences in the womb.

15. J. Yang et al., "Common SNPs Explain a Large Proportion of the Heritability for Human Height," *Nature Genetics* 42 (2010): 565–69, https://www.nature.com/articles/ng.608; P. Wainschtein et al., "Recovery of Trait Heritability from Whole Genome Sequence Data," www.biorxiv.org/content/10.1101/588020v1/.

16. Y. Erlich et al., "Identity Inference of Genomic Data Using Long-range Familial Searches," *Science* 362, no. 6415 (2018): 690–94.

17. Dorothy Roberts, Fatal Invention (New York: New Press, 2011). See also Angela Salini, *Superior: The Return of Race Science* (Boston: Beacon Press, 2019); and Adam Rutherford, *How to Argue with a Racist: What Our Genes Do (and Don't) Say About Human Differences* (New York: The Experiment, 2020).

18. L. Ouyang et al., "Three-Dimensional Bioprinting of Embryonic Stem Cells Directs Highly Uniform Embryoid Body Formation," *Biofabrication* 7 (2015).

19. H. Niederholtmeyer, C. Chaggan, and N. K. Devaraj, "Communication and Quorum Sensing in Non-Living Mimics of Eukaryotic Cells," *Nature Communications* (2018), https://www.nature.com/articles/s41467-018-07473-7.

20. S. Kriegman, D. Blackiston, M. Levin, and J. Bongard, "Kinematic Self-Replication in Reconfigurable Organisms," *Proceedings of the National Academy of Sciences* 118, no. 49 (2021), www.pnas.org/content/118/49/e2112672118/.

21. Tom McKay, "Tiny 'Living' Robots Figured Out How to Reproduce," Gizmodo, November 30, 2021, https://gizmodo.com/tiny-living-robots-figured-out -how-to-reproduce-1848139260/.

22. Laura Sanders, "Living Robots Can Self-Replicate," *Science News*, January 2022.

23. K. Murakami, N. Hamazaki, and N. Hamada et al., "Generation of Functional Oocytes from Male Mice in Vitro," *Nature* 615 (2023): 900–906, https:// doi.org/10.1038/s41586-023-05834-x; https://apnews.com/article/mice-sperm-eggs -embryos-male-cded47c4333a7176ef5c47d6b5ad7e63/.

CHAPTER 11. ANTHROPOCENE

1. In 2024, geologists voted that there is not yet sufficient evidence that humans have changed geology and rejected the formal declaration of an Anthropocene epoch, but I have eyes and I'm sticking with the Anthropocene. See Raymond Zhong, "Geologists Make It Official: We're Not in an 'Anthropocene' Epoch," *New York Times*, March 20, 2024, https://www.nytimes.com/2024/03/20/climate /anthropocene-vote-upheld.html/.

2. https://jembendell.com/.

3. www.ipcc.ch/.

4. https://nca2018.globalchange.gov/.

5. Brady Dennis and Chris Mooney, "Global Carbon Emissions Reach a New Record High in 2018," *Los Angeles Times*, December 5, 2018, www.latimes.com /world/la-fg-carbon-emissions-climate-change-20181205-story.html/.

6. James Gleick, *Chaos: Making a New Science* (New York: Viking Books,1987).

7. The phrase *butterfly effect* had many fathers in the community of meteorology but was popularized by Edward N. Lorenz, the pioneer of the application of chaos theory to studies of the weather. See E. N. Lorenz, *The Essence of Chaos* (Seattle: University of Washington Press, 1993).

8. A. T. Nottingham, P. Meir, and E. Velasquez et al., "Soil Carbon Loss by Experimental Warming in a Tropical Forest," *Nature* 584 (2020): 234–37, www .nature.com/articles/s41586-020-2566-4/.

9. https://svs.gsfc.nasa.gov/3915/; https://svs.gsfc.nasa.gov/4251/; https://svs .gsfc.nasa.gov/4860/.

10. K. Schulz et al., "An Improved and Observationally-Constrained Melt Rate Parameterization for Vertical Ice Fronts of Marine Terminating Glaciers," *Geophysical Research Letters* (2022), https://www.researchgate.net/publication/363701864_An_improved_and_observationally-constrained_melt_rate_parameterization_for_vertical_ice_fronts_of_marine_terminating_glaciers.

11. www.cftc.gov/sites/default/files/2020-09/9-9-20%20Report%20of%20the%20Subcommittee%20on%20Climate-Related%20Market%20Risk%20-%20Managing%20Climate%20Risk%20in%20the%20U.S.%20Financial%20System%20for%20posting.pdf/.

12. www.esd.whs.mil/Portals/54/Documents/DD/issuances/dodd/471521p.pdf/.

13. https://media.defense.gov/2019/Jan/29/2002084200/-1/-1/1/CLIMATE-CHANGE-REPORT-2019.PDF/.

14. Lauren Hirsch, "Large Insurers Are Hatching a Plan to Take Down Coal," *New York Times*, November 23, 2021, www.nytimes.com/2021/11/23/business/dealbook/insurance-companies-coal.html?searchResultPosition=85/.

15. A. R. Siders, M. Hino, and K. J. Mach, "The Case for Strategic and Managed Climate Retreat," *Science* 365, no. 6455:761–63, https://science.sciencemag.org/content/365/6455/761.full/.

16. Elizabeth Kolbert, "From Obama's Top Scientist, Words of Caution on Climate," E360, https://e360.yale.edu/features/obama_top_scientist_words_of_caution_climate_john_holdren/.

17. D. McKay et al., "Exceeding 1.5°C Global Warming Could Trigger Multiple Climate Tipping Points," *Science* 377, no. 6611, www.science.org/doi/10.1126/science.abn7950/.

18. https://cleanpower.org/.

19. This control proved grossly inadequate, however, during an extended freeze in February 2021 induced in part by climate change and shifting jet stream patterns. People died when they lost power for an extended time.

20. Y. Tao, C. D. Rahn, L. A. Archer, and F. You, "Second Life and Recycling: Energy and Environmental Sustainability Perspectives for High-Performance Lithium-Ion Batteries," *Science Advances* 7 no. 45 (2021).

21. www.energy.gov/articles/doe-launches-new-energy-earthshot-slash-cost-geothermal-power?utm_medium=email&utm_source=FYI&dm_i=1ZJN,80MJ0,E29F44,WSJI1,1/.

22. Karen Kwon, "Hydrogen: Coming to an Aircraft Near You," Aerospace America, https://aerospaceamerica.aiaa.org/features/hydrogen-coming-to-an-aircraft-near-you/.

23. Jennifer Chu, "Sugar Rush," *MIT Technology Review*, www.technologyreview.com/2022/08/24/1057254/sugar-rush/.

24. https://usfusionenergy.org/.

25. H. Abu-Shawareb et al., "Lawson Criterion for Ignition Exceeded in an Inertial Fusion Experiment," *Physical Review Letters* 129 (2022).

26. Andrea Peterson, "White House Sets Sights on Commercial Fusion Energy," AIP, www.aip.org/fyi/2022/white-house-sets-sights-commercial-fusion-energy/.

27. www.cfs.energy/.

28. www.whitehouse.gov/ostp/news-updates/2022/04/19/readout-of-the-white -house-summit-on-developing-a-bold-decadal-vision-for-commercial-fusion-energy/.

29. www.energy.gov/science/articles/department-energy-announces-50-mil lion-milestone-based-fusion-development-program?utm_medium=email&utm _source=FYI&dm_i=1ZJN,81S9G,E29F44,WXQQJ,1/.

30. The Department of Energy released an "Industrial Decarbonization Road-map" in 2022.

31. https://environmenthalfcentury.princeton.edu/.

32. www.nationalacademies.org/news/2021/02/new-report-charts-path-to-net -zero-carbon-emissions-by-2050-recommends-near-term-policies-to-ensure-fair -and-equitable-economic-transition-and-revitalization-of-manufacturing-industry/.

33. www.census.gov/popclock/world. For a thorough summary of population is-sues see Jennifer D. Scuibba, *8 Billion and Counting: How Sex, Death, and Migra-tion Shape Our World* (New York: W.W. Norton & Company, 2022).

34. Buckminster Fuller, *Critical Path* (New York: St. Martin's Press, 1981).

35. Paul R. Erlich and Anne H. Erlich, *The Population Bomb* (New York: Bal-lantine Books, 1968).

36. www.mauldineconomics.com/frontlinethoughts/light-in-the-covid-tunnel/.

37. https://databank.worldbank.org/home/.

38. www.footprintnetwork.org/.

39. An old-fashioned but powerful lecture on population issues titled "Arithme-tic, Population, and Energy" was given by deceased University of Colorado physi-cist Albert Bartlett. See https://www.albartlett.org/presentations/arithmetic_popu lation_energy_video_full_length.html/.

40. C. J. L. Murray et al., "Population and Fertility by Age and Sex for 195 Countries and Territories, 1950–2017: A Systematic Analysis for the Global Bur-den of Disease Study 2017," *Lancet* 392 (November 10, 2017), www.thelancet .com/journals/lancet/article/PIIS0140-6736(18)32278-5/fulltext/; S. E. Vollset et al., "Fertility, Mortality, Migration, and Population Scenarios for 195 Countries and Territories from 2017 to 2100: A Forecasting Analysis for the Global Burden of Disease Study," *Lancet* 396, no. 10258 (October 17, 2020): 1285–306.

41. Kohei Saito, *Capital in the Anthropocene (Hitoshinsei no Shihonron)*, Shuei-sha, Tokyo, 2020 (in Japanese); *Marx in the Anthropocene: Towards the Idea of De-growth Communism* (Cambridge: Cambridge University Press, 2023), doi:10.1017 /9781108933544.

42. N. Georgescu-Roegen, *The Entropy Law and the Economic Process* (Cam-bridge, MA: Harvard University Press, 1971).

43. https://cusp.ac.uk/.

44. See Mary-Jane Rubenstein, *Astrotopia: The Dangerous Religion of the Corporate Space Race* (Chicago: University of Chicago Press, 2022); and Erika

Nesvold, *Off Earth: Ethical Questions and Quandaries for Living in Outer Space* (Cambridge, MA: MIT Press, 2023).

45. E. A. DeBoy et al., "Familial Clonal Hematopoiesis in a Long Telomere Syndrome," *New England Journal of Medicine* 388 (June 2023): 2422–33, https://www.nejm.org/doi/full/10.1056/NEJMoa2300503/.

46. Y. Lu et al., "Reprogramming to Recover Youthful Epigenetic Information and Restore Vision," *Nature* 588 (2020), www.nature.com/articles/s41586-020-2975-4/.

47. S. Tarazi et al., "Post-Gastriculation Synthetic Embryos Generated *ex utero* from Mouse Naïve ESCs," *Cell* 185 no. 18:3290–96, www.cell.com/cell/fulltext/S0092-8674(22)00981-3/; www.statnews.com/2022/08/01/synthetic-mouse-embryos-created-from-stem-cells-without-sperm-eggs-uterus/.

48. G. Amadei, C. E. Handford, and C. Qiu et al., "Synthetic Embryos Complete Gastrulation to Neurulation and Organogenesis," *Nature* (2022), https://doi.org/10.1038/s41586-022-05246-3/.

49. F. S. Loffredo et. al., "Growth Differentiation Factor 11 Is a Circulating Factor That Reverses Age-Related Cardiac Hypertrophy," *Cell* 4 (2013), www.cell.com/fulltext/S0092-8674(13)00456-X/.

50. D. Andrijevic at al. "Cellular Recovery after Prolonged Warm Ischaemia of the Whole Body," *Nature* 608 (2022): 405–12, www.nature.com/articles/s41586-022-05016-1/; see also Gina Kolata, "A 'Reversible' Form of Death? Scientists Revive Cells in Dead Pigs' Organs," *New York Times*, August 3, 2022, www.nytimes.com/2022/08/03/science/pigs-organs-death.html/.

51. https://en.wikipedia.org/wiki/Frankenstein_(1931_film)/.

52. Peter H. Diamandis, "How Long Will You Live?" The Tech Blog, March 28, 2021.

CHAPTER 12. ECONOMICS

1. Azeem Azhar, *Exponential: How Accelerating Technology Is Leaving Us Behind and What to Do About It* (New York: Random House Business, 2021).

2. Jaron Lanier, *Who Owns the Future?* (New York: Simon & Schuster, 2013).

3. Thomas Piketty, *Capital and Ideology* (Cambridge, MA: Harvard University Press, 2020).

4. www.nbcnews.com/business/corporations/ceos-public-u-s-firms-earn-320-times-much-workers-n1263195/.

5. From a conversation between Stewart Brand and Steve Wozniak at the first Hackers Conference in 1984 recorded by Steve Levy (https://digitopoly.org/2015/10/25/information-wants-to-be-free-the-history-of-that-quote/).

6. Sheera Frenkel and Cecilia Kang, *An Ugly Truth: Inside Facebook's Battle for Domination* (New York: Harper, 2021).

7. Shoshana Zuboff, *The Age of Surveillance Capitalism* (New York: Hachette Book Group, 2019).

8. Lanier, *Who Owns the Future?*

9. Nobel Laureate Herbert A. Simon said, "What information consumes is rather obvious: It consumes the attention of its recipients." See also Michael H. Goldhaber, "Attention Shoppers," *Wired*, December 1, 1997, www.wired.com /1997/12/es-attention/.

10. Jamelle Bouie, "Facebook Has Been a Disaster for the World," *New York Times*, September 18, 2020, www.nytimes.com/2020/09/18/opinion/facebook-de mocracy.html/.

11. The name is a Romanization of the word meaning "voice" in languages such as Hindi आवाज़ and Urdu آواز.

12. https://secure.avaaz.org/campaign/en/facebook_threat_health/.

13. Lanier, *Who Owns the Future?*

14. Zuboff, *The Age of Surveillance Capitalism*.

15. Charlie Warzel and Stuart A. Thompson, "How Your Phone Betrays Democracy," *New York Times*, December 21, 2019, www.nytimes.com/interac tive/2019/12/21/opinion/location-data-democracy-protests.html?action=click&mod ule=RelatedLinks&pgtype=Article/; and Charlie Warzel and Stuart A. Thompson, "They Stormed the Capitol: Their Apps Tracked Them." *New York Times*, February 5, 2021, www.nytimes.com/2021/02/05/opinion/capitol-attack-cellphone-data.html/.

16. Kevin Roose, "Goodbye to the Wild Wild Web," *New York Times*, July 2, 2020, www.nytimes.com/2020/07/02/technology/goodbye-to-the-wild-wild-web .html?searchResultPosition=5/.

17. Adam Satariano, "What the G.D.P.R., Europe's Tough New Data Law, Means for You," *New York Times*, May 6, 2018, www.nytimes.com/2018/05/06 /technology/gdpr-european-privacy-law.html/.

18. Adam Satariano, "E.U. Takes Aim at Social Media's Harms with Land-mark New Law," *New York Times*, April 22, 2022, www.nytimes.com/2022/04/22 /technology/european-union-social-media-law.html?searchResultPosition=1/; and Cecilia Kang, "As Europe Approves New Tech Laws, the US Falls Further Be-hind," *New York Times*, April 22, 2022, www.nytimes.com/2022/04/22/technology /tech-regulation-europe-us.html?searchResultPosition=1/.

19. *New York Times*, "The House Antitrust Report on Big Tech," October 6, 2022, www.nytimes.com/interactive/2020/10/06/technology/house-antitrust-report -big-tech.html/.

20. Tim Wu, *The Curse of Bigness: Antitrust in the New Gilded Age* (New York: Columbia Global Reports, 2018).

21. https://ainowinstitute.org/.

22. Paul Mozur, "Forget TikTok. China's Powerhouse App is WeChat, and Its Power is Sweeping." *New York Times*, September 4, 2020, www.nytimes .com/2020/09/04/technology/wechat-china-united-states.html/.

23. www.fhi.ox.ac.uk/windfallclause/.

24. https://kpmg.com/us/en/podcasts/2022/twist-102422.html/.

25. Steve Lohr, "He Created the Web: Now He's Out to Remake the Digital World." *New York Times*, January 10, 2021, www.nytimes.com/2021/01/10/technol ogy/tim-berners-lee-privacy-internet.html?searchResultPosition=1/.

26. Kaveh Waddell, "Would Your Let Companies Monitor You for Money?" *Atlantic*, April 1, 2016, www.theatlantic.com/technology/archive/2016/04/would -you-let-companies-monitor-you-for-money/476298/.

27. www.odeinfinity.com/.

28. https://gener8ads.com/.

29. https://mitpress.mit.edu/books/technoprecarious/.

30. Kate Crawford, *Atlas of AI: Power, Politics, and the Planetary Costs of Artificial Intelligence* (New Haven, CT: Yale University Press, 2021).

31. Byron Reese, *The Fourth Age: Smart Robots, Conscious Computers, and the Future of Humanity* (New York: Atria Books, 2018).

32. Sigal Samuel, "Everywhere Basic Income Has Been Tried, in One Map," *Vox Future Perfect*, October 20, 2020, www.vox.com/future-perfect/2020/2/19/21112570 /universal-basic-income-ubi-map/.

33. Kurtis Lee, "Guaranteed Income Programs Spread, City by City," *New York Times*, September 10, 2022, www.nytimes.com/2022/09/10/business/economy /guaranteed-income.html?searchResultPosition=1/.

34. www.givedirectly.org/about/.

35. Suresh Naidu, "Eudaimonic Jobs," *Daedelus* (Winter 2023): 119.

36. Anikket Kittur et al., "The Future of Crowd Work," *Proceedings of the 2013 Conference on Computer Supported Cooperative Work* (February 2013): 1301–18.

37. Alex Gladstein, "Why Bitcoin Matters for Freedom," *Time*, December 28, 2018, https://time.com/5486673/bitcoin-venezuela-authoritarian/.

38. https://bitcoin.org/bitcoin.pdf/.

39. "First US Bitcoin ATMs to Open Soon in Seattle, Austin," Reuters, February 18, 2014, https://web.archive.org/web/20151019092802/www.reuters.com /article/2014/02/18/us-bitcoin-robocoin-idUSBREA1H05F20140218/.

40. Fred Imbert, "BlackRock CEO Larry Fink Calls Bitcoin an 'Index of Money Laundering,'" CNBC, October 13, 2017, www.cnbc.com/2017/10/13/blackrock -ceo-larry-fink-calls-bitcoin-an-index-of-money-laundering.html/.

41. Jeanna Smialek, "The Federal Reserve Takes a Closer Look at Digital Currency as Stablecoins Loom Large," *New York Times*, May 24, 2021, www.nytimes .com/live/2021/05/24/business/economy-stock-market-news#the-federal-reserve -takes-a-closer-look-at-digital-currency-as-stablecoins-loom-large/.

42. https://v.cent.co/tweet/20/.

43. Robert Sanders, "First-Ever Auction of NFT Based on Nobel Prize Nets UC Berkeley $50,000," *Berkeley News*, June 8, 2021, https://news.berkeley.edu /2021/06/08/first-ever-auction-of-nft-based-on-nobel-prize-nets-uc-berkeley-50000/.

44. Jason Farago, "Beeple Has Won: Here's What We've Lost." *New York Times*, March 12, 2021, www.nytimes.com/2021/03/12/arts/design/beeple-nonfungible-nft-review.html/.

45. https://workofthefuture.mit.edu/research-post/the-work-of-the-future-building-better-jobs-in-an-age-of-intelligent-machines/.

46. www.workerinfoexchange.org/post/managed-by-bots-a-report-by-worker-info-exchange/.

47. Peter H. Diamandis, "Future of Money (Crypto and DeFi)," The Tech Blog, June 20, 2021.

48. Miguel Arduengo and Luis Sentis, "The Robot Economy: Here It Comes," *International Journal of Social Robotics* 13 (August 1, 2020): 937–47.

49. Yuval Noah Harari, *Homo Deus: A Brief History of Tomorrow* (New York: Harper Perennial, 2017).

CHAPTER 13. DEMOCRACY

1. Carl Becker, *The Dilemma of Modern Democracy*, www.vqronline.org/essay/dilemma-modern-democracy/.

2. Winston Churchill, speech in the House of Commons, November 11, 1947, quoting an unknown originator of the aphorism.

3. Siva Vaidhyanathan, *Anti-Social Media: How Facebook Disconnects Us and Undermines Democracy* (Oxford: Oxford University Press, 2018).

4. Zeynep Tufekci, "How Social Media Took Us from Tahir Square to Donald Trump," *MIT Technology Review*, August 14, 2018, www.technologyreview.com/2018/08/14/240325/how-social-media-took-us-from-tahrir-square-to-donald-trump/.

5. Yuval Noah Harari, *Homo Deus: A Brief History of Tomorrow* (New York: HarperCollins, 2017).

6. Shoshana Zuboff, *The Knowledge Coup* (New York: PublicAffairs, 2020); and Shoshana Zuboff, "The Coup We Are Not Talking About," *New York Times*, January 29, 2021, www.nytimes.com/2021/01/29/opinion/sunday/facebook-surveillance-society-technology.html/.

7. Tufekci, "How Social Media Took Us from Tahir Square."

8. Michael Goldhaber, "Attention Shoppers!" *Wired,* December 1, 1997, www.wired.com/1997/12/es-attention/; and Michael Goldhaber, "I Talked to the Cassandra of the Internet Age," *New York Times*, February 24, 2021, www.nytimes.com/2021/02/04/opinion/michael-goldhaber-internet.html?searchResultPosition=6/.

9. Andrew Marantz, "How a Pro-Trump Islamophobe Who Just Won a Congressional Primary Got Famous on the Internet," *New Yorker*, August 20, 2020, www.newyorker.com/news/daily-comment/how-pro-trump-islamophobe-laura-loomer-got-famous-on-the-internet/.

10. Tiffany Hsu, "On TikTok, Election Misinformation Thrives Ahead of Midterms," *New York Times*, August 14, 2022, www.nytimes.com/2022/08/14/business /media/on-tiktok-election-misinformation.html?searchResultPosition=1/.

11. Britt Paris and Joan Donovan, "Deepfakes and Cheap Fakes," *Data & Society*, September 18, 2019, https://datasociety.net/library/deepfakes-and-cheap -fakes/; Tiffany Hsu and Stuart A. Thompson, "AI's Ease at Spinning Deception Raises Alarm," *New York Times*, February 9, 2023, https://www.nytimes .com/2023/02/08/technology/ai-chatbots-disinformation.html; and Adam Statariano and Paul Mozur, "The People Onscreen Are Fake," *New York Times*, February 9, 2023, www.nytimes.com/2023/02/07/technology/artificial-intelligence-training -deepfake.html/.

12. Robert Chesney and Danielle Keats Citron, "Deep Fakes: A Looming Challenge for Privacy, Democracy, and National Security," *California Law Review* 107 (2019), https://papers.ssrn.com/sol3/papers.cfm?abstract_id=3213954/.

13. Charlie Warzel, "Can Tech Giants Help Our Democracy?" *New York Times*, September 22, 2020, www.nytimes.com/2020/09/22/opinion/sunday/2020-election -security-tech.html?searchResultPosition=1/.

14. Charlie Warzel, "Believable: The Terrifying Future of Fake News," Buzzfeed, February 11, 2018, www.buzzfeednews.com/article/charliewarzel/the -terrifying-future-of-fake-news/.

15. www.deeptracetech.com/; https://medium.com/sensity/newsletter/home/.

16. https://truepic.com/.

17. Chesney and Citron, "Deep Fakes."

18. In 2022, the Texas legislature passed HB20 that would expressly remove these protections. The issue is in the courts (Chuck Lindell, "Appeals Court Lets Texas Enforce, For Now, Social Media Law Sought by Conservatives," *Austin American Statesman*, May 11, 2022, www.statesman.com/story/news/2022/05/11/texas-can -enforce-social-media-law-bans-facebook-twitter-censoring-users/9738480002/).

19. John Mauldin, "Thoughts from the Frontline," https://ggc-mauldin-images .s3.amazonaws.com/uploads/pdf/TFTF_Jul_03_2021.pdf/.

20. Christina Zhao, "'Black Mirror' in China? 1.4 Billion Citizens to Be Monitored through Social Credit System," *Newsweek*, May 1, 2018, www.newsweek .com/china-social-credit-system-906865/.

21. Center for Humane Technology, "Digital Democracy Is Within Reach," Your Undivided Attention Podcast, www.humanetech.com/podcast/23-digital -democracy-is-within-reach/.

22. Lawrence Lessig, *Code and Other Laws of Cyberspace* (New York: Basic Books, 1999).

23. Ephrat Livni, "For Rules in Technology, the Challenge Is to Balance Code and Law," *New York Times*, November 23, 2021, www.nytimes.com/2021/11/23 /business/dealbook/cryptocurrency-code-law-technology.html/.

24. Anand Giridharadas, *Winners Take All: The Winners Charade of Changing the World* (New York: Vintage, 2019).

25. Charlie Warzel, "America's Tech Billionaires Could Help Protect the Election: If They Wanted To," *New York Times*, September 22, 2020, www.nytimes.com/2020/09/22/opinion/sunday/2020-election-security-tech.html/.

26. Yuval Noah Harari, *Sapiens: A Brief History of Humankind* (New York: HarperCollins, 2014).

27. Mark Montague and Daniel Basov, "Patent Law Alert—AI Machines Are Not Human Inventors," Cowan, Liebowitz & Latman, June 2, 2020, www.cll.com/newsroom-news-172873/.

28. www.caidp.org/.

29. Amanda Gorman, www.cnn.com/2021/01/20/politics/amanda-gorman-inaugural-poem-transcript/index.html/.

CHAPTER 14. SPACE

1. Brian Greene, *The Elegant Universe: Superstrings, Hidden Dimensions, and the Quest for the Ultimate Theory* (New York: W.W. Norton & Company, 1999); J. Craig Wheeler, *Cosmic Catastrophes: Exploding Stars, Black Holes, and Mapping the Universe* (Cambridge: Cambridge University Press, 2007).

2. Lisa Randall, *Warped Passages: Unravelling the Mysteries of the Universe's Hidden Dimensions* (New York: HarperCollins, 2005).

3. Lee Smolin, *The Life of the Cosmos* (Oxford: Oxford University Press, 1997).

4. About 186,000 miles per second or 300,000 kilometers per second.

5. Douglas Brinkley, *American Moonshot: John F. Kennedy and the Great Space Race* (New York: HarperCollins, 2019).

6. Jeff Foust, "Griffin's Commercialization Legacy," Space Review, Monday, December 8, 2008, www.thespacereview.com/article/1266/1/.

7. Lori Garver, *Escaping Gravity: My Quest to Transform NASA and Launch a New Space Age* (New York: Diversion Books, 2022).

8. www.innovationnewsnetwork.com/american-commercial-space-sector/8734/.

9. Ashley Vance, *When the Heavens Went on Sale: The Misfits and Geniuses Racing to Put Space Within Reach* (New York: Ecco, 2023).

10. Gerard O'Neill, *The High Frontier: Human Colonies in Space* (North Hollywood, CA: Space Studies Institute Press, 1976); Larry Niven, *Ringworld* (New York: Random House, 1970).

11. Chris Boshuizen, cofounder of Planet Labs; Glen de Vries, cofounder of Mediadata; and Audrey Powers, a vice president of Blue Origin.

12. https://science.slashdot.org/story/12/03/20/1922248/elon-musk-future-round-trip-to-mars-could-cost-under-500000/.

13. David Mackay and Michael Masucci.

14. Beth Moses, the company's chief astronaut instructor; Colin Bennett, lead operations engineer; and Sirisha Bandla, vice president of government affairs and research operations.

15. https://www.nasa.gov/multimedia/nasatv/#public/.

16. https://www.youtube.com/watch?v=mhJRzQsLZGg/.

17. Hayley Arceneaux, a physician's assistant and cancer survivor; Christopher Sembroski, a data engineer; and Sian Proctor, a community college geology professor who had earlier just missed being chosen for the NASA astronaut corps. Proctor was pilot for the mission.

18. Michael López-Alegría commander of the flight and real estate magnate Larry Connorission the pilot. Mission specialists Israeli businessman Eytan Stibbe and shipping CEO Mark Pathy each paid $65 million.

19. Former NASA astronaut and current Axiom director of human space flight Commander Peggy Whitson, private astronaut Pilot John Shoffner, and Saudi mission specialists Ali Alqarni and Rayyanah Barnawi, the first female Saudi astronaut.

20. https://www.rocketlabusa.com/.

21. https://www.statista.com/statistics/264472/number-of-satellites-in-orbit-by-operating-country/.

22. https://www.universetoday.com/17754/explore-earths-satellites-with-google-earth/.

23. https://maps.esri.com/rc/sat2/index.html/.

24. "Celestial palace" in Chinese.

25. "Heavenly questions" in Chinese.

26. A Kazachok is an iconic Russian folk dancer who performs high kicks from a squatting position.

27. About 10^{23}, roughly the number of atoms in a sugar cube.

28. Kip Thorne, *Black Holes and Time Warps: Einstein's Outrageous Legacy* (New York: W.W. Norton & Company, 1994).

29. Not *that* easy; I did not do it!

30. Formally known as *polarization*.

31. The satellite was named after Chinese philosopher Mozi or Mengzi, Latinized as *Micius*, who lived in the fifth century BC and interpreted Confucianism for the emperor.

32. P. T. Dumitrescu et al. "Dynamical Topological Phase Realized in a Trapped-Ion Simulator," *Nature* 607 (2022), www.livescience.com/fibonacci-material-with-two-dimensions-of-time?utm_campaign=0CFA9D5B-9285-47DF-A018-5ED9E96C0C0B/.

33. Lawrence M. Krauss, *The Physics of Star Trek* (New York: HarperPerennial, 1995).

34. The exact wording of Fermi's query was debated in recollections among his colleagues.

35. For a contemporary realization see https://sites.psu.edu/astrowright/2021/06/11/how-a-species-can-fill-the-galaxy/.

36. J. T. Wright, et al., "The Infrared Search for Extraterrestrial Civilizations with Large Energy Supplies: I. Background and Justification," *Astrophysical Journal* (2014).

37. The Hawaiian name approximately translates as "first distant messenger."

38. www.seti.org/.

39. Sarah Scoles, *They Are Already Here: UFO Culture and Why We See Saucers* (New York: Pegasus Books, 2020).

40. www.dni.gov/files/ODNI/documents/assessments/Prelimary-Assessment -UAP-20210625.pdf/.

41. Edward C. Condon, *The Scientific Study of Unidentified Flying Objects* (New York: Bantam Books, 1969).

42. Carl Sagan and Ann Druyan, *The Demon-Haunted World: Science as a Candle in the Dark* (New York: Penguin Random House, 1995).

CHAPTER 15. THE FUTURE

1. Ray Kurzweil, *The Singularity Is Near* (New York: Penguin Group, 2005).

2. M. Alfonseca, M. Cebrian, A. F. Anta, L. Coviello, A. Abeliuk, and I. Rahwan, "Superintelligence Cannot be Contained: Lessons from Computability Theory," *Journal of Artificial Intelligence Research* 70 (2021): 65–76, https://doi.org/10.1613 /jair.1.12202; www.mpib-berlin.mpg.de/computer-science-superintelligent-machines/.

3. I used an AI–based program to check the grammar of this book; it was not always correct.

4. Sally Adee, "Zap Your Brain Into the Zone: Fast Track to Pure Focus," New Scientist, February 1, 2012, www.newscientist.com/article/mg21328501-600-zap -your-brain-into-the-zone-fast-track-to-pure-focus/.

5. *Engineering and Technology*, "Carbon Capture System Sequesters Record Amounts of CO2 for Direct Air Capture," May 30, 2022, https://eandt.theiet .org/content/articles/2022/05/carbon-capture-system-sequesters-record-amounts -of-co2-for-direct-air-capture/; Tanya Weaver, "£200m Carbon Capture Project in North Wales Could Remove 235,000 Tonnes of CO2 Annually," *Engineering and Technology*, April 11, 2024, https://eandt.theiet.org/2024/04/11/ps200m-carbon -capture-project-north-wales-could-remove-235000-tonnes-co2-annually.

6. M. Stillings, Z. K. Shipton, and R. J. Lunn, "Mechanochemical Processing of Silicate Rocks to Trap CO2," *Nature Sustainability* (2023), doi:10.1038/s41893 -023-01083-y/.

7. DNA analysis by Harvard evolutionary biologist David Reich and colleagues suggests farming started in the Anatolia region of the Middle East after separate migrations from the Mediterranean coast and Iraq mixed those peoples with local hunter-gatherers (I. Lazaridis et al., "Ancient DNA from Mesopotamia Suggests Distinct Pre-Pottery and Pottery Neolithic Migrations into Anatolia," *Science* 377, no. 6609 (2022): 982–87, www.science.org/doi/10.1126/science.abq0762/.

8. David Graber and David Wengrow, *The Dawn of Everything: A New History of Humanity* (New York: Farrar, Straus & Giroux, 2021).

9. Yuval Noah Harari, *Homo Deus: A Brief History of Tomorrow* (New York: HarperCollins, 2017).

10. For an interesting personal meditation on contemporary aspects of this issue see Kashmir Hall, "Your Memories, Their Cloud." *New York Times*, December 31, 2022, www.nytimes.com/2022/12/31/technology/cloud-data-storage-google-apple -meta.html?searchResultPosition=1/.

11. www.archmission.org/nanofiche/.

12. Thomas Dietterich, "Robust Artificial Intelligence and Robust Human Organizations," https://arxiv.org/abs/1811.10840/.

13. K. E. Weick, K. M. Sutcliffe, and D. Obstfeld, "Organizing for High Reliability: Processes of Collective Mindfulness," in R. S. Sutton and B. M. Staw, eds., *Research in Organizational Behavior*, vol. 1:81–123 (Stanford, CA: Jai Press, 1999).

14. Stuart Russell, *Human Compatible: Artificial Intelligence and the Problem of Control* (New York: Viking, 2019).

15. Kurzweil, *The Singularity Is Near*.

16. The gray goo scenario is the notion that runaway, self-replicating machines consume all the biomass of the Earth. See K. Eric Drexler, *Engines of Creation* (New York: Doubleday, 1986).

17. Quoted by Will Douglas Heaven, "A Mind of Its Own," *MIT Technology Review* (September/October 2021).

18. Luciano Floridi, "Digital Ethics Online and Off," *American Scientist* (July/August 2021): 218.

19. https://ai100.stanford.edu/.

20. https://bridgingbarriers.utexas.edu/good-systems/.

21. https://news.utexas.edu/2022/07/21/new-partnership-will-scale-up-investment-in-ethical-ai-research-and-innovation/.

22. https://dl.acm.org/journal/jrc/.

23. www.responsible.ai/news/raii-scc-pilot-press-release/.

24. https://partnershiponai.org/.

25. https://ai4good.org/.

26. www.safe.ai/.

27. https://c4tt.org/.

28. https://centres.weforum.org/centre-for-the-fourth-industrial-revolution /home/.

29. Junfeng Jiao, "Highlights: Our Week with the World Economic Forum," University of Texas, https://bridgingbarriers.utexas.edu/news/highlights-our-week -world-economic-forum/.

30. www.nsf.gov/pubs/2020/nsf20604/nsf20604.htm/.

31. www.amazon.science/blog/amazon-supports-nsf-research-in-human-ai-interaction-collaboration/.

32. https://blog.google/technology/ai/partnering-nsf-human-ai-collaboration/.

33. www.fedscoop.com/ai-bill-of-rights-teeth/?utm_source=Stanford +HAI&utm_campaign=2d5b0129e0-Mailchimp_HAI_NewsletterNovember +2021_2&utm_medium=email&utm_term=0_aaf04f4a4b-2d5b0129e0-2140 23414/.

34. Katie M. Palmer, "Why the Gene Editors of Tomorrow Need to Study Ethics Today," *Wired,* September 18, 2018, www.wired.com/story/wired25-jennifer -doudna-jiwoo-lee-crispr-gene-editing-ethics/.

35. William MacAskill, *What We Owe the Future* (New York: Basic Books, 2022); and William MacAskill, "The Case for Longtermerism," *New York Times,* August 5, 2022, www.nytimes.com/2022/08/05/opinion/the-case-for-longtermism .html/. In an odd twist, MacAskill was a mentor to Sam Bankman-Fried, who was an active proponent of longtermism until he was caught up in the cryptocurrency scandal discussed in chapter 12.

36. www.vhemt.org/; Cary Buckley, "Earth Now Has 8 Billion Humans: This Man Wishes There Were None," *New York Times,* November 23, 2022, www.nytimes.com/2022/11/23/climate/voluntary-human-extinction.html?search ResultPosition=1/.

37. Sherry Turkle, *Alone Together: Why We Expect More from Technology and Less from Each Other* (New York: Basic Books, 2017).

38. Yuval Noah Harari, "'Homo sapiens Is an Obsolete Algorithm': Yuval Noah Harari on How Data Could Eat the World," *Wired,* September 1, 2016, www .wired.com/story/yuval-noah-harari-dataism/.

39. G. René and D. Mapes, "The Spatial Web: How Web 3.0 Will Connect Humans, Machines, and AI to Transform the World."

40. *Transmission control protocol/internet protocol* (TCP/IP) is a suite of communication protocols used to connect network devices on the Internet.

41. *Hypertext transfer protocol* (http) is the foundation of data communication for the World Wide Web.

42. https://spatialwebfoundation.org/.

43. Active inference is a framework for explaining, simulating, and understanding the mechanisms that enable perception, decision making, and action. Active inference presumes that self-organizing biological, physical, and computational systems are driven by an imperative to minimize surprise, which can be quantified as the degree to which future predictions based on past history differ from actual received sensory data.

44. The free energy principle is a variational principle of information physics based on Bayesian statistical approaches to brain functions that registers the difference between predictions and sensory data. It was first introduced in the context of computational neuroscience and biology by Friston. See Karl J. Friston, "A Theory of Cortical Responses," *Philosophical Transactions of the Royal Society B: Biological Sciences* 360, no. 1456:815–836, https://doi.org/10.1098/rstb.

45. www.verses.ai/.

46. https://21624003.fs1.hubspotusercontent-na1.net/hubfs/21624003/white papers/Designing%20explainable%20artificial%20intelligence%20with%20ac tive%20inference.pdf/.

47. The number factorial *n*, written n! = n(n-1)(n-2)(n-3) . . . 3, 2, 1, is a huge number if *n* is large.

48. Sonia Orwell and Ian Angus, eds., *George Orwell: In Front of Your Nose 1946 to 1950* (Boston: Nonpareil Books, 2000).

49. Zeynep Tufekci, "We Need to Take Back Our Privacy," *New York Times*, May 19, 2022, www.nytimes.com/2022/05/19/opinion/privacy-technology-data .html?searchResultPosition=1/.

50. Thomas L. Friedman, "Our New Promethean Moment," *New York Times*, March 21, 2023, www.nytimes.com/2023/03/21/opinion/artificial-intelligence-chat gpt.html?searchResultPosition=1/; see also Yuval Harari, Tristan Harris, and Aza Raskin, "You Can Have the Blue Pill or the Red Pill, and We're Out of Blue Pills," *New York Times*, March 24, 2023, www.nytimes.com/2023/03/24/opinion/yuval-ha rari-ai-chatgpt.html?smid=nytcore-ios-share&referringSource=articleShare/; Ezra Klein, "The Imminent Danger of A.I. Is One We're Not Talking About," *New York Times*, February 26, 2023, www.nytimes.com/2023/02/26/opinion/microsoft-bing -sydney-artificial-intelligence.html?action=click&module=RelatedLinks&pgtype =Article/; and Jaron Lanier, "There Is No A.I." *New Yorker*, April 20, 2023, www .newyorker.com/science/annals-of-artificial-intelligence/there-is-no-ai/.

51. Aspen Institute Commission on Information Disorder, "Final Report November 2021," www.aspeninstitute.org/wp-content/uploads/2021/11/Aspen-Institute _Commission-on-Information-Disorder_Final-Report.pdf/; Global Disinformation Lab, https://gdil.org/.

APPENDIX

1. Strictly speaking, each step corresponds to multiplying by a factor of e = 2.71828.

2. David Roodman, "Modeling the Human Trajectory," Open Philanthropy, June 15, 2020, www.openphilanthropy.org/research/modeling-the-human-trajectory/.

3. https://en.wikipedia.org/wiki/ito_calculus.

INDEX

UPS, 61
Upwork, 223
uracil, 159
Urnov, Fyodor, 166
US Navy, 86, 281–82
US Patent and Trademark Office
 (USPTO), 250
Utah array, 103
UTM. *See* Unmanned Aircraft Traffic
 Management

vaccines, 161–62
Valkyrie, 59
velocity, of money, 233
Venmo, 224
Venter, Craig, 173
VERSES, 305
vertical take-off and landing (VTOL), 61
video games, 112–13
Virgin Galactic, 257, 262, 263, 265, 266,
 268
Virgin Orbit, 266
virtual private networks (VPNs), 228
virtual reality (VR), 111–12, 124, 304, 307
viruses, 160, 163, 167
Vision Pro, 304
visual processing, 97–98
Vollrath, Dietrich, 201
Voluntary Human Extinction, 303
voluntary population optimization, 199
VOX Space, 266
VPNs. *See* virtual private networks
VR. *See* virtual reality
VTOL. *See* vertical take-off and landing

Waal, Frans de, 153
Walking Beast, 53
WALL-E, 40, 57
Wallet, 224
Walmart, 72, 112
Warzel, Charlie, 213
Watson, James, 156
Waymo, 72

weak AI, 27–28, 129
weather, 185
Web 3.0, 304–8
WeChat, 216–17
Weizenbaum, J., 330n10
Wengrow, David, 293
West Nile, 172
WhatsApp, 210
What We Owe the Future (MacAskill),
 303
Wheeler, John Archibald, 272
White Knight, 263
Whitson, Peggy, 343n19
Whole Foods, 210
Windfall Clause, 217–19
wind power, 191–92
Woese, Carl, 149
Work, Robert O., 86–87
Worker Info Exchange, 232
wormholes, 272
Wozniak, Steven, 42, 88
Wright brothers, 255
Wu, Tim, 216

X-47B, 86
XAI. *See* explainable AI
X-chromosome, 157
xenobots, 177–78
xenotransplantation, 167
Xplore, 268

Yandex, 72
Yang, Andrew, 28, 42, 220, 222, 312
Y-chromosome, 150, 157
YouTube, 32, 210, 238–39, 245, 265
ytterbium, 274–75
Yuste, Rafael, 110–11

Zeilinger, Anton, 275
Zero Gravity Corporation, 265
Zika, 172
Zuboff, Shoshana, 210, 213, 239
Zuckerberg, Mark, 278